Equivalence and Priority: Newton versus Leibniz

Equivalence and Priority: Newton versus Leibniz

Including Leibniz's Unpublished Manuscripts on the Principia

DOMENICO BERTOLONI MELI

CLARENDON PRESS · OXFORD
1993

Oxford University Press, Walton Street, Oxford OX2 6DP
Oxford New York Toronto
Delhi Bombay Calcutta Madras Karachi
Kuala Lumpur Singapore Hong Kong Tokyo
Nairobi Dar es Salaam Cape Town
Melbourne Auckland Madrid
and associated companies in
Berlin Ibadan

Oxford is a trade mark of Oxford University Press

Published in the United States
by Oxford University Press Inc., New York

A catalogue record for this book is available from the British Library

Library of Congress Cataloging in Publication Data
Bertoloni Meli, Domenico.
Equivalence and priority: Newton versus Leibniz. Including Leibniz's
unpublished manuscripts on the Principia / Domenico Bertoloni Meli.
Includes index.
1. Newton, Isaac, Sir, 1642–1727. Principia. 2. Leibniz,
Gottfried Wilhelm, Freiherr von, 1646–1716. Tentamen de motuum
coelestium causis. 3. Mathematics–History–17th century.
4. Mechanics, Celestial–History–17th Century. I. Title,
QA803.B47 1992 510'.9'032–dc20 92-18967
ISBN 0-19-853945-2 (hardback)

Typeset by Cotswold Typesetting Ltd, Gloucester
Printed in Great Britain by
Biddles Ltd,
Guildford & King's Lynn.

ACKNOWLEDGEMENTS

Completion of this work has been made possible by the generous assistance of several institutions. An exchange programme between Ghislieri College, Pavia, and Rheinland-Pfalz allowed me to spend the year 1984–5 in Mainz and from there to make several trips to Hanover. Subsequent scholarships from the British Council and St John's College, Cambridge, allowed me to complete my M.Phil. and Ph.D., respectively. Lastly, a Research Fellowship from Jesus College, Cambridge, gave me the opportunity to revise my work for publication in a most pleasant environment.

During my stays in Hanover I benefited from several conversations with Heinz-Jürgen Hess, Herbert Breger, and the late Albert Heinekamp. His kindness and hospitality made my time there particularly pleasant.

Eberhard Knobloch, more than anyone else, taught me how to decipher Leibniz's manuscripts and has constantly been a source of encouragement. In the UK I wish to thank for their advice my supervisor Simon Schaffer, Nick Jardine, Tom Whiteside, and my Ph.D. examiners A. Rupert Hall and the late Eric Aiton. In addition, Tom Whiteside has generously suggested many improvements to my translation of the *Tentamen.* I am also grateful to the editors of *Studies in History and Philosophy of Science* and Pergamon Press for permission to use part of my article 'Public claims, private worries' (Vol. **22**, 1991, 415–49). The readers for the Oxford University Press offered much valuable advice. I thank especially Henk Bos, who kindly agreed to read a draft of Chapters 3 and 4, and Michael Mahoney. Finally, I am grateful for the expert and kind assistance of the staffs of the Niedersächsische Landesbibliothek, Cambridge University Library, and Whipple Library. All manuscripts are reproduced by courtesy of the Niedersächsische Landesbibliothek. My most outstanding debt is to my wife for her help, support, and patience.

I claim sole responsibility for all remaining errors and inaccuracies, and especially for the selection, presentation, and dating of the manuscripts.

Cambridge   D. B. M.
November 1992

CONTENTS

GENERAL INTRODUCTION

This book examines the competing world systems put forward by Newton and Leibniz in the late 1680s and their reception up to the beginning of the eighteenth century. The two main published texts of my story are *Philosophiae Naturalis Principia Mathematica* of 1687, and *Tentamen de Motuum Coelestium Causis* of 1689, namely Leibniz's article in response to Newton's theories about the motion of celestial bodies.

Over the last few decades Newton's itinerary to the *Principia* has been meticulously reconstructed through the study of his private manuscripts and correspondence; the new picture which has emerged has enriched enormously our understanding of his work in many aspects, ranging from the interaction between mathematics and mechanics to the role of alchemy and theology. The formation of Leibniz's techniques and ideas about planetary motion is extremely well documented in an exciting series of previously unknown manuscripts which are published and analysed here for the first time. Leibniz's manuscripts, dating from the late 1680s, include his first annotations to the *Principia* and his first attempts to construct an alternative theory; they are among the most extensive documents we possess on the early reception of Newton's masterpiece, and while providing us with deeper insights into Leibniz's thought and strategy, allow us to gain a better understanding of Newton's theory as well. They constitute stimulating material for reconsidering the issues of equivalence between rival theories, mathematical formulations and physical interpretations, private research versus publication, intellectual property, and priority in the late seventeenth century. These issues are inextricably interwoven with the dispute about the world system between the Cambridge Lucasian Professor and the Court Librarian and Councillor in Hanover.

The names of the protagonists of this war are immediately associated with the controversy over the invention of the calculus, of which they are now considered to be joint and independent inventors, Newton in 1665–6 and Leibniz in 1675. With regard to the theory of planetary motion, however, the question of priority and plagiarism requires a different answer. In order to introduce this issue, I start from a curious prehistory.

Leibniz took charge of his duties as librarian to the Duke of Hanover at the end of 1676. He reached Germany from Paris via London, having spent the previous four years in the French capital in contact with the most advanced philosophical and mathematical circles in Europe. Soon after his arrival, in the first few months of 1677, he wrote an important letter to his friend the Jesuit Honoré Fabri containing an essay on cosmology, the role and propagation of light, elasticity, cohesion, gravity, and planetary motion. In the letter Leibniz stressed the fertility of a way of philosophizing without hypotheses, in which experiments, geometry, and laws of mechanics are conjoined. In his view this approach was about to bring sweeping and breathtaking advancements of knowledge: 'And I doubt not that if a certain number of selected people pursued the matter in earnest, so many things are in our power, that with the work of one decade it would be possible to obscure the efforts of all past centuries.'[1]

Exactly one decade after Leibniz's letter to Fabri, Newton's *Principia* appeared in print. Despite the superficial similarity in the language with regard to experiments, mathematics, and mechanics, Leibniz had good reasons for being enthusiastic and worried at the same time. His early study of Newton's masterpiece can be reconstructed on the basis of three documents: his annotations or *Marginalia* in his own copy of the *Principia*, first made available in 1973; two sets of *Excerpts* from several propositions, including some comments, first published in 1988;[2] his working sheets or *Notes*, published here for the first time. The *Notes* are the most important of the three: they complement and clarify several cryptic remarks in the *Marginalia*, and constitute at the same time the starting point for a series of manuscript essays on planetary motion. Although Leibniz admired the wealth of results attained by Newton, he also strongly objected to the rejection of vortices carrying the planets and to the lack of a physical explanation for gravity. These two features affected the very criteria for studying nature in a mathematical fashion and were in contradiction with some fundamental pillars of Leibniz's philosophy. Hence the need of a reply in which Newton's extraordinary mathematical achievements could be reconciled with a physical theory based on vortices. That which he had taken for granted in the letter to Fabri, needed now to be fully spelt out and developed. Crucially, Leibniz did not present his essay as a reinterpretation of Newton's masterpiece; rather, he claimed that he had read only a review of the *Principia*, not the book itself, and that the review had stimulated him to publish some ideas he had conceived previously. His double claim will be examined presently.

---

[1] *LSB*, III, 2, pp. 119–48, esp. pp. 142–3, Leibniz to Fabri, dated May 1677: 'Neque dubito si homines aliquot lecti serio agerent, quae in nostra potestate sunt, unius decennii opera omnium retro seculorum labores obscurari posse.'
[2] G. W. Leibniz, *Marginalia in Newtoni Principia Mathematica*, ed. E. A. Fellmann (Paris, 1973); D. Bertoloni Meli, 'Leibniz's Excerpts from the *Principia Mathematica*', *AS*, *45*, 1988, pp. 477–504; hereafter *Marginalia* and *Excerpts*, respectively.

After an initial response, interest in the *Tentamen* vanished for longer than a century. In the 1850s the French physicist and historian Jean-Baptiste Biot published a long essay-review of John Edleston, *Correspondence of Sir Isaac Newton and Professor Cotes* (London, 1850). The book contained Newton's attack on the *Tentamen*, which was fully endorsed by Biot. He severely criticized the essay on the assumption that universal gravity was a 'simple fact' and vortices were superfluous. However, while stating that Newton had completely demolished Leibniz's essay, he did not examine either the contents of the *Tentamen* or any of Newton's criticisms. Biot also claimed that there was not sufficient evidence to accuse Leibniz of plagiarism. The same position was taken, not without some regret, by Lord Brougham and Edward Routh in their *Analytical View of Sir Isaac Newton's Principia*, again without discussing the matter in any detail.[3] More recently, René Dugas has examined Leibniz's text and defended the originality of his reasoning. This, however, did not prevent the French historian from expressing some doubts about Leibniz's 'bonne foi' with respect to Newton. By far the most perceptive and accurate account of Leibniz's theory has been provided by Eric Aiton. Among other recent interpreters I wish to mention Alexandre Koyré, who has defended Leibniz's 'good faith', claiming that if Leibniz had seen the *Principia*, he would not have made so many mistakes. Apart from the fact that some of these mistakes are Koyré's and not Leibniz's, as Aiton has convincingly shown, the argument does not hold: Leibniz became aware of one of his errors more than 15 years after having seen the *Pincipia*, in his correspondence with Pierre Varignon.[4]

In order to appreciate Leibniz's theory, strategy, and the dispute with Newton in the light of the new manuscripts which were not available to previous commentators, it is essential to start from a detailed investigation of the circumstances of composition of Leibniz's essay and of the code of behaviour concerning priority, intellectual property, and publication.

---

[3] The papers by Biot were published in the *Journal des Savants*, 1852, pp. 133–47; 217–32; 269–83 (pp. 136–7 and 273–5 are on the *Tentamen*); see also pp. 400–3; 458–70; 522–35. H. Brougham and E. J. Routh, *Analytical View of Sir Isaac Newton's Principia* (London, 1855), pp. 437–42.

[4] R. Dugas, *La Mécanique au XVIIe Siècle* (Neuchâtel, 1950), p. 492. A. Koyré, *Newtonian Studies* (Cambridge, Mass., 1965), appendix A, to be seen together with E. J. Aiton, 'An Imaginary Error in the Celestial Mechanics of Leibniz', *AS*, *21*, 1965, pp. 169–73. I. B. Cohen, *Introduction to Newton's 'Principia'* (Cambridge, Mass., 1971), p. 154; R. S. Westfall, *Force in Newton's Physics. The science of dynamics in the seventeenth century* (London and New York, 1971), ch. 6.

Leibniz versus Hooke: establishing the code of behaviour

Priority in the seventeenth century involved a complex series of factors and had different connotations from those with which we are familiar. Partly as a result of such different conventions, the second part of the century has been recently aptly characterized as 'the golden age of the mud-slinging priority disputes'. The usage of anagrams and sealed depositions to secure property over an invention without revealing it, for example, was common practice in the age of Galileo and Newton.[5] At a time when journals were just beginning to appear, correspondence, and even the disclosure in front of reliable witnesses of private manuscripts or instruments, had considerable weight: a letter to Marin Mersenne or to the Secretary of the Royal Society Henry Oldenburg, for example, often had a status closer to that of a published article than of a private communication. Today, by contrast, we tend to consider the criterion of printed publication enshrined in the copyright act as decisive. Priority also had a subtle relation with the prestige and credibility of an interpretation: the ingenuity of the 'first inventor' went hand in hand with the favourable reception of his work. The 'second inventor', by contrast, had to protect himself from the shade of doubt about his integrity and from the suspicion of having stolen a secret not from Nature, but from a colleague. Intellectual priority also related to legal matters concerning lucrative patents of technological inventions and machines, as Henry Oldenburg, Christiaan Huygens, and Robert Hooke were well aware with regard to the spring balance watch. But whether instruments or ideas were involved, a central issue in disputes was the identification of the invention or discovery: what one party presented as a major innovation was often perceived by the rival party as a mere modification, involving little or no skills, of well known ideas and techniques.[6]

[5] N. Jardine, *The Birth of History and Philosophy of Science* (Cambridge, 1984), ch. 2, especially pp. 32–4. O. Gingerich and R. S. Westman, 'The Wittich Connection: Conflict and Priority in Late Sixteenth-Century Cosmology', *Transactions of the American Philosophical Society*, Part 7, 1988, *78*, on pp. 50–69, and the essay-review by N. Jardine, 'How to appropriate a world system', *JHA*, *21*, 1990, pp. 353–8, quot. from p. 353. W. Eamon, 'From the secrets of nature to public knowledge', in D. C. Lindberg and R. S. Westman, eds., *Reappraisals of the Scientific Revolution* (Cambridge, 1990), pp. 333–66. On priority, patronage, and anagrams see R. S. Westfall, 'Science and patronage: Galileo and the Telescope', *Isis*, *76*, 1985, pp. 11–30.

[6] *The Correspondence of Henry Oldenburg*, ed. and transl. by A. R. Hall and M. Boas Hall, 13 vols. (Madison and London, 1965–86), vol. *11*, passim. C. MacLeod, *Inventing the Industrial Revolution. The English patent system, 1660–1800,* (Cambridge, 1988). R. K. Merton, *The Sociology of Science* (Chicago, 1973), 'Priorities in Scientific Discovery', pp. 286–324; 'Behaviour Patterns of Scientists', pp. 325–42; 'The Ambivalence of Scientists', pp. 383–412, and the critical observations by J. A. Secord, *Controversy in Victorian Geology: The Cambrian–Silurian Dispute* (Princeton, 1986), ch. 7, esp. pp. 230 and 239–40. L. R. Patterson, *Copyright in historical perspective* (Nashville, 1968).

In our case we are particularly lucky because Leibniz himself spelt out the criteria he deemed appropriate in priority disputes. During his first stay in London in the winter of 1673 he became involved in two difficulties concerning plagiarism. The first case concerned a method for the interpolation of series by constructing series of differences. As a guest of Robert Boyle, in the presence of the respected mathematician John Pell, Leibniz presented the method as his own, but Pell promptly remarked that the discovery had already been made by François Regnaud, as reported in Gabriel Mouton, *Observationes diametrorum Solis et Lunae apparentium* (Lyons, 1670). Reacting to the initial embarrassment, on the following day Leibniz hastily composed a letter to Oldenburg explaining what had happened. On looking at Mouton's book, he found that 'what Pell had said was perfectly true', but felt justified in defending himself from Pell's implicit reproach of plagiarism:[7]

I will vindicate the honesty of my conduct by two arguments: first by displaying my actual disordered notes in which not only my discovery appears but the occasion and manner of making it; and secondly by adding certain points of great importance not stated by Regnaud and Mouton, which is not likely that I could have huddled together since yesterday evening, nor are they to be readily expected of a copyist.

Leibniz's strategy involved the public display of private notes and 'important' additions to Regnaud's discovery; this double move was meant to safeguard his honesty and originality at the same time. Although Leibniz's honour seemed to have been saved, this embarrassing event was to be referred to later and used against him during the priority dispute.

The second case concerned the calculating machine presented by Leibniz at the meeting of the Royal Society on 1 February 1673. The curator of experiments Robert Hooke examined with extraordinary care Leibniz's model and at the following meeting of the Society a fortnight later, when Leibniz was not present, he criticized it and claimed he could produce a simpler model. On hearing of Hooke's criticisms, Leibniz was understandably incensed and shortly afterwards, back in Paris, he spelt out his case very forcefully in a letter to Oldenburg. Leibniz claimed that the basis of the construction of Hooke's machine was the same as his own:[8]

Nor could he say that this basic idea would have entered his head but for me,

[7] *The Correspondence of Henry Oldenburg*, vol. 9, p. 444, 3 Febr. 1673 o.s. J. E. Hofmann, *Leibniz in Paris* (Cambridge, 1974), pp. 25–7.

[8] *The Correspondence of Henry Oldenburg*, vol. 9, p. 493, 8 March 1673. L. von Mackensen, 'Zur Vorgeschichte und Entstehung der ersten digitalen 4-spezies-Rechenmaschine von Gottfried Wilhelm Leibniz. *SL*, Supplementa 2, 1969, pp. 34–68.

since two things are obvious, (1) he had never spoken to anyone of such a thing before I came to England with my machine; (2) my machine was very thoroughly and carefully examined by him at close quarters. For when I explained that [machine of mine] before the Royal Society he was certainly very well forward; he removed the back plate which covered it, and absorbed every word I said; and so, such being his familiarity with mechanics and his skill in them, it cannot be said that he did not observe my machine. That he did not distinctly trace out all its wheels I readily admit. But in such cases it is enough for a man who is clever and mechanically-minded to have once perceived a rough idea of the design, indeed the external manner of operation, and then for him afterwards to add a little of his own, consisting only of some involvement of the wheels which can be effected by different people in different ways.

Hooke could not display private notes nor refer to reliable witnesses in order to vindicate his honesty, unlike Leibniz with respect to Regnaud. The final part of this quotation concerns the problem of originality of and equivalence between inventions. Priority affected the definition of 'invention', the basic idea and the detailed arrangement of the parts, or, as we shall see, the final result and the method of demonstration. Leibniz continued his attack in his forceful letter by establishing a code of behaviour based on recent history:[9]

We know that right-minded and decent men have preferred when they understood something that was relevant to the improvement of other person's discoveries to ascribe their [own] improvements and additions to the [original] discoverers, rather than incur the suspicion of intellectual dishonesty and want of true magnanimity should they chase after falsehood with an unworthy kind of greed.

Nicole Fabri de Peiresc and Pierre Gassendi, Leibniz continued, embarked on the observations of the periods of the Jovian satellites and the surface of the Moon respectively. But when they found out that Galileo Galilei and Johann Hevelius had already taken up a similar task, they spontaneously desisted and made their partial investigations available to Galileo and Hevelius respectively. 'Conversely, it is for the discoverer also to admit before the public his obligation to anyone whose advice has caused his own ideas to prosper.' This was the code of conduct Leibniz deemed appropriate: the curator of experiments was expected to desist from his attempts and claims, and to impart his advice to Leibniz who, in his turn, would publicly acknowledge his obligations. If Hooke 'does this I shall praise his good spirit in public; if he does not, he will do something unworthy of his own estimate of himself, unworthy of his nation, and unworthy of the Royal Society.'[10]

[9] Ibid.
[10] Ibid., pp. 493–4 and 497, notes 12 and 13. On Hooke see M. Hunter and S. Schaffer, eds., *Robert Hooke. New Studies* (Woodbridge, Boydell, 1989), esp. S. Shapin, 'Who was Robert Hooke?', pp. 253–85.

Leibniz versus Newton: violating the code of behaviour

Fifteen years later Leibniz found himself in an analogous situation with respect to Isaac Newton. Following the appearance of the *Principia* in 1687, Leibniz rushed to print in the *Acta Eruditorum* three essays on optics, motion in a resisting medium, and the causes of planetary motion, namely *De lineis opticis, Schediasma de resistentia medii*, and *Tentamen*, respectively. The first piece is little more than an introduction to the two following memoirs; Leibniz's main aim was to explain his situation at the time and set the scene for the following essays:[11]

While I was engaged on a mission that I had undertaken at the command of my prince, His Most Serene Highness, and had to spend much time in travelling to quite distant parts, in the course of my regular sifting of records in Archives and Libraries I was offered by a friend certain monthly parts of the *Acta* from which to discover what was afoot in the Republic of Letters, since I had long been out of touch with new publications. So, when I was examining the Proceedings for June of this year I came across an account of the celebrated Isaac Newton's Mathematical Principles of Nature. This account I have read eagerly and with much enjoyment . . .

Indeed we know that between November 1687 and June 1690 Leibniz was travelling through Southern Germany, Austria, and Italy on behalf of his Duke on a diplomatic mission aimed at establishing the origins of the House of Hanover and its links with the Este House. Leibniz's implicit claim in the quotation above—later repeated in the *Tentamen*—that he had seen only the review of the *Principia* by Christoph Pfautz in the *Acta Eruditorum*,[12] not the book itself, sounds somewhat clumsy. A partial justification for Leibniz would be that book reviews were a new literary genre in the last quarter of the seventeenth century, and their status with regard to priority was dubious. However, the idea of publishing a dozen pages on planetary motion two years after Newton had devoted several

---

[11] G. W. Leibniz, 'De lineis opticis et alia', *AE* Jan. 1689, pp. 36–8, *LMG, 7,* pp. 329–31; partial transl. in *NC, 3,* p. 3f. 'Schediasma de Resistentia Medii et Motu Projectorum', *AE* Jan. 1689, pp. 38–47, *LMG, 6,* pp. 135–47. 'Tentamen de Motuum Coelestium Causis'. *AE* Febr. 1689, pp. 82–96, *LMG, 6,* pp. 144–61, apart from some later corrections published by Leibniz in 'Excerptum ex Epistola Autoris, Quam pro Sua Hypothesi Physica Motus Planetarii olim (Febr. 1689) His Actis Inserta, ad Amicum Scripsit', *AE* Oct. 1706, pp. 446–51, on p. 450, and inserted by Gerhardt in the text. On Leibniz's journey see K. Müller and G. Krönert, *Leben und Werk von G. W. Leibniz. Eine Chronik* (Frankfurt a.M., 1969), and A. Robinet, *G. W. Leibniz, Iter Italicum; mars 1689–mars 1690* (Florence: Olschki, 1988), with the review by D. Bertoloni Meli, *SL, 22,* 1989, pp. 211–13.

[12] For the identification of the reviewer as Christoph Pfautz, friend and correspondent of Leibniz, and professor of mathematics in Leipzig, see *Excerpts*, p. 478, n. 8. Among Leibniz's papers at the NLB, Hanover, there is a manuscript in his hand with his excerpts from Pfautz's review in the *Acta*: LH 35, 10, 7, f. 13–14. The text contains no commentary and shows no sign of particular interest by Leibniz. On Pfautz's review see Cohen, *Introduction*, pp. 150–2.

hundred pages to the same topic without having allegedly looked at his work is open to several obvious objections. One of them is that the reader is left in doubt as to Leibniz's definitive views on the matter, since he could have changed his mind on reading the *Principia*. Christiaan Huygens was concerned precisely with this issue: in the letter to Leibniz of 8 February 1690 he inquired whether Leibniz had changed anything in his theory, since before composing the *Tentamen* he had seen only 'un extrait de son livre et non pas le livre mesme'. Leibniz did not receive this letter until late September, after a further letter of 24 August in which Huygens reiterated the point, namely whether Leibniz had changed his mind after having seen the *Principia*. Pressed twice by Huygens, Leibniz drafted a reply—which he never sent—dealing with planetary motion in which he evaded the question; he stated that he had first seen the *Principia* in Rome, where he resided from April to December 1689. Since by then the *Tentamen* was already published, Leibniz was confirming the substance of what he had already put in print. Implicitly he admitted that passing from Pfautz's review to Newton's masterpiece he had found nothing to change in his own work. Indeed, Leibniz's later formulations of his theory show virtually no sign of an increased influence by Newton.[13]

At the end of *De lineis opticis* he made the following claim stating, if not priority, at least independent discovery:[14]

Conclusions about the resistance of the medium, which I have put on a special sheet, I had reached to a considerable extent twelve years ago in Paris, and I communicated some of them to the famous Royal Academy. Then, when I too had chanced to reflect on the physical cause of celestial motions, I thought it worth while to bring before the public some of these ideas in a hasty extemporization of my own, although I had decided to suppress them until I had the chance to make a more careful comparison of the geometrical laws with the most recent observations of astronomers. But (apart from the fact that I am tied by occupations of quite another sort) Newton's work stimulated me to allow these notes, for what they are worth, to appear, so that sparks of truth should be struck out by the clash and sifting of arguments, and that we should have the penetration of a very talented man to assist us.

It is immediately clear from this quotation that the two areas are not treated symmetrically. With regard to motion in a resisting medium Leibniz could point to a specific essay he had sent to the Paris Academy. Indeed manuscripts on this topic dating from his stay in Paris have been

---

[13] See for example the 'zweite Bearbeitung' of the *Tentamen*, in *LMG*, 6, pp. 161–87, and D. Bertoloni Meli, 'Leibniz on the censorship of the Copernican system', *SL*, 20, 1988, pp. 19–42, section 4.

[14] *LMG*, 7, pp. 330–1, transl. in *NC*, 3, p. 5.

traced.[15] His strategy resembles that adopted in the Regnaud–Pell case only in part. Leibniz referred to a text which, strictly speaking, was not public in order to safeguard his honesty. However, he could not present any improvement upon a book which—in his account—he had not seen. His case concerning planetary motion was considerably weaker. The statement that he had withheld publication in order to attain a better agreement between theory and observations implied that he had found the results contained in the *Tentamen* some time earlier. However, Leibniz was unable to refer to any letter, manuscript, or other evidence supporting his statement.

Thus Leibniz had put forward two claims concerning priority in *De lineis opticis*. Although one of them was apparently made more precise in the letter intended for Huygens, the context of their correspondence shows that the evidence of this later statement, though at first sight compelling, is deceptive: both claims have a comparable status. From my reading of Leibniz's manuscripts I have reached the conclusion that they have to be rejected: Leibniz formulated his theory in autumn 1688, and the *Tentamen* was based on direct knowledge of Newton's *Principia*, not only of Pfautz's review. Evidence supporting my conclusion is displayed in Part 2 and in Appendix 1. But setting aside for the moment this issue, I wish to draw attention to the analogies between Hooke's position with respect to Leibniz's calculating machine, and Leibniz's public position with respect to Newton's *Principia*. The analogy becomes closer if one considers that a few years later Newton attacked Leibniz on the issue of priority and claimed that Leibniz had merely rearranged Newton's own propositions in a new manner. Having affirmed that immediately after publication a copy of the *Principia* had been dispatched to Leibniz via Fatio de Duillier, Newton added:[16]

If such unrestrained licence is allowed, any author can easily be robbed of his discoveries ... Leibniz had seen the epitome in the Leipzig *Acta*. Through the wide exchange of letters which he had with learned men, he could have learned the principal propositions contained in that book [that is, the *Principia*] and indeed have obtained the book itself. But even if he had not seen the book itself, he ought nevertheless to have seen it before he published his own thoughts concerning these same matters, and this so that he might not err through haste in

---

[15] H.-J. Hess, 'Die unveröffentlichten naturwissenschaftlichen und technischen Arbeiten von G. W. Leibniz aus der Zeit seines Parisaufenthaltes. Eine Kurzcharakteristik', *SL*, Supplementa *17*, 1978, 2 vols., vol. *1*, pp. 183–217, on pp. 206–10, 'Du Frottement', dated 'Hyeme, 1675'. Compare also E. J. Aiton, 'The Application of the Infinitesimal Calculus to some Physical Problems by Leibniz and His Friends', *SL*, Sonderheft *14*, 1986, pp. 133–43, on p. 134, n. 6.

[16] Cohen, *Introduction*, p. 154, translated from Edleston, *Correspondence*, pp. 308–9; the original draft is in ULC, Lucasian professorship papers, Res. 1893(a), Packet 8 (the manuscripts are not numbered).

a new and difficult subject, or by stealing unjustly from Newton what he had discovered, or by annoyingly repeating what Newton had already said before.

In the 'Account of the *Commercium Epistolicum*', an anonymous review published in the *Philosophical Transactions* for 1715, Newton stated:[17]

In the *Commercium Epistolicum*, mention is made of three tracts written by Mr. Leibnitz, after a copy of Mr. Newton's *Principia Philosophiae* had been sent to Hanover for him, and after he had seen an account of that book published in the *Acta Eruditorum* for January and February 1689. And in those tracts the principal propositions of that book are composed in a new manner, and claimed by Mr. Leibnitz as if he had found them himself before the publishing of the said book. But Mr. Leibnitz cannot be a witness in his own cause. It lies upon him either to prove that he found them before Mr. Newton, or to quit his claim.

In other words, Leibniz had to name reliable witnesses testifying that he had worked on the topics of the *Principia* before 1687. Referring to the three 1689 papers, Newton went on to claim that 'the Propositions contained in them (Errors and Trifles excepted) are Mr Newton's (or easy Corollaries from them)'. If in the previous passages he had focused on scholarly ethics and on the equivalence between the results published by Leibniz and those contained in the *Principia*, in the following text Newton adopts an openly sarcastic tone:[18]

Galileo began to consider the effect of Gravity upon Projectiles. Mr Newton in his Principia Philosophiae improved that consideration into a large science. Mr Leibnitz christened the child by a new name as if it had been his own, calling it *Dynamica*. Mr Huygens gave the name of vis centrifuga to the force by which revolving bodies recede from the centre of their motion. Mr Newton, in honour of that author, retained the name and called the contrary force vis centripeta. Mr Leibnitz to explode this name calls it sollicitatio Paracentrica, a name much more improper than that of Mr Newton. But his mark must be set upon all new inventions. And if one may judge by the multitude of new names and characters invented by him, he would go for a great inventor.

In a related passage he explained that Leibniz 'changed the name of vis centripeta used by Newton into that of sollicitatio paracentrica, not because it is a fitter name, but to avoid being thought to build upon Mr Newton's foundations.'[19] The eruption of Newton's anger corresponds closely to Leibniz's: if Hooke had removed the back plate of the calculating machine, Leibniz had seen the review in the *Acta*, and it was 'enough for a man who is clever . . . to have once perceived a rough idea

[17] The 'Account of the Book Entituled *Commercium Epistolicum*', *PT*, *29*, 1715, pp. 173–224, is reproduced in A. R. Hall, *Philosophers at War* (Cambridge, 1980); see pp. 208–9 and 263–314. Notice that Pfautz's review was published in June 1688.
[18] ULC, Ms Add 3968, f. 415v.; Cohen, *Introduction*, p. 296.
[19] ULC, Ms Add 3968, f. 412v.; Cohen, *Introduction*, p. 297.

of the design . . . and then for him afterwards to add a little of his own . . . which can be effected by different people in different ways.' Although the line between true inventor and mere imitator was at times difficult to discern, the consequences of falling on one or the other side were considerable in terms of prestige and credibility. Several years later Leibniz conceived the project of a *Machina Coelestis*, a planetarium based on his own theory of planetary motion: the analogy between intellectual and material inventions is indeed remarkable.[20]

Leibniz's failure to meet the very standards he had set in the controversy with Hooke convincingly shows that the appearance of the *Principia* had left him in an awkward position. Newton's extraordinary success in providing a new world system and his 'bad' philosophy required a prompt reply. Setting aside questions of personal pride, the danger was that if Leibniz had not intervened, Newton's interpretation might have gone unchallenged and might have been accepted on the basis of its empirical success. The code of behaviour followed by Peiresc and Gassendi, and proposed by Leibniz to Hooke, looked utterly inadequate in those circumstances. An intervention claiming priority, however, might have incurred 'the suspicion of intellectual dishonesty and want of true magnanimity', as Leibniz had written about Hooke; unlike the case when he had been accused by John Pell, this time he had no manuscripts or other documents to present in his defence. Until today no evidence has emerged that Leibniz had formulated any significant part of his theory of planetary motion before the autumn of 1688. A purely philosophical response would have left Newton as the only master of celestial mechanics; Leibniz, though, wanted to compete in mathematics, mechanics, and astronomy as well. However, had he stated that his essay depended on Newton's *Principia*, the authority of his own theory—which Leibniz knew was problematic in several respects, as he had to admit in the conclusion of the *Tentamen*—would have been undermined. In order to raise its profile and gain higher credit, he thought he had to claim independent and prior discovery: thus priority was used as a philosophical weapon against Newton. Pressed by the circumstances, Leibniz opted for action. In doing so he was prompted by an unexpected finding that allowed him to transform Newton's explanation of the area law, and then to create a different theory and interpretation. The unfolding of his strategy, the development of his alternative theory, and the ensuing war with Newton, illuminate a crucial episode in the history of science.

[20] E. Gerland, *Leibnizens nachgelassene Schriften physikalischen, mechanischen und technischen Inhaltes* (Leipzig, 1906), pp. 134–41. The planets were supposed to move along a rotating stick. In this General Introduction I have relied on my 'Public Claims, Private Worries: Newton's *Principia* and Leibniz's theory of planetary motion', *SHPS*, 22, 1991, pp. 415–49.

Plan of the work and historiographic observations

This book consists of three parts with appendices; each part fulfils different aims, hence the style of the exposition varies accordingly. Part 1 provides a historical background for the interpretation of Leibniz's manuscripts. While concentrating primarily on his philosophical and mathematical works, I have found it useful to compare and contrast his views with Newton's. Chapter 1 is devoted to Leibniz's deployment of the Keplerian programme and to astronomy. His attempt to retrieve the theme of the integration of mathematical and physical astronomy constitutes a major aspect of my interpretation. Leibniz selected Kepler as an ally against Newton: the German astronomer was an adroit and natural choice because of his prestige, of the crucial role of his three laws of planetary motion, and because Kepler had made of the integration of mathematical and physical astronomy one of the central themes of his work. Following Kepler, Leibniz wanted to show that mathematical astronomy could and indeed ought to be developed together with a physical interpretation based on vortices. Leibniz's double gambit of claiming priority and summoning Kepler's authority was all the more necessary since he knew that this theory was problematic both with regard to mathematics and to mechanical causes. Chapter 2 deals with vortex theories and explanations for celestial motions, gravity, and elasticity. A set of annotations in Leibniz's hand on the third part of the *Principia Philosophiae* reveals the extent of Leibniz's debt to Descartes and establishes a correlation between the theories of celestial motions and of motion in resisting media. I also examine Huygens's mathematical and physical accounts of centrifugal force and gravity and conclude with a survey of seventeenth-century, and especially Leibniz's, views on elasticity. In Chapter 3 some fundamental features of Leibnizian and Newtonian mathematics are outlined, with special emphasis on the issues of representations of curves, the role of infinitesimals, and physical dimensions. This highly selective account sets the scene for Chapter 4, which is devoted to mechanics. Whilst the current historiographic tradition tends to consider Leibnizian mechanics and especially dynamics together with his metaphysics, I emphasize the links between mechanics and mathematics. These links are crucial to appreciate the characteristic features of Leibniz's notions, such as impetus and solicitation, dead and living force, and their mutual relations. Once again it is useful to contrast Leibnizian and Newtonian notions. Although the specialist in the seventeenth century may find much that is already known in Part 1, my presentation should pave the way, I hope, for a better understanding of the material presented and discussed below. The degree of success of these introductory chapters is measured not by the extension of the survey of Leibniz's and Newton's sources, but by their role in the rest of the work.

Part 2 presents a close examination of the main stages of the formation of Leibniz's theory from his reading of the *Principia* to the composition of the published *Tentamen*. Chapter 5 charts the private development of Leibniz's itinerary and includes large excerpts of the manuscripts in English translation. Despite the initial dependence on the *Principia*, later Leibniz attained original results involving a sophisticated analysis of orbital motion. A full translation of the *Tentamen* with closely textual annotations can be found in Chapter 6.

Part 3 consists of three interpretive essays providing an exegesis of Leibniz's texts, a reappraisal of Newtonian mechanics in the light of his war with Leibniz, and a study of the reception of their world systems, respectively. The common aim of the essays is to investigate how Leibniz's manuscripts affect our interpretation of mathematical theories of celestial motions around 1700. Chapter 7 is devoted to a comparative study of Leibniz's private and public work presented in Part 2, to the analysis of his style of writing in the *Tentamen*, and to the development of his theory. I also draw some general conclusions on the importance of circumstances of composition and audience in the interpretation of Leibniz's texts. Chapter 8 focuses on Newton: I present a fresh reading of the problems in mathematics, mechanics, and physics encountered in the correspondence with Hooke in 1679–80 and in the formulation of the tracts *De motu* in the mid-1680s. At the end of the chapter I analyse Newton's onslaught on the *Tentamen* and especially his interpretation of centrifugal force in terms of the law of action and reaction. This crucial aspect in Newton's line of attack was later deployed by his champion, the Savilian Professor of Astronomy John Keill. The final chapter surveys the reception of the competing theories and the reasons for the final defeat of Leibniz's alternative by focusing on the practice of celestial mechanics at the beginning of the eighteenth century. In my interpretation the defeat of the *Tentamen* has to be attributed primarily to the inconsistency between Kepler's laws and vortex theory. However, other factors were also important, such as Newton's interpretation of centrifugal force in terms of the third law of motion, and Leibniz's claim that centrifugal force in orbital motion is inversely proportional to the third power of the distance. If the first reason is perfectly consonant with our understanding, the others are alien to our perception and can only be appreciated by an investigation of the generation of mathematicians following Leibniz and Newton. Moreover, while the *Principia* has opened up a wide field of research, the *Tentamen* had never been able to overcome the problems already present in its original formulation.

02

Indeed, in later years even Leibniz was forced in practice to abandon the Keplerian programme and stress the theological and philosophical issues typical of his correspondence with the divine Samuel Clarke in 1715–16. Despite the victory of the *Principia*, however, at the beginning of the eighteenth century celestial mechanics was beginning to show autonomous features with respect to Newton's work. Lastly, Appendix 1 contains transcriptions of the most important manuscripts with textual analyses and commentaries.

In my investigation I have found it helpful to take into account some historiographic tenets which require a brief explanation. They can be described as the 'controversy thesis' and the 'symmetry thesis'. Controversies are particularly instructive for the historian because the protagonists are often induced to spell out assumptions which otherwise remain tacit. We have now come to realize that, far from being pathological manifestations, controversies constitute an integral part of scientific life; the virulent priority dispute over the invention of the calculus, for example, exerted a profound influence on the community of mathematicians, stimulated them to produce new results and to read avidly publications appearing on both sides of the Channel. It is helpful in the study of controversies to adopt a symmetric treatment of winning and losing theories. Only in this way is it possible to analyse which factors counted in the resolution of a dispute. Indeed, unless this investigation of the actual historical developments and practice of the protagonists is carried out, no conclusion can be drawn about the reasons why an interpretation prevailed or was defeated. Far from following these tenets in a dogmatic way, I employ them selectively and in a very personal fashion depending on the historical material on which I am working. The great number of excellent works and editions concerning Newtonian mechanics has induced me to give more weight to the comparatively less studied and unpublished Leibnizian manuscripts. A further example may help to clarify my approach. Leibniz's theory of planetary motion was defeated and interest in it was eclipsed only to re-emerge in comparatively recent times. Newton's theory, on the other hand, has been considered as the dominant world-view and constantly reformulated over three centuries. Like a venerated painting where successive waves of restorers have emphasized chiaroscuro and fading colours in a personal and sometimes fanciful way, the *Principia Mathematica* bears the marks of successive reinterpretations alien to a seventeenth century context. Hence the importance of such a detailed and extensive contemporary analysis as that provided by Leibniz. His controversy with Newton is particularly helpful in understanding contemporary approaches to centrifugal force and a wealth of subtleties concerning mathematics and mechanics. I hope that the juxtaposition and contrast of competing explanations will result in a much needed defamiliarization from the *Principia*. A large portion of Newton's masterpiece was based on a conceptual framework which was considerably different from that of later reformulations improperly called 'Newtonian'. The identification of some of those differences prepares the ground for a fresh look at the intellectual horizon and practice of mechanics around 1700. A further problem resulting from Newton's victory concerns our difficulties in understanding Leibniz's ideas and techniques in all their diversity and complexity. In cases like his reading of Newton's text and the developing of an alternative theory it is tempting to appeal to the lazy and unhistorical category of confusion rather than taking seriously an alien-looking corpus of texts. However, it is precisely those cases, when at first sight it was tempting to use the notion of confusion, which revealed themselves on second and sometimes third reading as the most original and stimulating. With all their inadequacies and mistakes, Leibniz's analyses and interpretations deserve the historian's full attention.

A general and more philosophical theme running through the pages of this study is the notion of equivalence between rival theories. As A. Rupert Hall has observed in the conclusion of his study of the priority dispute, the word 'equivalent' as opposed to 'identical' implies some differences 'of a more than symbolic character'.[21] A careful handling of the problem of equivalence in the theories of celestial motions, as in the formulations of the calculus, reveals a complex web of connections between mathematical techniques, reflections on nature, and philosophical doctrines. These considerations can be extended to the works by Pierre Varignon, Jakob Hermann, and Johann Bernoulli at the beginning of the eighteenth century. The translation of the eminently geometrical language of the *Principia* into the algebraic form of the calculus presents a set of problems which can be profitably compared with those related to the *Tentamen*: the issues of translation and equivalence are closely linked. Newton's champion John Keill, for example, claimed that the analysis of the inverse problem of central forces by Johann Bernoulli differed from that in the *Principia* as Latin and Greek versions of the same passage would. The adequacy of this comparison with Indo-European languages, however, can be questioned since their structure is too close for our purposes. Other comparisons are probably more appropriate: the syntaxes of two computer languages, for example, often present deeper differences and the skills required to operate with them may vary considerably. Thus later developments may

---

[21] Hall, *Philosophers at War*, pp. 257–8; C. Truesdell, 'A program toward rediscovering the rational mechanics of the Age of Reason', *AHES*, *1*, 1960, pp. 3–36; M. S. Mahoney, 'Algebraic vs. geometric techniques in Newton's determination of planetary orbits' (forthcoming).

follow different routes suggested by the specific structure and style of a language. Underlying these observations is a definition of 'theory' which is worth making explicit. The current understanding of this notion involves a set of definitions, axioms, propositions, and demonstrations. As such, a theory is a corpus of written interrelated statements. Besides this set of statements it is useful to consider the skills necessary to manipulate them. This extension of the notion of theory has several implications: it adds a human dimension by placing alongside doctrine and practitioners; it emphasizes the operative and hence the dynamic aspect of theories, as opposed to the static aspect. Therefore, the specific fashion in which they are formulated takes on an important role. Hence my scrupulous handling of the actors' notation is not merely a tribute to philological accuracy, but is an integral part of my story: nuances in terminology and notation often have wide implications for their conceptual significance and role in the evolution of a theory. Only by contrasting the specific features of the interaction between mathematics, mechanics, and natural philosophy in Leibniz, Newton, and their successors, can we gain a deeper understanding of the intellectual subtleties and practice of the theories about the system of the world around 1700.

PART 1
THE BACKGROUND OF THE
NEWTON–LEIBNIZ DISPUTE

1
ASTRONOMY AND THE KEPLERIAN PROGRAMME

1.1 The Cassirer thesis: Kepler and Leibniz

Although Leibniz explicitly mentions Kepler as one of his masters, the role of Keplerian themes on Leibniz's system has been little studied and often altogether ignored. One of the few exceptions is represented by Ernst Cassirer. In his classic monograph Cassirer claims that Kepler's concept of *vera hypothesis* developed in the *Mysterium Cosmographicum* and *Apologia pro Tychone contra Ursum* was very influential on Leibniz. Cassirer identifies two astronomical traditions whose aims were to provide a description of phenomena and to explain them causally, respectively. Purely descriptive accounts were incapable of going beyond sensory data and establishing a link between experience and the general laws of knowledge. Causal accounts were merely speculative and considered phenomena in a negative way, as something contradicting the course of thought. According to Cassirer, the central theme of the Keplerian reform of astronomy was the view that phenomena were a positive challenge influencing the direction of knowledge. The primary tool for accomplishing this task is identified in the notion of 'vera hypothesis', which is capable of establishing the crucial link between isolated sensory data and the laws of knowledge. Thus a hypothesis is properly called 'true' not simply by virtue of its immediate agreement with single observations, but because of its correlation with a system of general physicomathematical principles. Similarly, in mathematics the role and value of hypotheses has to be established not *per se*, but in relation to a system of definitions. Following Cassirer's interpretation, the legitimacy of hypotheses in natural philosophy and mathematics was defended by Leibniz exactly as Kepler had done in astronomy. In their philosophical systems phenomena assume a new dignity and the true hypothesis becomes the instrument for binding them to the laws of knowledge.[1]

[1] E. Cassirer, *Leibniz' System in seinen wissenschaftlichen Grundlagen* (Marburg, 1902), pp. 362–3 and 503; *Das Erkenntnisproblem in der Philosophie und in der Wissenschaft der neueren Zeit*, vol. 1 (Berlin, 1902), pp. 328–52. See also my preliminary study 'Kepler and Leibniz', *Leibniz. Tradition und Aktualität*. II. Teil (Hannover, 1989), pp. 88–94, where I

I call this association of Kepler with Leibniz, so beautifully expressed in the style of history of ideas at its best, the 'Cassirer thesis'. This section shows that the thesis put forward by Cassirer is a powerful tool for studying Leibniz's theory of planetary motion. My argument follows two main steps. First, I introduce Kepler's notion of hypothesis, paying attention to his emphasis on the links between mathematics and physics. Then I contrast Leibniz's statements about the mathematization of nature before and after he studied Newton's *Principia* and published his own essay. This analysis reveals a change of attitudes, if not of personal convictions: before the late 1680s Leibniz took for granted that mathematization could be reconciled with sound philosophy. From the 1680s onwards, however, he became more and more concerned that the mathematization of nature ought to be constrained within clear philosophical and physical boundaries, lest the proper system of knowledge be subverted. Although this shift was probably part of a more complex evolution of Leibniz's philosophical thought, it seems to me that the appearance of the *Principia* represented at least a significant factor in this evolution. A visible result of this change of preoccupations was the 1688–9 theory of planetary motion: from then onwards Kepler assumed a new importance for Leibniz. At the end of this section I examine Leibniz's views on laws of nature and their analogies with Keplerian true hypotheses. This introductory discussion will help us to understand why Keplerian themes became relevant to Leibniz's work after the appearance of Newton's *Principia* in 1687.

In the *Mysterium* Kepler expresses his disagreement with those who believe that since true conclusions can be drawn from false premises, the astronomical hypothesis of Copernicus may well lead to true phenomena even though it is false. More generally, this sceptical stance implies that the truth of astronomical hypotheses cannot be proved by their capacity to save the phenomena. Kepler replies that the success of false premises is fortuitous and fails as soon as they are applied to related matters:[2]

I have never been able to agree with those who, relying on the example of an accidental demonstration, which with syllogistic necessity yields something true

---

mention several areas in which Kepler was an important source for Leibniz. Collections of essays on various aspects of Kepler's activities are in F. Krafft, K. Mayer, B. Sticker, eds., *Internationales Kepler-Symposium* (Hildesheim, 1973); A. and P. Beer, *Kepler, four hundred years* (*Vistas in Astronomy, 18*) (Oxford, 1975).

[2] Quoted from Jardine, *Birth*, p. 215 f.; see also p. 140. J. Kepler, *Mysterium Cosmographicum* (Tübingen, 1596) = *KGW, 1*, p. 15; (Frankfurt/M, 1621, 2nd edn) = *KGW, 8*, p. 8. English translation by A. M. Duncan, *Mysterium Cosmographicum. The secret of the universe* (New York, 1981). P. Duhem, *'To save the phenomena'. An essay on the idea of physical theory from Plato to Galileo* (Chicago, 1969; transl. from the French, Paris, 1908).

from false premises ... used to maintain that it could be that the hypotheses which Copernicus adopted are false, but nevertheless the true phenomena follow from them as if from genuine principles. In fact the example is inappropriate. For this outcome of false premises is fortuitous, and that which is false by nature betrays itself as soon as it is applied to another related matter; unless you gratuitously allow him who argues to adopt infinitely many other false propositions and never, as he goes backwards and forwards [in his reasonings], to stand his ground.

In other words, false hypotheses have to be constantly modified by *ad hoc* assumptions in order to prevent them from being refuted when they are applied to similar problems. If one accepts heliocentrism, though, Kepler claims that everything follows most directly and there is no need for *ad hoc* assumptions. However, another objection can be raised. In the case of astronomical hypotheses Kepler is confronted with world systems which are observationally equivalent but inconsistent with each other. Thus no hypothesis can be lightly dismissed and if one is rejected as false, by the same logic any other can suffer the same fate. Kepler's reply to this further difficulty is twofold. First, he claims that the Copernican hypothesis is superior to its rivals because it can provide the causes of phenomena, such as the numbers, extents, and duration of retrograda-tions, which can be determined from the motion of the earth. Secondly, the rival world systems are not completely irreconcileable, since they agree in ascribing relative motions to the celestial bodies. On the basis of observations alone none of the astronomical hypotheses can be rejected— at the time when Kepler was writing. Thus the Copernican hypothesis does not refute the predications of its rivals, but expands and explains them from a different standpoint. These themes were expanded and refined in later years.

The *Apologia*, which was first published in 1858, was composed around 1600 in connection with the priority dispute between Tycho Brahe and Raimarus Ursus. Their controversy hinged on the trigono-metric rules of prostaphacresis and on the world system known as Tychonic. The astronomer and mathematician Ursus denied the capacity of astronomical hypotheses to 'portray the form of the world'. Kepler's defence of true hypotheses in the *Apologia* is based on a number of arguments countering the attack by Ursus point by point. Ursus had claimed that the falsity of hypotheses could be inferred even from the original connotation of the word. Moreover, he argued that predictive accuracy alone guarantees the adequacy of hypotheses, but 'accurate prediction and retrodiction of apparent celestical coordinates, does not guarantee the truth of astronomical hypotheses.' Indeed, he concluded that all hypotheses involve blatant absurdities. Kepler's reply ranged from a historical survey of the etymology of 'hypothesis' to the claim that astronomy has attained some truths about the world and that philosophers can make it progress further. As in the passage from the *Mysterium* quoted above, Kepler claimed that true conclusions follow from false premises only accidentally. Concerning the choice between rival hypotheses, he invoked the role of physical considerations as a powerful tool for the selection. In addition, Kepler refined his argument by drawing a distinction between true and merely geometrical hypotheses. True hypotheses concern the accurate description of phenomena as well as their causal explanation. Geometrical hypotheses, however, concern the specific ways in which a certain orbit can be represented for the sake of calculation. Ptolemy's equant point, for example, has to be considered as a purely geometrical construction rather than as a true hypothesis. A distinctive feature of geometrical hypotheses is that often several of them can account for the same observations. From the existence of this plurality of geometrical representations it does not follow that one of them is correct and the others false. Thus geometrical hypotheses could differ not substantively, but merely in the manner of exposition, since the same relative motions of celestial bodies could be represented by equivalent geometrical constructions. Although mathematics was linked to physics and other disciplines, at the same time it retained a certain degree of autonomy because the same phenomena could be legitimately represented in a variety of geometrical ways.[3]

The themes of the *Apologia* permeate Kepler's thought and were developed in his mature works, *Astronomia Nova*, *Harmonice Mundi* and *Epitome Astronomiae Copernicanae*. At the beginning of the *Epitome*, in a passage that Leibniz probably found very appealing, Kepler wrote:[4]

Physics is popularly deemed unnecessary for the astronomer, but truly it is in the highest degree relevant to the purpose of this branch of philosophy, and cannot, indeed, be dispensed with by the astronomer. For astronomers should not have absolute freedom to think up anything they please without reason; on the contrary, you should give *causas probabiles* for your hypotheses which you propose as the true cause of the appearances, and thus establish in advance the principles of your astronomy in a higher science, namely physics or metaphysics.

This quotation expresses a major aspect of Kepler's reform of astronomy

---

[3] Jardine, *Birth*, pp. 153–4 and 212–13, provides also an extensive and detailed analysis of the controversy between Ursus and Kepler; see also N. Jardine, 'Scepticism in Renaissance Astronomy: a preliminary study', in R. H. Popkin and C. B. Schmitt, *Scepticism from the Renaissance to the Enlightenment* (Wolfenbüttel, 1987), pp. 83–102.

[4] The following passage from *Epitome Astronomiae Copernicanae* = *KGW* 7, p. 25, is translated in Jardine, *Birth*, p. 250. See also *KGW*, *13*, pp. 140–4, esp. 141, Kepler to Michael Maestlin, Oct. 1597. E. J. Aiton, 'Johannes Kepler and the Astronomy without Hypotheses', *Japanese Studies in the History of Science*, 14, 1975, pp. 49–71.

and constitutes a central point in my interpretation of Leibniz. Kepler thought that a choice among astronomical hypotheses should be based not just on mathematical considerations or on the agreement between theory and observations, but also on physical and philosophical grounds. The *causae probabiles* he invoked had to be adequately justified in a broader disciplinary context. The true hypothesis emerges from the intersection between two or more disciplines, from considerations ranging from scripture to mathematical harmony and simplicity, and from the search for causes. Indeed, for Kepler these issues were mutually related because the discovery of mathematical harmony and simplicity was not an end in itself, but was instrumental in identifying the purpose and order of the creation. Therefore causal explanations involved both specific phenomena and the cosmic order of things. The choice of the Copernican hypothesis was a matter of observational accuracy and simplicity as well as of theological and physical considerations about the Sun and the *aetiologia* of planetary motion.

Leibniz's essay *Tentamen de Motuum Coelestium Causis* shows an analogy with Kepler's programme from the very title. But the analogy is more complex and can be pushed further if one considers that the *Tentamen* is Leibniz's response to Newton's *Principia Mathematica*. According to Leibniz's reading, Newton was spectacularly successful in providing mathematical imaginary constructs to account for planetary motions, but physical explanations were either lacking or unsound. In particular the rejection of vortices and of celestial matter seemed to suggest that Newton was not merely employing imaginary hypotheses in order to save the phenomena, but was pretending to explain nature by means of 'occult qualities', such as attraction. Whatever his intentions, the task of the *Tentamen* was to attain a theory mathematically equivalent to Newton's in accounting for planetary motion and especially for the inverse-square law and Kepler's laws, but physically sound and capable of explaining the causes of phenomena. The physical cause of planetary motion had to be ascribed to a fluid rotating around the Sun. Leibniz's praise of Kepler in the introduction to the *Tentamen* is largely based, certainly not by chance, on this aspect. Thus Leibniz found in Kepler a natural and authoritative ally against Newton; natural because both Kepler and Leibniz believed in a theory which brings together mathematical representations of phenomena and physical explanations, and authoritative because of the prestige of the German astronomer, and in particular because Kepler's three laws occupy a central position in Newton's theory.

Other texts by Leibniz show different preoccupations, as we have seen in the letter to Fabri referred to in the General Introduction. In a letter of 1676 to Claude Perrault—a member of the Paris Academy—concerning the cause of gravity, Leibniz concludes his analysis with the following words:[5]

So much so that concerning the laws of motion, I believe that at present I can satisfy myself with entirely geometrical demonstrations, without employing any suppositions or principles of experience; and that from now on what one could say on this matter will be nothing but *res calculi et geometriae*. Thus I believe that at present we are in a condition to pretend to a true physics without hypotheses.

In *Demonstrationes Novae de Resistentia Solidorum*, for example, an essay of 1684 on the resistance of materials, Leibniz states:[6] 'These few things having been considered, the whole matter is reduced to pure geometry, which is the one aim of physics and mechanics.' Similar pronouncements on the role of mathematics can be found in several texts preceding Newton's *Principia*, such as *De Arcanis Motus et Mechanica ad Puram Geometriam Reducenda* of 1676 and the correspondence with Henry Oldenburg, where Leibniz makes the reduction of mechanics to geometry dependent on the metaphysical principle of conservation of *vis viva*.[7]

After having read the *Principia* Leibniz would not take for granted the relationships between mathematical representations and natural philosophy and would sharpen his views on this problem. In a number of texts such as the *Tentamen, Specimen Dynamicum, Antibarbarus Physicus* and in all principal works up to the correspondence with Samuel Clarke, he stressed more forcefully the insufficiency of purely mathematical laws, the need for physical explanations and once again for metaphysical principles: 'Because we cannot derive all truths concerning corporeal things from logical and geometrical axioms alone, ... we must admit something metaphysical, something perceptible by the mind alone over and above that which is purely mathematical and subject to the imagination.'[8] Although Leibniz did not change his philosophical views

---

[5] *LSB*, II, *1*, pp. 262–8, on p. 267: 'D'autant que je croy me pouvoir satisfaire à présent sur les loix de mouvement, par des demonstrations entièrement geometriques, sans me servir de suppositions aucunes, ny des principes d'expérience; et que ce qu'on pourra dire la dessus doresnavant ne sera que *res calculi et geometriae*. Ainsi je tiens que nous sommes en estat à présent de prétendre à une physique véritable, et sans hypothèse.'

[6] *AE* July 1684, pp. 319–25 = *LMG*, *6*, pp. 106–12, on p. 112. Translation from C. Truesdell, *Essays in the History of Mechanics* (Berlin, 1968), pp. 181–2.

[7] 'De Arcanis Motus et Mechanica ad Puram Geometriam Reducenda', in Hess, 'Kurzcharakteristik', pp. 202–5; *NC*, *2*, pp. 57–75, on pp. 64 and 71, 17 Aug. 1676. Compare also Leibniz's 1686 *Discours de Métaphysique*, in G. Le Roy, ed., *Discours de Métaphysique et Correspondance avec Arnauld* (Paris, 1957), paragraphs 18 and 21.

[8] G. W. Leibniz, 'Specimen Dynamicum', *AE* April 1695, pp. 145–57; in *LMG*, *6*, pp. 234–46; a second part was first published in *LMG*, *6*, pp. 246–54. A critical edition was published in Hamburg, 1982 (pp. 22–4). Transl. by R. Ariew and D. Garber in G. W. Leibniz, *Philosophical Essays* (Indianapolis, 1989), pp. 117–38, on p. 125. 'Antibarbarus Physicus', *LPG*, *7*, pp. 336–44, on p. 343, dating from 1706 according to Müller and Krönert, *Chronik*, p. 203; transl. in Ariew and Garber, *Essays*, pp. 312–20, on pp. 318–19. Compare also the letter to Newton of 7 March 1693, *NC*, *3*, pp. 257–60.

as a result of his reading of the *Principia Mathematica*, the new emphasis on physical explanations led him to the deployment of Keplerian arguments in the *Tentamen*. Hence the Cassirer thesis needs to be set in a historical context where the theory of planetary motion occupies a central position.[9] In order to carry out this task it is helpful to examine Leibniz's notion of law of nature in greater detail. I refer to his expositions in several texts dating from the 1680s up to the *Essais de Théodicée* of 1710.

Leibniz believed that mathematical or logical principles alone are not sufficient to identify the laws of nature observed in our world, such as conservation of force, continuity, equality between cause and effect, the impact laws, and the law of inertia. Those principles narrow down the choice to a broad set of propositions satisfying the criterion of non-contradiction. A further selection is needed to move from a multiplicity of non-contradictory propositions to laws of nature. This second step cannot be carried out by mathematical or logical means, nor by the data of experience, which inevitably lack generality. In other words, laws of nature are not absolutely necessary or demonstrable from such principles as identity or non-contradiction, nor can they be established simply by induction. Observations and experiments can lead to a single statement about this or that phenomenon, namely to truths of fact. They can also instantiate and corroborate a general law by specific examples, yet they cannot prove or refute it. In general, every time that an experiment or the observation of a phenomenon seemed to contradict a law, Leibniz did not take the outcome at face value; rather, he tried to work out alternative explanations saving the generality of the law. An example of this strategy is briefly discussed in Section 2.4 in relation to impacts and conservation of force. If on the one hand laws are not demonstrable, on the other they are not arbitrary; their certainty is not similar to that of a mathematical proposition, but descends from the wisdom and perfection of God and from his choice for the best. Theological arguments via the principle of perfection and harmony, together with the agreement between predictions and data, guarantee the moral certainty of laws of nature. This justification for mechanical laws and the mathematization of nature was related to broader debates and concerns. At a time when mechanical explanations were becoming more and more successful, Leibniz reports that some theologians, 'shocked at the corpuscular philosophy', deny that all phenomena can be explained in mechanical terms. His own solution is that although mechanical explanations can be given to all phenomena, the laws of mechanics 'depend on more sublime principles which show the wisdom of the Author in the order and

[9] D. Bertoloni Meli, 'Some Aspects of the Interaction between Mathematics and Natural Philosophy in Leibniz', *The Leibniz Renaissance* (Florence, 1989), pp. 9–22.

perfection of his work.' This solution satisfied a double purpose, since it paved the way to a mechanical explanation of nature while overcoming theological and philosophical objections against those who destroy contingency and freedom, and subject everything to geometrical necessity and mathematical laws alone.[10] Like Kepler's true hypotheses, Leibniz's laws of nature emerge from a plurality of disciplines, go beyond a simple description of nature, and involve a causal and physical explanation of phenomena. Although they are often formulated in a mathematical fashion, they are not derived by mathematical means alone, but are selected from the broad set of non-contradictory statements with the help of a variety of disciplines.

The similarity between Keplerian and Leibnizian notions should not be treated as an identity, since the problems they addressed and the debates in which they were involved at the beginning and at the end of the century varied considerably. With regard to the issues we have considered thus far, it seems to me that Kepler paid more attention to the observational adequacy of a hypothesis, Leibniz focused on the logical status of laws. Similarly, Kepler's theological preoccupations concerned the order and structure of the universe, Leibniz's the necessity or contingency of propositions about nature. Further, the notion of harmony is strictly related to geometry for Kepler, whereas for Leibniz it involves more general principles such as combinatorial arguments, continuity, and conservation. In conclusion, even if several important differences should not be disregarded, true hypotheses and laws of nature show more than a superficial analogy: Leibniz certainly had ample ground for using Kepler in his response to the *Principia Mathematica*.

The discussion in this section is largely philosophical. The issues raised so far can be seen and hopefully clarified in a different context,

---

[10] G. W. Leibniz, *Théodicée* (Amsterdam, 1710); in *LPG, 6*, paragraphs 345–9, esp. 346, and 371, where he attacked Hobbes and Spinoza. Leibniz developed similar views much earlier: 'Elementa Physicae', early 1680s, Gerland, *Schriften*, pp. 110–13, on p. 111; 'Cogitationes de Physica Nova Instauranda', *Vorausedition zur Reihe VI—Philosophische Schriften*, Faszikel 3 (Münster, 1984), pp. 625–43; both texts are translated in L. E. Loemker (ed.), *G. W. Leibniz. Philosophical Papers and Letters* (Dordrecht: Reidel, second edition, 1969), pp. 277–80 and 280–9. 'Animadversiones in Partem Generalem Principiorum Cartesianorum', *LPG, 4*, pp. 354–92, on p. 391 (dated 1691–2); 'Tentamen *Anagogicum*', *LPG, 7*, pp. 270–9, on pp. 271–3, transl. in Loemker *Papers and Letters*, p. 478. This topic is related to a major area in Leibniz's philosophy, namely the notion of contingency. The literature on this topic is vast; with regard to the more specific problem of laws of nature see: M. D. Wilson, 'Leibniz's Dynamics and Contingency in Nature', in P. K. Machamer and R. G. Turnbull, eds., *Motion and Time, Space and Matter* (Ohio State University Press, 1976), pp. 264–89; H. Poser, 'Apriorismus der Prinzipien und Kontingenz der Naturgesetze. Das Leibniz-Paradigma der Naturwissenschaft', *SL* Sonderheft *13*, 1984, pp. 164–79; K. Okruhlik, 'The Status of Scientific Laws in the Leibnizian System', in K. Okruhlik and J. R. Brown, eds., *The Natural Philosophy of Leibniz* (Dordrecht, 1985), pp. 183–206.

paying attention to the historical circumstances of their emergence. This is the aim of the following section, which outlines Leibniz's deployment of Keplerian themes and in particular the account of Kepler's physical explanations in the *Tentamen*. The last section of this chapter contrasts Leibniz's and Newton's interests in astronomy and examines their respective competence and sources for the three laws of planetary motion. This task is made easier because over the last few decades the diffusion of Kepler's laws in the seventeenth century has been investigated by several commentators, especially in an attempt to trace the origin of Newton's ideas on mechanics and universal gravity. Thus while my itinerary develops from philosophical themes, to physical explanations, and lastly to astronomy, I constantly look at Kepler as the unifying element.

1.2   Kepler's framework and his physical explanations

Eric Aiton correctly pointed out that Leibniz's theory of planetary motion reflects the framework of Kepler's ideas with regard to both mathematics and physics. If in some cases one can talk of Kepler's influence on Leibniz, in others it is more correct to emphasize Leibniz's usage of Kepler for a variety of purposes. At the end of this section we are going to see several instances of this usage. Let us consider first the analysis of planetary trajectories. Kepler decomposed orbital motion into a circular motion of revolution around a centre and a libration. The former is due to a *species motrix immateriata* emitted from the Sun, similar to light and decreasing according to the law of the inverse distance;[11] the latter is composed of an *accessus* and *recessus* along the radius to the Sun and is due to magnetic attraction or repulsion. Kepler often talked of libration as taking place along a rigid rotating stick, or an arm of a *libra*, which is moved by the rotation of the Sun. The composition of these two motions produces planetary ellipses.[12]

In the *Tentamen* Leibniz decomposed orbital motion into a *circulatio harmonica* and a *motus paracentricus*. The former is a circular motion

[11] E. Aiton, *The Vortex Theory of Planetary Motion* (London and New York, 1972), ch. 6, esp. p. 128. J. Kepler, *Mysterium*, *KGW*, *1*, p. 71 and *KGW*, *8*, p. 113; *Astronomia Nova* (Prague, 1609) = *KGW*, *3*, pp. 239–40, 248–51; *Epitome*, *KGW*, *7*, pp. 298, 304–5, 333.

[12] *Astronomia Nova*, *KGW*, *3*, p. 237 and 376; *Epitome*, *KGW*, *7*, pp. 295–6, 301 (where Kepler explains that without attraction or repulsion the orbit would be circular), 332, 367–70 (compare the figure with the diagram from the *Tentamen*). Both texts were known to Leibniz: the *Epitome* is mentioned in the *Tentamen* (see below); the manuscript LH 35, 9, 2, f. 80, contains excerpts from the *Astronomia Nova*, pp. 252–6 (original), namely p. 182, lines 6–13; p. 184, lines 15–20; p. 186, lines 15–17. Compare also A. Koyré, *La Revolution Astronomique* (Paris, 1961), part 2.

around a centre in which the velocity of rotation is inversely proportional
to the radius; this condition is equivalent to the area law. The latter is the
radial motion towards or from a centre, due to gravity and centrifugal
force. Leibniz also talked of this component of motion as taking place
along a rigid rotating ruler. The combination of the two motions
generates planetary ellipses. The similarity between the two representations is striking. However, it should not be overlooked that for Kepler the
outward motion of a planet was due to magnetic repulsion, whereas for
Leibniz it was an effect of the planet's own orbital motion: no repulsion
was needed in his account.

In the introduction to his essay Leibniz reveals first-hand knowledge
of the *Epitome*: he refers almost verbatim to a passage where Kepler
discusses the place of the Earth in the Universe, and whether heavy
bodies tend towards the centre of the world. Kepler proposes the
following argument, in which he examines the case of circular violent
motion, only to refute it immediately afterwards. The reasoning is based
on Aristotle's physics. If some stalks or straws float in a vortex of water,
the water, being denser than the floating bodies, pushes them towards the
centre. In Kepler's account this mechanism cannot be valid for the
motion of the Earth, because the aether supposedly keeping the Earth in
its orbit is certainly more tenuous than the Earth and cannot counterbalance its tendency to fly away. Surprisingly, this passage is not in Book
4 of the *Epitome*, in the section on the causes of planetary motion where
Kepler explained his theory of the *species motrix*, but in Book 1. The
example of the bodies floating in water was to be analysed by several
philosophers in the seventeenth century, including Descartes and
Huygens. Further, Kepler compared the *species motrix* emitted from the
Sun with the rotation of a vortex both in the *Astronomia Nova* and
Epitome.[13]

In the *Tentamen* Leibniz praises Kepler for having first made known to
the mortals 'jura poli, rerumque fidem, legesque Deorum'.[14] In this
quotation from Claudian, which originally referred to Archimedes,
Leibniz significantly mentions the laws of the heavens, the order of
nature, the precepts of the Gods, namely three central areas of intersection with Kepler's philosophy. Among the reasons for this praise are
the explanation of gravity and the three laws of planetary motion: planets
move in ellipses where the Sun occupies one of the foci (first law); the
radius from the Sun to the planet sweeps out equal areas in equal times
(second law); the squares of the times of revolution are proportional to

---

[13] *KGW*, 7, p. 75–6, section 4 of Book 1, *De loco telluris in mundo, eiusque proportione
ad mundum*; quoted in Bertoloni Meli, 'Public Claims', p. 427, n. 22. *KGW*, 3, p. 34; *KGW*,
7, p. 299.
[14] Claudian, *Carminum Minorum Corpusculum*, LI: 'Jura poli rerumque fidem legesque
Deorum | ecce Syracusius transtulit arte senex.'

the third power of the major axes of planetary ellipses (third law).[15] Nevertheless, Leibniz claims that Kepler would not have found the causes of his discoveries, both because he still believed in angelic intelligences moving the planets and in the doctrine of sympathies, and because the most advanced mathematics and the science of motion had not progressed sufficiently in his time. However, following Leibniz's account, Kepler gave the first indication of the true cause of gravity with the example of the bodies floating in water, but he was hesitant and did not draw full consequences from it. As we have seen, Kepler was not hesitant at all and rejected altogether the explanation which Leibniz attributed to him.[16]

In the introduction to the *Tentamen* Leibniz also credited Kepler with the discovery that rotating bodies tend to escape along the tangent. He was certainly referring to the passage immediately following that on the vortex of water discussed above. Indeed, Kepler does say that rotating bodies tend to fly away along the tangent; if a body is kept on a rotating wheel and then is released, it will fly away in a straight line, as it is shown in the accompanying figure in the *Epitome*.[17] However, in the letter of 1677 to Fabri, Leibniz criticized both Kepler and Galileo for failing to realize that bodies moving along a circumference, regardless of its size, tend to escape along the tangent. It is well known that in some passages Kepler and Galileo considered motion in very large circles—of the size of planetary orbits—to be natural. Leibniz's statement reveals his understanding of what we call the principle of rectilinear inertia and his sympathetic reading of Kepler in the *Tentamen*.[18]

[15] The first two laws appeared in *Astronomia Nova*, *KGW*, *3*, p. 265 (first law) and 366 (second law); compare C. Wilson, 'Kepler's derivation of the elliptical path', *Isis*, *59*, 1968, pp. 5–25; E. J. Aiton, 'Kepler's Second Law of Planetary Motion', *Isis*, *60*, 1969, pp. 75–90. The third law appeared in the *Harmonice Mundi*, *KGW*, *6*, p. 302.

[16] Similar statements are in Leibniz, 'De Causa Gravitatis', *AE* May 1690, pp. 228–39; in *LMG*, *6*, pp. 193–203, on p. 195.

[17] *KGW* 7, p. 76: 'Ex adverso, solet motus violentus, horizonti parallelus, cum gravia corripuit, incitare illa, si soluta a rota fuerint et in lineam rectam a circumferentia circuli excutere.' Compare *Tentamen de Systemate Universi*, LH 35, 9, 9, f. 1–2:1r.: 'Hanc vim [Leibniz means centrifugal] primus quod sciam observavit Keplerus, notans rotae circulanti imposita rejici per tangentem circuli.' Throughout the seventeenth century, and for several decades in the eighteenth century, centrifugal force was not seen as a fictitious force arising in a non-inertial reference system, but as a real force due to the tendency of rotating bodies to escape along the tangent. D. Bertoloni Meli, 'The Relativization of Centrifugal Force', *Isis*, *81*, 1990, pp. 23–43.

[18] *LSB*, III, *2*, pp. 136–7 (Proposition 14). Leibniz mentioned the law of inertia already in the 1669 'De Rationibus Motus', *LSB*, VI, *2*, pp. 161–2. See also Koyré, *Revolution*, p. 215, n. 1, and *Études Galiléennes* (Paris, 1966). H. R. Bernstein, 'Passivity and Inertia in Leibniz's Dynamics', *SL*, *13*, 1981, pp. 97–113. In the 'Animadversiones' Leibniz attributes the law of nature according to which a body, '*quantum in se est*', persists in the same state, to Galileo and Pierre Gassendi, *LPG*, *4*, p. 372. I. B. Cohen, '*Quantum in se est*: Newton's Concept of Inertia in Relation to Descartes and Lucretius'. *NRRS*, *19*, 1964, pp. 131–55. For Gassendi's influence on the young Leibniz through his professor of mathematics at Jena Erhard Weigel compare K. Moll, *Der junge Leibniz*, vol. 2 (Stuttgart, 1982).

Another issue relevant to Leibniz's reading of Keplerian inertia and to his relations with Newton is worth discussing at this point. Although my analysis focuses on physical problems, one should not forget the existence also of a metaphysical dimension. On a small piece of paper pasted in his own copy of the second edition of the *Principia* (1713), Newton inserted an addition to his definition 3. The definition reads: 'The *vis insita*, or innate force of matter, is a power of resisting, by which every body, as much as in it lies, continues in its present state, whether it be of rest, or of moving uniformly forwards in a right line.' The addition follows the claim, a few lines after the passage just quoted, that the *vis insita* can be called *vis inertiae*, and states: 'I do not mean the Keplerian force of inertia by which bodies incline to rest, but a force maintaining them in the same state of rest or of motion.' I. Bernard Cohen successfully identified the addressee of this clarification as Leibniz, who refers to Keplerian inertia in paragraphs 30 and 380 of the *Théodicée*. Kepler defended a neo-Aristotelian view according to which bodies tend to lose their velocity because of their own inertia or resistance to motion, and force produces speed rather than acceleration.[19] Leibniz tried to clarify his reference to Keplerian inertia by the following passage, which led Newton and some modern interpreters to believe that the German philosopher shared with Kepler pre-Cartesian views about motion:[20]

Let us suppose that the current of one and the same river carried along with it various boats, which differ among themselves only in the cargo, some being laden with wood, others with stone, and some more, the others less. That being so, it will come about that the boats most heavily laden will go more slowly than the others, provided it be assumed that the wind or the oar, or some other similar means, assist them not at all.

The situation described by Leibniz is quite complex, and at first sight it is not immediately clear how it relates to the problem of inertia. Before tackling this issue, following Bernard Cohen I wish to notice the similarity between boats carried by the river and planets moved by a vortex. The reference to Kepler and the notion that motion is in some

[19] I. B. Cohen, 'Newton and Keplerian Inertia: an Echo of Newton's Controversy with Leibniz', in A. G. Debus, ed., *Science, Medicine and Society in the Renaissance*, 2 vols. (London, 1972), 2, pp. 199–211. *Isaac Newton's 'Philosophiae Naturalis Principia Mathematica', the Third Edition (1726) with Variant Readings*, ed. A. Koyré, I. B. Cohen and A. Whitman (Cambridge, 1972), pp. 40–1. Transl. by Motte and Cajori, p. 2. I. B. Cohen, *Newtonian Revolution*, sect. 4.5; 'Newton's Copy of Leibniz's *Théodicée*', *Isis*, 73, 1982, pp. 410–14. A. Gabbey, 'Force and Inertia in Seventeenth-Century Dynamics', *SHPS*, 2, 1971, pp. 1–67. Bernstein, 'Passivity'. Catherine Wilson, *Leibniz's metaphysics* (Manchester, 1989), p. 141. On the links of inertia or 'passivity' with metaphysical problems see G. Buchdahl, *Metaphysics and the Philosophy of Science* (Oxford, 1969), ch. 7.

[20] For the quotations from Leibniz, *Théodicée*, pars. 30 and 380, I use the transl. by E. M. Huggard, *Theodicy* (Routledge and Kegan Paul, London, 1952), pp. 140–1 and 353.

inverse proportion to the resistance or inertia may sound surprising to the reader. However, Leibniz frequently used terms and concepts from various traditions and reinterpreted them within his own philosophy. In this circumstance he clarified the difference between his own concept of inertia and Kepler's in the very paragraph 30 of the *Théodicée*, soon after the passage quoted above:

It is therefore matter itself which originally is inclined to slowness or privation of speed; not indeed of itself to lessen this speed, having once received it, since that would be action, but to moderate by its receptivity the effect of the impression when it is to receive it. Consequently, since more matter is moved by the same force of the current when the boat is more laden, it is necessary that it goes more slowly; and experiments on the impact of bodies, as well as reason, show that twice as much force must be employed to give equal speed to a body of the same matter but of twice the size.

Of course, here 'force' means living force or mass times the square of velocity, rather than Newtonian force. The last claim in the quotation is not very accurate, because 'twice the force' could be obtained by doubling the mass or the square of the velocity, and the result of these operations is not the same. From the context, however, it appears that Leibniz meant the former alternative. His interpretation of 'inertia' can be further grasped from paragraph 380 of the *Théodicée*, where he claims: 'Kepler, one of the most excellent of modern mathematicians, recognized a species of imperfection in matter, even when there is no irregular motion: he calls it its "natural inertia", which gives it a resistance to motion, whereby a greater mass receives less speed from one and the same force.' As Ernst Cassirer observed and as these quotations show, Leibniz's concept of inertia is defined in terms of impacts, and in impacts, if the living force of the impelling body is given, the velocity of the impelled body is 'moderated' by its own inertia. Leibnizian inertia is resistance to impressed motion rather than *vis insita* or the tendency to continue motion uniformly in a straight line. While illustrating the differences between Leibnizian and Newtonian notions, this discussion shows how Leibniz reinterpreted terminology and concepts from several traditions with a sense of philological accuracy that the modern reader may often find baffling.[21] He was very quick in finding prestigious

---

[21] Cassirer, *Leibniz' System*, p. 339. Leibniz, 'Discours de Métaphysique', in G. Le Roy, ed., paragraph 21; *Phoranomus*, in C. I. Gerhardt, 'Zu Leibniz' Dynamik', *Archiv für Geschichte der Philosophie*, *1*, 1888, pp. 566–81, on pp. 575–81, and Bertoloni Meli, 'Censorship', section 2; *LPG*, *7*, pp. 446–9, on p. 447, Leibniz to Antonio Alberti, 1690(?); *LPG*, *4*, pp. 464–7, 1691; *Specimen Dynamicum* (Hamburg, 1982), pp. 8–11; 'De Ipsa Natura', *AE* Sept. 1698, pp. 427–40, in *LPG*, *4*, pp. 504–16, on p. 510, where the reference to Keplerian inertia is first mentioned in print; *Leibniz–Clarke Correspondence*, ed. G. M. Alexander (Manchester, 1956), Leibniz's fifth paper, par. 102. See also the valuable edition by A. Robinet, *Correspondence Leibniz–Clarke* (Paris, 1957). Leibniz's fifth letter to Clarke, paragraph 102.

predecessors to his own ideas, even if, as in this case with Kepler, those predecessors seem to us to be defending rather different views.

Before concluding this section I wish to discuss a short essay first published by Gerhardt, the *Tentamen de Physicis Motuum Coelestium Rationibus*, where Leibniz gives an account of the development of astronomy in which Kepler appears in a different light. The circumstances of composition, I shall suggest, induced Leibniz to write yet another different history. Since the manuscript is on Italian paper, we can be sure that the essay was composed in 1689 or shortly afterwards, therefore after the published *Tentamen*.[22] The *Tentamen de Physicis Motuum Coelestium Rationibus* is a retrospective gloss, an account of the events as Leibniz wanted to present them when his essay on planetary motion had already appeared in the *Acta*. The first striking feature is that Kepler is no longer mentioned in the context of the vortex theory. In his place Leibniz mentions Galileo, Torricelli, and Descartes, who was previously accused of having plagiarized Kepler's vortices.[23] These three 'restauratores philosophiae', as Leibniz calls them, were not able to explain from which kind of vortical motion the laws of planetary motion arose. On the other hand Kepler, allegedly having compared the accuracy of several explanations of phenomena, found that planets move in ellipses where the Sun occupies one of the foci, and that the areas swept out by the radii from the Sun to the planet are proportional to the times. The third law, which Leibniz had never been able to include satisfactorily in his theory, is ignored. Leibniz claims that Ismael Boulliau, Seth Ward, and others who followed Kepler tried to explain the area law by means of several devices, such as the composition of various circles, but it was not clear whence the proportion involving the areas could be generated. Hence he presents two separate and complementary traditions: on the one hand there are those providing physical causes without being able to account for observations; on the other hand there are those saving the phenomena without being able to give physical explanations. These two traditions have to be brought together, and this is the role Leibniz assigns himself with the discovery of the harmonic circulation, which accounts for the motion of the vortex carrying the planets and for the area law.

---

[22] *LBG*, pp. 612–13. Gerhardt erroneously identified this essay with a work mentioned by Leibniz in a letter intended for Huygens of October 1690. Compare *HOC*, 9, p. 528, n. 16. G. W. Leibniz, *Vorausedition zur Reihe VI—Philosophische Schriften—*, Faszikel 8 (Münster, 1989), pp. 1751–3.

[23] Galileo is mentioned, though with some reservations, also in the *Tentamen*. He referred to a subtle celestial aether in *Il Saggiatore* (Roma, 1623); in *GOF*, 6, p. 317. The reference to Torricelli is identical to that in the *Tentamen* and is taken from Balthasar de Monconys, *Journal des Voyages* (Lyons, 1665–6), pp. 130–1, first part. Leibniz neglects a host of believers in vortex theories; in Britain, for example, they were adopted by Vincent Wing and Thomas Streete. A detailed study of this area is in C. Donahue, *The Dissolution of the Celestial Spheres* (New York, 1981).

Although Kepler is confined to the role of a mathematical astronomer, Leibniz's dialectical ploy is typical of the German astronomer. It is conceivable that Leibniz assigned a different role to Kepler because his works were in the index in Italy, where the essay was composed.

1.3   Astronomy and the reception of Kepler's laws

After having surveyed some philosophical and physical themes intersecting Kepler's and Leibniz's writings, we move to an area which turned out to be of central importance in the Newton–Leibniz dispute, namely astronomy. In fact, the incorporation of Kepler's laws into their theories became a decisive factor in the course of the controversy.

Judging from the available sources, Leibniz's interest and competence in astronomy seem to have been weak. I am not aware of any observations he made, nor of any extensive and accurate study of the astronomical literature. Among his manuscripts at the Niedersächsische Landesbibliothek catalogued under the heading 'Astronomica', one finds a rather thin corpus of excerpts and remarks, the most extensive being from Robert Hooke, *Animadversiones on the first part of the Machina Coelestis* (London, 1674). Elsewhere among his papers I traced a manuscript containing excerpts largely based on Giandomenico Cassini's observations and bearing the title 'Observationes Astronomicae Novissimae'. The text contains references to telescopes, double and multiple stars, sunspots, comets, the Earth and the Moon, the planets including Jupiter with its satellites, and Roemer's observations on the speed of light, Saturn and its three satellites known before 1684. Some notes which have been added later refer to mean distances and apparent diameters of planets as reported by Newton in the *Principia*.[24]

By contrast, Newton is known to have studied major astronomical works and to have made observations. If the reflecting telescope he presented at the Royal Society at the end of 1671 was more related to his optical researches, other sources point to an unambiguous interest in astronomy. We know for certain that he studied and annotated the monumental compendium by Vincent Wing, *Astronomia Britannica* (London, 1669), and Thomas Streete, *Astronomia Carolina* (London, 1661). In his correspondence with the Astronomer Royal John Flamsteed, Newton revealed his competence on several occasions both in cometography and lunar theory, a topic about which the eminent mathematician and astronomer Nicolaus Mercator had already praised

---

[24] LH 35, 15, 6, 'Astronomica', esp. f. 10–16. LH 35, 10, 1, f. 4. Watermark 1794 in the catalogue at the NLB, 'crown, symbol' and year '1680', datable from 1680 to 1687.

him in 1676.[25] Indeed, as Curtis Wilson has convincingly argued, Mercator is Newton's most likely source for Kepler's second law. The German had publicly attacked the newly appointed Royal Astronomer in Paris, Giandomenico Cassini, precisely on the issue of the exact formulation of the area law. The first and third laws were more widely known in the seventeenth century and are clearly expressed in Streete, *Astronomia Carolina*. Newton employed the third law in a manuscript of the late 1660s, in the attempt to estimate the dependence on distance of a planet's centrifugal force.[26]

In the three main sources referred to by Leibniz which I have mentioned so far, namely the *Epitome* and the works by Boulliau and Ward, the first law is clearly stated; the second, however, requires a more careful study.[27]

In the manuscript draft of the *Tentamen* there is a marginal note, later crossed out, which alters paragraph 6 to read:[28]

With the radii drawn from the centre of the circulation [planets and satellites] describe areas proportional to the times, which Kepler discovered first, then Boulliau in the *Astronomia Philolaica* (rather *Kepleriana*) and Seth Ward beautifully explained and respectfully approved.

Interestingly Leibniz refers to the same authors mentioned in the *Tentamen de Physicis Motuum Coelestium Rationibus*, namely Boulliau and Ward. They were among the leading astronomers around the middle of the century, and their controversy which I am about to outline was widely known in Europe. The reference to them concerning the second

---

[25] Westfall, *Never at Rest*, pp. 232–7 and 258. D. T. Whiteside, 'Newton's Early Thoughts on Planetary Motion: a Fresh Look', *BJHS, 2*, 1964, pp. 117–37. J. E. McGuire and M. Tamny, 'Newton's Astronomical Apprenticeship: Notes of 1664/5', *Isis*, 1985, *76*, pp. 349–65. N. Mercator, *Institutiones Astronomicae* (London, 1676), p. 286: 'Harum . . . Librationum causas Hypothesi elegantissima explicavit nobis vir Cl. Isaac Newton', quoted in Edleston, *Correspondence*, p. li, n. 49.

[26] N. Mercator, 'Some considerations concerning the geometrick and direct method of Signior Cassini for finding the Apogees, Excentricities, and Anomalies of the Planets, as that was printed in the Journal des Sçavants of Septemb. 2 1669', *PT, 5*, 1670, pp. 1168–75. CUL, Ms Add 3958(5), f. 87, in Herivel, *Background*, pp. 192–8. On the reception of Kepler's laws in the seventeenth century see: J. L. Russell, 'Kepler's Laws of Planetary Motion', *BJHS, 2*, 1964, pp. 1–24; Whiteside, 'Early Thoughts', esp. pp. 122–4 on Newton's sources. V. E. Thoren, 'Kepler's Second Law in England', *BJHS, 7*, 1974, pp. 243–56; C. A. Wilson, 'From Kepler's Laws, so Called, to Universal Gravitation: Empirical Factors', *AHES, 6*, 1970, pp. 89–170, esp. pp. 129–34 and 141–2. C. A. Wilson, 'Horrocks, Harmonies, and the Exactitude of Kepler's Third Law', *Studia Copernicana, 15*, 1978, *Science and History, Essays in Honor of Edward Rosen*, pp. 235–59.

[27] I. Boulliau, *Astronomia Philolaica* (Paris, 1645), p. 4; S. Ward, *Astronomia Geometrica* (London, 1656), p. 1.

[28] LH 35, 9, 2, f. 56–9; f. 57r.: 'Radiis enim ex centro circulationis ductis, areas describunt temporibus proportionales [(quod primus invenit Keplerus, deinde Bullialdus in Astronomia Philolaica (potius Kepleriana) et Sethus Wardus pulchre illustrarunt, et observantes comprobaberunt)]'. I enclose the text of the marginal note in square brackets.

law is very puzzling. In *Astronomia Philolaica* Boulliau did not accept the area law, but rather claimed that the elliptical motion of the planet takes place along the surface of a cone: the points on the axis of the cone are those of mean motion, or with respect to which angles are proportional to times. Boulliau explicitly denied that his theory involved a *punctum equans*, since he believed that the points on the axis of his cone constituted an alternative to equant theories.[29] A few years later Seth Ward pointed out that the intersection of the plane of the planetary ellipse with the axis of Boulliau's cone is indeed an equant point of the ellipse. This observation highlighted a serious contradiction in Boulliau's theory, whose presupposition was to use only circular uniform motions: Ward had shown that Boulliau's theory was equivalent to the theories the French astronomer had criticized. In *Astronomia Geometrica* Ward adopted again the hypothesis that the second focus of planetary ellipses is an equant point.[30] In a later version of his theory, Boulliau adopted a modified form which saved the phenomena better, but was in contradiction with the assumptions of the *Astronomia Philolaica*; despite the better agreement between theory and observations, Boulliau's construction was different from the area law. His was one of the many approximations used at that time as a computational device.[31] In the passage from the manuscript of the *Tentamen* quoted above, Leibniz draws no distinction between Kepler's, Boulliau's, and Ward's versions of the second law. It is not clear whether Leibniz crossed it out for this reason.

[29] *Astronomia Philolaica*, p. 26, point 13, and p. 286. The best account of Boulliau's positions and of the reception of Kepler's laws is in Wilson, 'Empirical factors'. The equant point has the property that arcs measured with respect to it are proportional to the times.

[30] S. Ward, *In Ismaele Bullialdi Astronomiae Philolaicae Fundamenta, Inquisitio Brevis* (Oxford, 1653), p. 3; *Astronomia Geometrica*, p. 1. In a letter to Leibniz written ten years after Newton's book, Jakob Bernoulli raises the problem of the difference between Kepler's second law and Ward's equant theory, pointing out that these were not equivalent and that although Newton used the former it was by no means certain which was the correct one; *LMG*, *3*, p. 50, 27 Jan. 1697.

[31] *Astronomiae Philolaicae Fundamenta Clarius Explicata et Asserta* (Paris, 1657), p. 37 (clear summary); see C. A. Wilson, 'Empirical factors', pp. 119–21. Compare also I. Newton, *Principia*, scholium to proposition 31, and D. T. Whiteside, 'Early Thoughts', pp. 122–4, n. 25. Y. Maeyama, *Hypothesen zur Planetentheorie des 17. Jahrhunderts*, Doktorarbeit, Frankfurt a.M., 1971; 'The Historical Development of Solar Theories in the Late Sixteenth and Seventeenth Centuries', *Vistas in Astronomy*, *16*, 1974, pp. 35–60. V. E. Thoren, 'Second law', p. 245. O. Gingerich and B. Welther, 'Notes on Flamsteed's Lunar Tables', *BJHS*, *7*, 1974, pp. 257–8. See also the letter by Henry Oldenburg to Leibniz of 5 Aug. 1676, with James Gregory's calculations on Kepler's problem: *LSB*, III, *1*, pp. 520–2, Leibniz's excerpts from James Gregory's papers, ib. pp. 499–503 and Newton's *epistola prior*, *NC*, *2*, pp. 110–61, on p. 126, 24 Oct. 1676. Kepler's problem, stated in *Astronomia Nova*, *KGW* 3, p. 381, consists of finding the division of the area of a semicircle in a given ratio by a line drawn from a point on the diameter to the circumference. This is strictly related to the area law.

In the introduction to the *Tentamen* Leibniz refers to the third law with respect to planets and to the satellites of Jupiter and Saturn. The reference to the satellites of Saturn is omitted in the *editio princeps* of the *Principia*, and cannot have been taken from Newton.[32] Indeed, I traced a manuscript fragment in Leibniz's hand referring to Cassini's article on the *Journal des Sçavants* of 22 April 1686, where the Italian astronomer announced the discovery of two more satellites of Saturn—fourth and fifth—and clearly stated that they obey Kepler's third law.[33]

These observations indicate that around 1688 Leibniz probably knew Kepler's first and third laws; concerning the second law, although this is correctly formulated and applied in the *Tentamen*, I am less confident of Leibniz's appreciation of the different formulations he quotes. Evidence of his full understanding of the matter before the *Tentamen* is lacking, and even afterwards his references to Boulliau and Ward are puzzling. As far as I can tell from my reading, until 1688 Leibniz had not paid attention to Kepler's laws, both with respect to their mathematical formulation and physical interpretation. I believe that before reading Newton's *Principia*, his understanding of their relation to the inverse-square law and central forces, or harmonic circulation, was non-existent. In a letter of 1679 Leibniz stated that the astronomy of Descartes—a telling denomination—was basically the same as that of Copernicus and Kepler, to which the French added a better explanation of vortices.[34] This statement suggests that at least around 1679 Leibniz did not consider the three laws to be a fundamental part of Kepler's heritage. Within the two traditions identified in *Tentamen de Physicis Motuum*

[32] *NMW*, 6, p. 40, n. 26. In the first edition Newton omitted this reference after consultations with Flamsteed, who could not detect the satellites; *NC*, 2, pp. 403–5, on p. 405, 27 Dec. 1684.

[33] LH 35, 9, 5, f. 29: 'Notat Cassinus diar. eruditorum 22 April. 1686 ab omnibus satellitibus tam Saturni quam Solis observari legem a Keplero praescriptam rationis temporum periodicorum est distantiarum nempe ut sint t^2 ut d^3.' Leibniz added some computations to check the validity of Kepler's law. Compare *Journal des Sçavants*, 1686, pp. 139–54:149 (Amsterdam). See also Leibniz's excerpts from *Harmonice Mundi* (Linz, 1619), in LH 35, 10, 1, f. 9, watermark 488 in the catalogue at the NLB, 'crossed keys' and letters 'F M', used by Leibniz in Vienna in 1688. The excerpts are from pp. 193, 195–6, 187–90. The sketch on the verso representing planetary distances from the Sun reproduces the figure facing page 186 of the *Harmonice*. The relevant portion of the text excerpted by Leibniz reads: 'Proportio quae est inter binorum quorumcumque planetarum tempora periodica est praecise sesquialtera proportione mediarum distantiarum.' The text was written in pencil and subsequently inked over.

[34] For a different opinion about Leibniz's reading of Boulliau and Ward see E. J. Aiton, 'The mathematical basis of Leibniz's theory of planetary motion', *SL* Sonderheft *13*, 1984, pp. 208–25, on p. 219. Compare also LH 35, 8, 30, f. 69, *Keplerus plurimus aliorum inventis principium et occasionem dedit*: 'Dixerat Keplerus ex foco superiore tanquam centro motum planetae fere aequabilem apparet.' The date of composition of this manuscript is not known. *LSB*, II, *1*, pp. 499–504, on p. 503. The addressee is unknown. Compare also *LPG*, *4*, pp. 343–4 and 348, datable around 1683–6.

Coelestium Rationibus, in this area he concentrated on physical accounts in terms of vortices rather than on mathematical attempts to save the phenomena. We shall see in Section 5.4 that in a manuscript of autumn 1688 Leibniz explicitly attributed the generalization of the area law to Newton, a clear indication that his own harmonic circulation had not yet been conceived.[35]

---

[35] In the 'Leibniz-Nachlass' I traced Leibniz's own copy of the *Epitome* with his marginal annotations (Leibniz Marg. 97: Frankofurt, 1635) and a set of excerpts which closely correspond to the underlinings in the book; LH 35, 15, 6, f. 28–9: *Kepleri Loca Sunt Multa Meae de Corporis Constitutione Philosophiae Consentanea. Tantum Quaedam ex Epitome Astronomiae Excerpere Placet.* As the title suggests, Leibniz is interested in the theory of matter and excerpts several passages on inertia (compare the commentary on his notes to the *Principia,* Definition 3). The excerpts from Kepler are followed by other excerpts from books printed in 1696 or after; among these are: the Latin edition of Richard Bentley's sermons, *Stultitia et Irrationalitas Atheismi,* transl. by D. E. Jablonski (Berlin, 1696), and Edward Stillingfleet, *Answer to Mr Locke's Second Letter* (London, 1698). After Huygens's death in 1695 Leibniz acquired part of his library, and in particular a copy of the *Epitome* which may well be the same he excerpted some time afterwards. See H.-J. Hess, 'Bücher aus dem Besitz von Christiaan Huygens (1629–1695) in der Niedersächsischen Landesbibliothek Hannover', *SL, 12,* 1980, pp. 1–51, on p. 10. I believe that marginalia and excerpts date from the late 1690s.

2

VORTICES AND FLUIDS: FROM GRAVITY TO ELASTICITY

2.1 Introduction

In the second half of the seventeenth century natural philosophy relied heavily on Cartesian explanations of phenomena in terms of the motion of particles of different shape and size. Even if the specific mechanism devised by Descartes was questioned, vortices and subtle fluids were the cornerstone for an intelligible explanation of the universe. The following two sections examine some aspects of the vortex theories framed by Descartes and Huygens in order to gain some familiarity with this material. The examples I have selected were well known to Leibniz and Newton, and were important in the formation of their views; we shall find several references to them in Leibniz's manuscripts.

During my surveys of the 'Leibniz-Nachlass' I traced a set of excerpts from the third part of Descartes' *Principia Philosophiae, De mundo adspectabili.* They cover the four manuscript sides of an entire folded sheet and one side of a separate half sheet. Luckily the separate sheet has a watermark with a crown, a bell, and the year '1678'. Leibniz used this kind of paper especially in the early 1680s.[1] Descartes is mentioned in the introduction to the *Tentamen*, and there is no more obvious a source than him for a vortex theory; there is no doubt that Leibniz knew his work well before the 1680s. However, these manuscripts are important because they testify to Leibniz's interest in the causes of celestial motion in those years, and help us to establish a link between two areas which appeared thus far to be separate, namely motion in a resisting medium and celestial motion. Further, these excerpts show how influential the *Principia Philosophiae* was as late as the early 1680s. In the following section I discuss Leibniz's excerpts from Descartes' *Principia* and conclude with some observations on the mathematization of nature.

Leibniz's sources certainly went beyond excerpts and annotations. A more direct form of communication is represented by the conversations he had with mathematicians and philosophers, especially during his stay

[1] LH 35, 15, 6, f. 25–7; the watermark corresponds to number 1765 in the catalogue at the NLB.

in Paris. Among the astronomers discussed in Chapter 1, for example, Leibniz referred to his talks with his friend Ismael Boulliau.[2] Moreover, he probably discussed several issues with his mentor Christiaan Huygens, to whom Section 2.3 is devoted. I examine in particular the theory of centrifugal force with the application of evolutes, and the explanation of the cause of gravity. Although the relevant portions of Huygens's theories were not available in print for many years after Leibniz left Paris, it is likely that they were discussed by them before 1676. The section ends with some observations on the relationship between Descartes and Huygens.

In the last section I discuss the role and development of the notion of elasticity in Leibniz's system, paying attention to its function in the relationships between mathematics, natural philosophy, and metaphysical principles. Gravity and elasticity were often related by Leibniz, and at one stage he even considered them to be identical. While providing a broader picture of the mathematization of nature in the seventeenth century and of the Leibnizian universe, these reflections will prove useful in the interpretation of some passages discussed below.

2.2   The excerpts from Descartes' *Principia Philosophiae*

The *Principia Philosophiae* consists of four parts on the principles of human knowledge, the principles of material things, the visible world, and the Earth, respectively. The excerpts I am about to discuss do not appear to be part of a larger corpus; it seems that Leibniz focused from the start on the third part, dealing with the objects of sensible experience, or the system of the world.[3] Descartes gives a description of the universe, its constitution and dimensions, and the types of matter filling the heavens. His qualitative account explains the origin of the world, of its elements, and of vortices carrying planets and comets.[4] Leibniz's excerpts cover most of the propositions 1–133. Before examining some of them in detail, I wish to mention proposition 16 on the inadequacies of the Ptolemaic hypothesis, proposition 30, where the word 'vortex' is

[2] *LMG*, *3*, p. 944, Leibniz to Johann Bernoulli, 1715; *Nouveaux Essais*, *LSB*, VI, 6, p. 489. See also Müller and Krönert, *Chronik*, p. 42.

[3] A collection of articles on several aspects of Descartes' work is in S. Gaukroger, ed., *Descartes: Philosophy, Mathematics and Physics* (Brighton, 1981). On Descartes' mechanics compare Westfall, *Force*, ch. 2; Dugas, *Mécanique*, ch. 7. On Descartes and Leibniz see the classic Y. Belaval, *Leibniz critique de Descartes* (Paris, 1960).

[4] At the end of proposition 46 Leibniz inserts the following note, f. 25v.: '(+ Leucippus apud Laertium vocabat Δίνην quanquam mihi ut dicam quod res est satius videatur quamlibet partem rursus concipere ut exiguum vorticem, eorumque innumeros statuere varietates in infinitum. +)' The reference is to Diogenes Laertius, IX, 31 and is repeated in the introduction to the *Tentamen*.

first mentioned and where Descartes refers to the example of the straws
floating in a whirlpool similar to the passage from Kepler's *Epitome*
which we have seen above, and proposition 52 on the three forms of
matter or elements, among which the first forms the Sun and the fixed
stars, the second fills the heavens, and the third constitutes planets and
comets. Proposition 82 is particularly interesting from our perspective
because Descartes provides a qualitative account of the motion of the
vortex and of the planets floating in it: he states that the particles of the
second element move swifter the further away from the centre of the
vortex, apart from a certain region around the centre containing the
planets where they follow the opposite law. No attempt is made,
however, to retrieve the precise quantitative formulations given by
Kepler.[5]

Notoriously, the ingredients of the Cartesian account are size, figure,
and motion of the particles filling up the universe. It is very helpful to
consider in detail the mechanism devised by Descartes in at least one
case. Although the relevant propositions have been excerpted but not
commented upon by Leibniz, they occupy an important position in the
present work. Proposition 120 treats the motion of a body in a vortex
having clear recourse to two opposite actions, gravity and a tendency to
escape along the tangent.[6] There and in the ensuing propositions
Descartes explains the details of the mechanism he has devised. A body's
tendency to fly off along the tangent depends on its surface, because if
this is larger, the body is pushed by a greater number of particles of the
vortex and acquires a larger velocity to fly away along the tangent.
Gravity, however, depends on the volume of the body, because only the
particles of the vortex which occupy the space left when the body moves
must be taken into account. For Descartes only matter of the third
element forming the body must be included while considering its
soliditas—which I translate as 'solidity'—because matter of the first
element moves through it freely without contributing to its motion.[7] All

---

[5] Leibniz's excerpt reads, f. 25v.–26r.: 'Quod ad globulos secundi elementi vorticem
componentes attinet. Eae excepto interiore quodam vorticis nucleo, concipiendae sint ut
aequales, tales enim initio supponimus, nec magna ratio apparet diversitatis, motus autem
remotiorum a centro celerior est, ita ut paucis forte in casibus circuitum absolvunt, nam
(licet vi suae circulationis circa vorticem, aequalem et mediocrem motum initio habuisse
intelligantur unde tardius est circulatio in circulis a centro valde remotis) . . .' Marked in
the margin with 'NB'.

[6] Leibniz's excerpt reads, f. 26v.: 'Quod ut accurate intelligatur, considerandum est
astrum in vorticem a quo abripitur delatum, ibi obstare globulis vorticis, inferioribus, qui a
centro recedere conantur vi circulationis, ideo ab ipsis depellitur versus centrum. Sed cum
interim simul cum vortice circulari incipiat, vim nanciscitur recedendi a centro, qui duo
conatus pugnant, et praevalet validior.' Compare also *Principia Philosophiae*, proposition
140, Part III, and propositions 20–7, Part IV, on gravity.

[7] Ib., Leibniz's rephrased excerpts read: 'Vis autem qua depellitur, oritur a quantitate et
celeritate materiae coelestis seu numero globulorum eius locum occupare et a centro

these considerations can be expressed in compact form by saying that a
body's tendency to move away from or towards the centre depends on
its solidity: the greater the solidity, the stronger the tendency to move
away from the centre, and vice versa. Propositions 121–125 define and
explain the notion of solidity; as we have seen, Descartes means the
quantity of matter of the third element of a body, in conjunction with its
volume and surface.[8]

The interest of the analysis of orbital motion in proposition 120 lies
also in the correlation established with the theory of motion in a resisting
medium: the inward tendency depends on the volume of a body; the
outward tendency on its surface. In Leibniz's excerpt the 'vis qua
depellitur', namely the inward tendency, depends on the number of
globes of the vortex which can occupy the volume of the body; the 'vis
qua circulatur', though, depends on the number of globes rubbing its
surface. An analogous reasoning is applied to proposition 126, where
Descartes explains how a comet begins to move assuming that particles
or globes of the second element filling the heavens are less solid than the
comet, which is carried by them. Leibniz writes three separate comments
to this proposition, the third of which is the most interesting:[9]

There is also this difficulty about the motion of a comet, that more globes can fill
its space than rub its surface. Therefore, the force of the pressing down globes is
greater; the greater the number of globes that rub its surface, the less the solidity
of the comet, and the weaker the force of recession from the centre, in spite of it

---

recedere conantium ducto in velocitatem eorum. Vis qua circulatur oritur a quantitate
superficiei seu numero globulorum radentium et secum circumagentium. Ipsa tamen
celeritas quam accipit astrum ducenda est in eius corpus seu soliditatem, a qua defungi
debet quicquid ei non cohaeret, ut materia primi vel secundi elementi in poris eius fluens
quippe quae aliorsum cursum suum flectit, nec agitationem impressam cum astro
conservat, aut potius determinationem potius motum, quam novum motum accessit.' This
quotation is discussed below in the text.

[8] Proposition 121 in the *Principia Philosophiae* states: 'Per soliditatem hic intelligo
quantitatem materiae tertii elementi, ex qua maculae hoc sidus involventes componuntur,
cum eius mole et superficie comparata.' We can also say that solidity depends on the
density of the matter of the third element in a body and on its surface. On some occasions
Leibniz seems to take 'solidity' as a synonym for 'quantity of matter', whereas at other
times he understands it in a proper Cartesian sense. Compare the letter to Claude Perrault
of 1676, *LSB*, II, *1*, pp. 262–8, on p. 264: '. . . celuy qui est plus solide ou qui contient plus
de materie . . .'

[9] Ib., f. 26v.: '(+ Est et haec circa cometae motum difficultas, quod plures globi locum
eius enplere quam superficiem radere possunt. Ergo major vis deprimentium, item quo
plures superficiem eius radunt, eo minus habet soliditatis; eoque minus vim habet a centro
recedendi, cum tamen a numero globorum superficiem radentium vim centrifugam habere
dicatur +)'. The previous two comments in ib., f. 26v., read: '(+ Difficultas quoad
cometam quod globuli vorticis extra nucleum suppositi sunt aequales. +)' '(+ Cum
cometa ad centrum accedat in spirali, fieri potest ut tandem impetu concepto perget in
tangente spiralis. +)'

being said that centrifugal force arises from the number of particles rubbing the surface.

Leibniz seems to take solidity to be proportional to the ratio between volume and surface. In this passage he finds some inconsistencies in the Cartesian account. However, despite some differences and criticisms, his analysis follows the main lines drawn by Descartes: orbital motion results from the imbalance between gravity and the tendency to recede from the centre, which Huygens named centrifugal force. These opposing tendencies are explained in terms of friction and density; thus the theory emerging from Leibniz's excerpts concerns orbital motion and is based on some concepts of the theory of motion in a resisting medium, two areas which so far seemed to be unrelated. In *Schediasma de Resistentia Medii* Leibniz introduced two types of resistance, called *absoluta* and *respectiva*:[10] '*Absolute resistance* is that which absorbs as much of the force of the moving body, whether it is moved with a large or small velocity, while it is moved, and depends on the *viscosity of the medium* . . . yet it is not of consequence what may be the velocity of striking.' Although each individual particle of fluid produces the same effect regardless of the speed with which its surface rubs the body, the overall effect of absolute resistance depends on the number of particles encountered and hence on velocity. For similar reasons the overall effect of respective resistance depends on the square of velocity, while the effect of each individual particle depends on simple velocity: '*Respective resistance* arises from the density of the medium, and is greater according as the velocity of the moving body is greater.' Leibniz clarifies his distinction further:

There is also this difference between the two kinds of resistance, that the absolute has relation in a certain manner to the surface of the moving body or of contact, the respective however to the solidity.

In this passage the reference to the *Principia Philosophiae* is very clear. In the *Schediasma* Leibniz was implicitly trying to transform the Cartesian qualitative account into a quantitative formulation. This mathematization blurred the link between the theories of motion in resisting media and celestial motions, because they had to be developed following contrasting requirements. Although in the *Tentamen* planetary motion was treated independently of the ideas in the *Schediasma*, for Leibniz the two theories were related in their common Cartesian origin. In Section 5.5 we shall find that a manuscript of 1688 on planetary

[10] Leibniz, 'Schediasma de Resistentia Medii', *LMG*, 6, p. 135–47, on p. 136. The following translations are from E. J. Aiton, 'Leibniz on motion in a resisting medium', *AHES*, 9, 1972, pp. 257–76, on pp. 260–1.

motion suggests that Leibniz was still trying to explain gravity and centrifugal force in terms of absolute and respective resistance.

Theories based on subtle fluids and Cartesian vortices proved extraordinarily influential on Newton too. As Tom Whiteside has shown, while studying Vincent Wing, *Astronomia Britannica*, Newton calculated that the terrestrial vortex is compressed by the solar vortex 'by about a 43rd of its width'. Whiteside has emphasized that 'it is still not realized how much such images coloured and structured his thinking on astronomy during the next fifteen years', thus approximately until the mid-1680s. As the *Géométrie* of Descartes 'had been Newton's entrance into higher mathematics in the late summer of 1664, so a few months later it was the beacon of Descartes' *Principia Philosophiae* that lit its way' to a world of vortices and to the analysis of curvilinear motion.[11] The Cartesian account of orbital motion based on the imbalance between opposing tendencies, for example, was closely examined by Newton. In *De gravitatione et aequipondio fluidorum*, of approximately 1670, he tried to refute the Cartesian theory of relativity of motion by arguing that the outward tendency was due to a real, as opposed to a relative, circulation. Referring to proposition 140 of the *Principia Philosophiae*, Part III, Newton paraphrased the Cartesian text: 'Later he attributes to the Earth and Planets a tendency to recede from the Sun as from a centre about which they are revolved, by which they are balanced at their [due] distances from the Sun by a similar tendency of the gyrating vortex.' A similar imbalance theory was also adopted by Giovanni Alfonso Borelli in his study of the Medicean Planets, *Theoricae Mediceorum Planetarum*; his book was owned and studied by Newton. Despite a certain similarity between Borelli's and Leibniz's theories, evidence that Leibniz studied the *Theoricae* before developing his own theory is lacking.[12]

It is worth pausing here to consider the implications of what has been said so far on the issue of the mathematization of nature. Cartesian and neo-Cartesian world-views were at the same time the springs driving forward the process of mathematization and the chains obstructing it.

[11] V. Wing, *Astronomia Britannica* (London, 1669), Trinity College Library, NQ.18.36. D. T. Whiteside, *The Preliminary manuscripts for Isaac Newton's 1687 'Principia' 1684–1686*, (Cambridge, 1989), p. x, n. 19–20.

[12] A. R. Hall and M. B. Hall, *Unpublished scientific papers of Isaac Newton*, (Cambridge, 1962), p. 124; Westfall, *Force*, ch. 7. Newton's later physical explanations of gravity are outlined in Chapter 8. G. A. Borelli, *Theoricae Mediceorum Planetarum ex Causis Physicis Deductae*, (Florence, 1666), discussed in Koyré, *Revolution*, section 3. For Newton's copy of the *Theoricae* see J. Harrison, *The library of Isaac Newton*, (Cambridge, 1978), with dog-earings. Borelli is also mentioned on p. 403 of the *Principia Mathematica*, first edition. On proposition 140 of *Principia Philosophiae* see Herivel, *Background*, p. 59, n. 4. A copy of Borelli's work with marginal annotations by Huygens is at the NLB (classmark 'Nm. A 104'); it was acquired by Leibniz after Huygens's death in 1965; see Hess, 'Bücher'.

This dichotomy was already present in Descartes, who conceived the universe in geometrical terms. His views seemed to pave the way for a mathematical description of the world; however, he was prevented from proceeding in this direction by the bewildering complexity resulting from huge numbers of colliding particles. His reaction to Galileo's analysis of parabolic trajectories in *Two New Sciences* is emblematic: Descartes despised Galileo's mathematical treatment not because it was fallacious, but because the void does not exist in nature. Therefore parabolic trajectories were merely an easy mathematical exercise inapplicable to the world of phenomena and unrelated to clear physical explanations. The fact that in the preface to the second edition of Newton's *Principia* Roger Cotes referred in a sarcastic tone to this problem shows that by 1713 the issues raised by Descartes' criticism of Galileo were still very much alive:[13]

Galileo has shown that when a stone projected moves in a parabola, its deflection into that curve from its rectilinear path is occasioned by the gravity of the stone towards the earth, that is, by an occult quality. But now somebody, more cunning than he, may come to explain the cause after this manner. He will suppose a certain subtle matter, not discernible by our sight, our touch, or any other of our senses, which fills the spaces which are near and contiguous to the surface of the earth, and that this matter is carried with different directions, and various, and often contrary, motions, describing parabolic curves. Then see how easily he may account for the deflection of the stone above spoken of. The stone, says he, floats in this subtle fluid, and following its motion, can't choose but describe the same figure. But the fluid moves in parabolic curves, and therefore the stone must move in a parabola, of course. Would not the acuteness of this philosopher be thought very extraordinary, who could deduce the appearances of Nature from mechanical causes, matter and motion, so clearly that the meanest man may understand it?

It is clear that in this passage Cotes had Leibniz in mind, since his attack paraphrases portions of the *Tentamen*. Despite his scornful irony, however, mechanical explanations and subtle fluids had proved valuable resources in the second half of the seventeenth century, especially when they were handled by such talented mathematicians as Christiaan Huygens.

2.3 Huygens: centrifugal force and the cause of gravity

The material in the present section is not based on a rich and relatively

---

[13] Newton, *Principia*, transl. by Motte and Cajori, p. xxix. W. Shea, 'Descartes as critic of Galileo', in R. E. Butts and J. C. Pitt, eds., *New perspectives on Galileo* (Dordrecht, 1978), pp.139–59.

unproblematic source, as in the previous section with the excerpts from Descartes. My attempts to trace Huygens's influence on Leibniz's theory are more conjectural and depend, at least in part, on some hints which can be found in the texts of 1688. By the time Leibniz arrived in the French capital in 1672, the Dutch mathematician was an established leading figure at the Académie Royale des Sciences. Following his discovery of the ring and main satellite of Saturn, as well as the invention of the pendulum clock in the 1650s, Huygens had become an authority in matters mathematical among Parisian circles. Three points deserve attention for the role they take in the rest of my work: the function of evolvents in the theory of centrifugal force; the belief that gravity depends on the centrifugal force of the vortex, and that the particles of the vortex differ only in velocity from common bodies on the Earth; the claim that vertical descent can be explained if the particles of the vortex move in all directions around the Earth. While presenting an account of Huygens's theory of gravity and centrifugal force, I concentrate on these aspects.

The recent work by Joella Yoder has convincingly reconstructed Huygens's intellectual itinerary to the *Horologium Oscillatorium*, showing its internal cohesion to a degree that was previously unrecognized. Since much of his production was probably available to Leibniz regardless of its date of publication, I shall treat the relevant works together. Stimulated by Marin Mersenne to determine the distance traversed by a body in the first second of its free fall—we would say the constant or gravitational attraction—Huygens undertook the investigation of an action which he believed was analogous to gravity, namely centrifugal force. The underlying idea was that of transforming a problem of motion into one of equilibrium. A heavy body attached to a rotating cord—the conical pendulum—was the simple tool employed for this transformation. If the cord has an inclination of 45 degrees to the vertical, gravity and centrifugal force are equal and one can be used to measure the other. Thus starting from gravity Huygens was led to the study of centrifugal force and the conical pendulum. By 15 November he had found that in the first second a body falls 15 6/10 feet. On 1 December 1659 he asked himself 'what ratio does the time of a very small oscillation of a pendulum have to the time of perpendicular fall through the height of the pendulum'. While working on this problem, Huygens unexpectedly found a more general result. Originally he had restricted his investigations to very small oscillations, since the period of the simple pendulum depends on the amplitude of the oscillations. While examining this problem, Huygens produced an intricate network of geometrical relations concealing a most precious result. Reviewing his proof, he determined a condition which would render his solution exact for

arbitrary arcs, or oscillations, not only for very small ones. This
condition identified the cycloid and proved that cycloidal oscillations are
isochronous. The question now was how to produce such oscillations.
With yet another brilliant result, Huygens found that if the thread is
constrained between two cycloidal arcs, the oscillations are also along a
cycloid and the pendulum is rigorously isochronous. The rolling and
unrolling of the thread on the constraining arcs led to the more general
mathematical problem of determining such related curves.[14] The curve
ABC, produced by the unrolling of the thread, was called evolvent or
involute, whilst the curve *AD* of the constraining arc was called evolute.

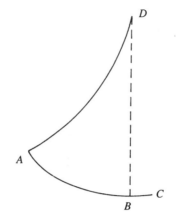

Fig. 2.1   Evolutes and evolvents in the study of pendular oscillations.

These studies were closely related to the problem of rectification of
curves. Lastly, the need to experiment with real pendulums differing
considerably from the abstract ones, where the thread has no weight and
the bob is considered to be a point, led Huygens to the study of centres
of oscillations and to the compound pendulum. The greatest part of
these researches was published in the *Horologium Oscillatorium* of 1673,
a work which exerted a considerable influence on Leibniz: the copy he
received 'ex dono autoris' in the year of publication is preserved in
Hanover.[15]

Although Huygens's first publication on gravity, the *Discours de la*

   [14] J. G. Yoder, *Unrolling Time. Christiaan Huygens and the Mathematization of Nature*
(Cambridge, 1989); quotation from p. 50.
   [15] C. Huygens, *Horologium Oscillatorium*, (Paris, 1673); in *HOC, 18*. English transl. by
R. J. Blackwell, *The Pendulum Clock*, (Iowa State University Press, 1986). Leibniz's copy
is at the NLB, classmark 'Leibn. Marg. 70'. A modern presentation of evolutes and
evolvents is in G. Loria, *Curve Piane Speciali Algebriche e Trascendenti*, (Milano, 1930), 2
vols.; *2*, p. 281f. J. Joder, *Unrolling time*, (Cambridge, 1988).

Cause de la Pesanteur, appeared in 1690 as an appendix to the *Traité de la Lumiere*, his ideas on the matter had been formulated long before, and had been presented at a debate on the cause of gravity at the Paris Academy in 1669.[16] Like Descartes, Huygens wanted to explain gravity in terms of the size and figure of particles in motion. Unlike his French predecessor, however, he assumed that all bodies are made of the same kind of matter and, crucially, introduced a quantitative analysis. His mathematics was largely based on a skilful use of geometry and proportions. Considering that a body moving along a circumference has a tendency to escape along the tangent and therefore away from the centre, he proved in the unpublished *De Vi Centrifuga* of 1659 that this tendency is proportional to the square of the velocity of rotation over the radius.

An observer placed at the centre *A* of a rotating wheel sees a body attached to the wheel and rotating with it along *BEFM*. If the body is released, it will move along the tangent *BS*. In the time it takes to travel along *BE*, *BF* and *BM*, it will reach *K*, *L*, and *N* respectively, because the circular and rectilinear motions are uniform. Rigorously, *EK*, *FL*, and *MN* are arcs of the evolvents of the circumference *BEFM*, namely arcs of

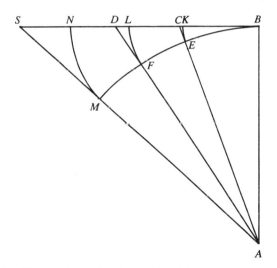

Fig. 2.2   Huygens's study of centrifugal force in circular motion.

[16]   C. Huygens, *Traité de la Lumiere*, (Leiden, 1690) = *HOC*, *19*, pp. 451–537; *Discours*, ib., *21*, pp. 427–99. For the debate at the Academy see *HOC*, *19*, pp. 625–45, and Aiton, *Vortex theory*, ch. 4. On Huygens compare: H. J. M. Bos, M. J. S. Rudwick, H. A. M. Snelders, R. P. W. Visser, eds., *Studies on Christiaan Huygens*, (Lisse, 1980). A. D'Elia, *Christiaan Huygens*, (Milano, 1985); Dugas, *Mécanique*, ch. 10; Westfall, *Force*, ch. 4. Joder, *Unrolling time*. H. J. M. Bos, 'The Influence of Huygens on the Formation of Leibniz's Ideas', *SL* Supplementa *17*, 1978, pp. 59–68.

the same curve occurring in the study of the isochronous pendulum. The arcs *BM*, *BF*, and *BE* can be conceived to unroll up to the tangent *BS* and represent the trajectories of the body seen by the rotating observer placed in *A*. Notice that this relativistic fashion of introducing the problem has to be interpreted as a rhetorical ploy inserted in order to make centrifugal force more easily understandable, rather than as a statement implying that centrifugal force is fictitious. For Huygens centrifugal force was real and depended on the circular motion of the body as opposed to the circular motion of the frame of reference. The arcs of the evolvents can be approximated by their tangents in *E*, *F*, and *M*, that is, *EC*, *FD*, and *MS* respectively, and their lengths can be calculated by means of simple geometry, on the assumption that the arc *BE* is very small. In conclusion, centrifugal force is proportional to EB^2:AB, and is represented by *CE*, with a constant factor $1/2$ because motion along *CE* is uniformly accelerated. Further, Huygens proved that setting the centrifugal force of a body moving along a circumference equal to the gravity retaining it in the orbit, its period must be equal to that of a pendulum whose length is half the radius. This allowed him to compare gravity and centrifugal force.[17] Before the publication of the theorems about centrifugal force in the *Horologium Oscillatorium*— where they appeared without proof—only Isaac Newton had been able to find similar results in a series of manuscripts dating from the 1660s.[18]

For Huygens the force of gravity acting on a body originated from the centrifugal force of a rotation fluid surrounding the body and was equal to the difference between the centrifugal force of the fluid and that of the body. Huygens believed that the Earth is surrounded by a fluid; on the assumption that matter is uniform and that a piece of lead on the Earth does not differ at all from the matter of the fluid, he calculated that its velocity of rotation must be 17 times swifter than that of a point on the equator. In order to explain why bodies tend to fall towards the centre of the Earth rather than towards the axis of rotation of the vortex, Huygens supposed that the particles of the fluid move in all directions around the Earth. A few years later, after Jean Richer's observations on the diminution of the length of the seconds pendulum at Cayenne, and after having read Newton's *Principia*, Huygens was able to account for the decrease in gravity from the pole to the equator on the basis of this

[17] 'De Vi Centrifuga', composed in 1659, was published posthumously (Leiden 1703); *HOC*, *16*, pp. 255–311, on pp. 260–1; see my 'Relativization', esp. sect. 1.

[18] Newton's manuscripts are: Waste Book, ULC, Add 4004, discussed by Herivel, *Background*, pp. 7–13, 45–8, 127–42; Vellum Manuscript, ULC, Add 3958, f. 45, discussed in Herivel, *Background*, pp. 145–50, 184–6; ULC, Add 3958(5), f. 87 and 89, discussed in Herivel, *Background*, pp. 192–8. Compare also *NMW*, *6*, p. 38, n. 22. The theorems on centrifugal force in the *Horologium*, *HOC*, *18*, pp. 366–8, are formulated using proportions.

theory; if the Earth is a perfect sphere the decrease is measured by a factor of $1/17^2$, as he explains in the *Discours*. Whereas Descartes explained gravity in terms of a difference in solidity, or density of the matter of the third element and surface of the body, Huygens initially employed only a difference in velocity. After the appearance of the *Principia Mathematica* he surmised that the particles of the aether could consist of a different and lighter matter with respect to planets; this would account for the apparent lack of resistance to the motion of planets and comets. Notice, however, that if the density of the aether changes, its velocity of rotation must vary accordingly in order to account for gravity. Huygens believed that the aether was necessary to explain gravity and the propagation of light.[19]

This observation leads to some general considerations on the relationships with Descartes, an important theme in the historiography on that period and especially in Leibnizian and Newtonian studies. In a passage dating from 1693, Huygens describes his change in attitude to Descartes from his first reading of *Principia Philosophiae*, aged fifteen or sixteen, to maturity:[20]

When I read the book of Principles the first time, it seemed to me that everything proceeded perfectly; and when I found some difficulty, I believed it was my fault in not fully understanding his thought. I was only fifteen or sixteen years old. But since then, having discovered in it from time to time things that are obviously false and others that are very improbable, I have rid myself entirely of the prepossession I had conceived, and I now find almost nothing in all his physics that I can accept as true, nor in his metaphysics and his meteorology.

Seen from our perspective 300 years later, Huygens's itinerary looks different. He appears to have moved away from Descartes on specific issues, such as the impact laws, while accepting his framework of general physical explanations. This is probably one of the reasons why Huygens's work appears to contain so little philosophical elaboration: the basic ingredients he needed for his great achievements in the mathematical disciplines and especially the very reasons for the mathematization of nature could be found in *Principia Philosophiae* and were well known to his contemporaries.

By contrast, Leibniz had much deeper philosophical preoccupations: his own reasons for the mathematization of nature resemble Descartes's and Huygens's only in part and on the level of phenomena. Leibniz had nothing to object to the *Horologium Oscillatorium*. Unlike the *Principia Mathematica*, Huygens's masterpiece did not challenge any Leibnizian

---

[19] *HOC, 21*, pp. 462–6 and *Addition*, pp. 466–88, on p. 473.
[20] *HOC, 10*, p. 403, transl. in Westfall, *Force*, p. 185, and quoted in R. S. Westman, 'Huygens and the problem of Cartesianism', in Bos et al., *Studies*, pp. 83–103, on p. 99.

philosophical tenet. However, Leibniz's own vision of nature in relation to mathematics and philosophy was extraordinarily more complex. In addition to mechanical explanations, he considered final causes, which gave rise to maxima and minima principles. In the 1682 *Unicum Opticae, Catoptricae et Dioptricae Principium* Leibniz criticized the Cartesians for rejecting finality in nature, and claimed that his principle, whereby light travels through the easiest path, served a double purpose: it allows the reduction of optical phenomena to geometry and the calculus, and it reveals the order and harmony of nature, and hence of its Creator. Further, he emphasized more strongly than ever before the conservation principle of living force. Take, for example, the case of colliding bodies. Although Huygens and Leibniz provided mathematically equivalent accounts of elastic impacts, their respective understanding of the phenomenon varied considerably. Huygens focused on conservation of momentum, whilst Leibniz's metaphysical commitments led him to emphasize conservation of living force. As we are about to see, he generalized this claim for all types of impact and phenomena. At an even deeper level, Leibniz conceived phenomena as the manifestation of metaphysical entities to which they were related in a way that recent scholarship has found highly problematic. We are going to see one aspect of this problem in the following section.[21]

2.4   The role of elasticity

I conclude this chapter with a brief discussion of the notion of elasticity. Not many years ago Herbert Breger emphatically stressed the importance of this notion occupying a nodal position in Leibniz's philosophy among metaphysical principles and laws of mechanics, phenomena and their mathematical representations.

From the middle of the seventeenth century elasticity began to take a progressively more important position in the investigations of nature. The idea that under certain conditions matter is capable of resuming its original shape, when deformed, emerged from several areas and had important applications. My immediate aim, before outlining the development of Leibniz's ideas on this subject, is to identify the main areas in which elasticity played a significant role rather than to summarize the main results. In acoustics the elasticity of vibrating strings

[21] G. W. Leibniz, *Unicum Opticae, Catoptricae et Dioptricae Principium*, *AE* June 1682, pp. 185–90. M. Gueroult, *Dynamique et Métaphysique Leibniziennes* (Paris, 1934; reprinted 1967), pp. 215–35; Buchdahl, *Metaphysics*, ch. 7; the collection of essays in *SL*, Sonderheft *13*, 1984, contains ample references to these themes. Bertoloni Meli, 'Some aspects'. Concerning Leibniz and the impact laws see the following section.

and of the air were discussed in such influential works as Marin Mersenne's *Harmonicorum Libri* and Galileo's *Two New Sciences*. Book II of *Harmonicorum Libri* contains a collection of observations on vibrating bodies. They include the statement that the vibrations of a string are isochronous regardless of the way the string is plucked and of the vibrations' decrease with time. Galileo's study of vibrating strings at the end of the first day of *Two New Sciences* is followed by an investigation of the resistance of materials on the second day. His work on the loaded beam led to further investigations by Leibniz in the 1684 *De Resistentia Solidorum*. Leibniz criticized Galileo's assumption that the beam is perfectly rigid; instead, he considered the beam as composed of elastic fibres acting as springs, so that the beam gives way considerably before it can be ruptured.[22] Another area related to the bodies' capacity to regain their original shape was that of collision. In the second part of *Principia Philosophiae* Descartes provided seven rules governing the impact of perfectly hard bodies. Although they were repeatedly criticized in the second half of the century, they set the agenda for the study of elastic and inelastic impacts. Huygens studied the problem of collision in the 1650s and devoted the essay *De Motu corporum ex percussione* of 1656 to this problem. Newton also studied the collision of bodies according to their *vis elastica* in several passages of the *Waste Book*. These investigations remained unpublished at the time, and it was only at the end of the 1660s that John Wallis, Christopher Wren, and Huygens published their studies, though only partially in the case of Huygens. Although their results were regarded to be in mutual agreement, the specific formulations provided by each mathematician varied. On the whole, however, impacts were classified according to the elasticity and hardness of matter.[23] A further relevant field of inquiry was the study of the elastic properties of air in several works on the barometer and the air-pump around the middle of the century. The

[22] M. Mersenne, *Harmonicorum Libri* (Paris, 1636; revised edition, 1648), esp. book 2, prop. 29; *Traité de l'Harmonie Universelle* (Paris, 1636-7). G. Galilei, *Discorsi e dimostrazioni matematiche intorno a due nuove scienze* (Leyden, 1638); in *GOF*, *8*, first and second day. On the mathematical theory of elasticity see K. Stiegler, 'Einige Probleme der Elastizitätstheorie im 17. Jahrhundert', *Janus*, *56*, 1969, pp. 107–22; C. Truesdell, *The rational mechanics of flexible or elastic bodies. 1638-1788* in L. Euler, *Opera Omnia*, Leipzig etc., 1912-ser. II, vol. *11*, pp. 28–64. S. Dostrovsky, 'Early vibration theory: Physics and Music in the seventeenth century', *AHES*, *14*, 1975, pp. 169–218.

[23] J. Wallis, 'A Summary Account of the General Laws of Motion', *PT*, *3*, 11 Jan. 1669, p. 864–6; C. Wren, 'Lex Naturae de Collisione Corporum', ib., pp. 867–8, Huygens, 'Règles du Mouvement dans la Rencontre des Corps', *Journal des Sçavants*, 18 March 1669; *PT*, *4*, 12 May 1669, pp. 925–8 and *HOC*, *16*, pp. 1–186. J. Wallis, *Mechanica sive De Motu Tractatus Geometricus*, (London, 1670-1), ch. 11 (*De Percussione*) and 13 (*De Elatere*). A. R. Hall, 'Mechanics and the Royal Society, 1668-70', *BJHS*, *3*, 1966, pp. 24–38. Westfall, *Force*, chs. 4 and 5. Herivel, *Background*, pp. 1–6 and 128–82, esp. pp. 133 and 142. J. A. Bennett, *The mathematical science of Christopher Wren* (Cambridge, 1982), pp. 71–3.

debates centred around Robert Boyle's experiments with the air-pump, the notion of 'spring of the air', and the prolonged controversy with the Jesuit Franciscus Linus and with Thomas Hobbes, contributed to projecting elasticity on to the centre-stage of philosophical debates. The most famous result of Boyle's work was the formulation of 'his' law in the 1662, 'A defence of the doctrine touching the spring and weight of the air'.[24]

At the time of Leibniz's stay in Paris, Huygens applied the empirically known isochronism of the vibrating string to horology. This application was based on the principle that if *incitation* is proportional to the displacement from equilibrium, oscillations are isochronous. *Incitation* was defined as 'the force acting on a body to set it in motion when it is at rest, or to increase or decrease its speed when it is in motion.' This seems to be one of the few areas in which the interesting notion of *incitation* was employed by Huygens. His reflections on it seem to have had no precedent in seventeenth-century mechanics, and were not greatly developed even by their author; as many other of his studies, they remained unpublished. The generalization to springs of the findings about harmonic oscillations led him to the invention of a watch regulated by a coiled spring. Interestingly, in a 1691 letter to Huygens, Leibniz observed that Newton had not treated elasticity in the *Principia*, adding that he could remember Huygens having told him of having demonstrated the isochronism of vibrations. In his reply Huygens confirmed Leibniz's recollection.[25] Lastly, in *De Potentia Restitutiva* (London, 1678), Robert Hooke formulated the law '*ut tensio sic vis*; That is, the power of any spring is in the same proportion with the tension thereof.' This survey provides convincing evidence that elasticity had already gained a prominent position within the conceptual and practical horizon of scholars in mechanics well before Hooke's celebrated work.[26]

Leibniz's reception of the ideas on elasticity dates from the early

[24] See also R. Boyle, *Nova Experimenta Physico-Mechanica de Vi Aeris Elastica*, (Oxford, 1661); T. Hobbes, *Problemata Physica*, (London, 1662); O. von Guericke, *Neue (sogenannte) Magdeburger Versuche über den leeren Raum*, (Amsterdam, 1672; reprinted and translated by H. Schimank, Düsseldorf, 1968). C. Webster, 'The discovery of Boyle's law, and the concept of the elasticity of the air in the seventeenth century', *AHES*, 2, 1965, pp. 441–502. J. Agassi, 'Who discovered Boyle's law?', *SHPS*, 8, 1977, pp. 189–250. S. Shapin and S. Schaffer, *Leviathan and the air-pump* (Princeton, 1985).

[25] *LMG*, 2, pp. 85 and 88. *LSB*, III, 1, pp. 181–216, esp. p. 206, March 1675. M. Mahoney, 'Christiaan Huygens: The measurement of time and longitude at sea', in Bos et al., *Studies*, pp. 234–70, esp. pp. 254–5. A. Gabbey, 'Huygens and mechanism', in ib., pp. 166–99, esp. pp. 176–7. Westfall, *Force*, ch. 4, esp. pp. 177–81 and 184. *HOC*, 18, pp. 479–498, esp. pp. 483–4. On Newton and elasticity see *Principia*, book I, proposition 38.

[26] R. Hooke, *Lectures de potentia restitutiva*, is reprinted in R. T. Gunther, *Early science in Oxford*, vol. 8, (Oxford, 1931), pp. 331–88, on p. 333. E. Williams, 'Hooke's law and the concept of elastic limit', *AS*, 12, 1956, pp. 74–83.

1670s and follows his reading of the works by Robert Boyle, Thomas Hobbes, and Otto von Guericke on the air-pump, and by Christopher Wren, John Wallis, Huygens, and Edme Mariotte on elastic impact. In the *Hypothesis Physica Nova* of 1671 Leibniz explained gravity and elasticity in terms of the interaction between the aether and matter. He claimed that elastic phenomena could be studied in a physicomathematical fashion by a new branch of mixed mathematics with optics, music, and statics. This belief was to lead to several quantitative studies in later years. From the *Hypothesis Physica Nova* onwards elasticity became a central theme for him. Although its role with respect to mathematics, natural philosophy, and metaphysics changed over the years, elasticity constantly kept a prominent position.[27]

In *Propositiones Quaedam Physicae* of 1672 Leibniz assumed that the whole universe is elastic and that elasticity and gravity differ only in name. A body lifted in the air would be heavy because the elasticity of the universe would tend to restore the original position of equilibrium through a series of impacts. Even though this radical solution was later modified, elasticity and gravity remained correlated because they were both the result of particles in motion and impacts.[28] Thus it is not surprising that several years later this connection reappeared in a different form. Commenting on proposition 10, book I of the *Principia*— where Newton proves that if a body moves along an ellipse the centripetal force towards the centre is proportional to distance—Leibniz immediately noticed the analogy between gravity and elasticity:[29]

[I question] whether it may be in agreement with the nature of things, that the further a body is from a centre, so the more strongly it tends to it, and is to be regarded almost an elastic substance [*elastrum*] receding more and more from its natural state. Indeed, this operation of magnetic attraction would be contrary to

[27] Huygens's paper in the *Philosophical Transactions* was excerpted by Leibniz in 1669, *LSB*, VI, *2*, pp. 157–9. Compare also the introduction, pp. xxxi–xxxiii. E. Mariotte, *Traité de la Percussion*, (Paris, 1673). G. W. Leibniz, *Hypothesis Physica Nova*, (Mainz, 1671); in *LSB*, VI, *2*, pp. 219–76; esp. pp. 225, 227, 229–31, 254–5. Hess, 'Kurzcharakteristik', pp. 211–17, 'Calculus Elasticus'; H. Breger in 'Elastizität als Strukturprinzip der Materie bei Leibniz', *SL* Sonderheft *13*, 1984, pp. 112–21, mentions many unpublished manuscripts on elasticity.

[28] *LSB*, VI, *3*, pp. 4–72, on p. 38. Compare the correspondence with Otto von Guericke of 1671–2 in *LSB*, II, *1*, and F. Krafft, *Otto von Guericke*, (Darmstadt, 1978), pp. 30–7; *LSB*, III, *2*, p. 134, Leibniz to Fabri, 1677. It is worth noticing that Otto von Guericke was led to his investigations by his reflections on planetary motion and on the Copernican system, which occupy the first portion of his work.

[29] *Marginalia*, M 48; transl. in Bertoloni Meli, 'Public Claims', p. 444, n. 51. Here Leibniz calls Newtonian central attraction a magnetic operation; see also paragraph 9 of the *Tentamen*. Concerning the notion of 'elastrum' compare T. Hobbes, *Problemata Physica* (London, 1662) = *Opera Philosophica Latina*, ed. W. Molesworth, 5. vols., (London, 1839–45); vol. 4, p. 335: 'Per elastrum intelligo partium internarum conatum restituendi se ad situm, a quo per tensionem abductae fuerat.'

such a notion, which would be strengthened if gravity arose not from the attraction of the central body but from the impulse of a vortex.

In the letter to Edme Mariotte of July 1673 Leibniz identified the cause of all impact laws in the 'grande principe du Ressort'. As Breger noticed, in the reception of the theories by Wallis and Mariotte one finds the source of the link between the principle of conservation of force and elasticity: the return of an elastic body to its original state after an impact or any other external action is intrinsically related to the idea of conservation. The law of continuity too is closely related to the elasticity of matter, because in the impact between two bodies elasticity guarantees that their change in velocity and direction is not instantaneous. The colliding bodies are compressed and deformed like two inflated balls, and later rebound with no sudden transitions.[30]

Together with the 1684 *De Resistentia Solidorum*, several manuscripts, only in part published, testify to Leibniz's mathematical treatment of elasticity since the 1670s. Acoustic phenomena in particular occupied an interesting position as a fertile area for the mathematical investigation of elasticity, a topic explored by Leibniz with the help of differential equations. Moreover, the study of sound had important conceptual implications derived from the vibration of rigid bodies. In *De Resistentia Solidorum* he claimed: 'that there is nothing so rigid but that it is bent a little by the lightest stroke follows from the nature of sound, which is a certain trembling or reciprocal bending of the parts of the sounding body. The more rigid and indiscernible is the restitution, the higher is the sound, since the tremulous parts are the shorter and the tenser, and they constitute the harder body.' Thus acoustic phenomena provided evidence for the elastic properties of matter.[31]

In his mature views Leibniz considered all matter of the universe as elastic, solid bodies as well as fluids and the aether. In spite of its pertaining to all bodies, elasticity was not treated as a primitive notion but was explicable in mechanical terms by means of subtle fluids, as in the example of the collision between two inflated balls. There are no bodies which are perfectly hard or perfectly fluid, but only different degrees of hardness and fluidity. Leibniz often had recourse to elasticity in order to reconcile his metaphysical principles with the mechanical explanation of specific phenomena, as we have seen in the case of the principle of continuity. In the *Essay de Dynamique*, for example, he considered those impacts where living force, or mass times the square of velocity, appears

---

[30] *LSB*, III, *1*, p. 102. See also *Specimen Dynamicum*, (Hamburg, 1982), pp. 44–9, in Loemker, *Papers and Letters*, pp. 446–8, and Ariew and Garber, *Essays*, pp. 132–3.

[31] Leibniz, *De resistentia solidorum*, transl. in Truesdell, *Flexible bodies*, p. 63. Gerland, ed., *Schriften*, 'De Vibrationibus Aeris Tensi', pp. 31–37, contains several differential equations (see also pp. 10–15).

not to be conserved, in contradiction to his most famous principle. According to his explanation, that portion of living force which seems to be lost is in fact absorbed by the little parts constituting the colliding bodies. Their elasticity guarantees the principle of conservation of force. Moreover, by taking into account the motion of the constituent parts of a body, Leibniz could claim that nothing in the universe is truly at rest. Living force pertains to all bodies and is a manifestation of their inner activity. Elasticity is also employed in the metaphysical analysis of impact, which led Leibniz to deny that bodies truly affect each other. In his view, not only the total living force is conserved in impact, but also the living force of each body taken in isolation. Thus any change in velocity would result from a corresponding change in the internal motion of a body: the sum of the living forces due to the global motion and the motion of the internal parts of a body is constant. These surprising views highlight Leibniz's metaphysical preoccupations and in particular his reluctance to accept that primitive substances influence each other; in his views each substance evolves according to its own internal principle. Such metaphysical beliefs, however, were only partially related to physical problems.[32]

In conclusion, all matter is composed of elastic fluids which in their turn are composed of particles and fluids: in the Leibnizian universe elasticity is a structural principle of matter. If elasticity was discussed by several philosophers and mathematicians in the late seventeenth century, no one gave it a more important and universal role in his system than Leibniz:[33]

I hold all the bodies of the universe to be elastic, not though in themselves, but because of the fluids flowing between them, which on the other hand consist of elastic parts, and this state of affairs proceeds *in infinitum*.

---

[32] Leibniz's solution to the metaphysical problem of impact involves the notion of pre-established harmony. Compare L. Couturat, *Opuscules et fragments inédits de Leibniz* (Paris, 1903), pp. 518–23, transl. in Loemker, *Papers and Letters*, esp. p. 269; Ariew and Garber, *Essay*, p. 33. 'De ipsa natura', *AE* Sept. 1698, pp. 427–40 = *LPG*, 4, pp. 504–16, in Loemker, *Papers*, p. 506; Ariew and Garber, *Essays*, pp. 165 and 254–5.

[33] *LMG, 3*, p. 81, Leibniz to Jakob Bernoulli, 3 Dec. 1703: 'Corpora omnia universi puto Elastica esse, non quidem per se, sed ob fuida interlabentia, quae rursus tamen partibus Elasticis constant, atque ea res procedit in infinitum.' Ib., pp. 536, 544–5, 616, Leibniz to Johann Bernoulli, 1698–99. A well-documented account of elasticity is in H. Stammel, *Der Kraftbegriff in Leibniz' Physik*, Doktorarbeit, (Mannheim, 1984), pp. 326–31. 'Essay de Dynamique', *LMG, 6*, pp. 218–19 and 228–31; Breger, 'Elastizität', pp. 120–21; *LPG, 4*, (dated 1702), pp. 397–8.

3

GEOMETRY AND THE CALCULUS

3.1 Introduction

In the seventeenth century mathematics underwent profound transformations affecting methods of demonstration and objects of investigation as well as the very notions of rigour and proof. The invention of the
calculus—namely of a method for finding tangents and quadratures
which identifies the reciprocity between these two operations—was the
culmination of a process involving several important advances. The
establishment of a new, highly abstract, and general form of algebra, the
formulation of analytic geometry, and the creation of a variety of
techniques for finding maxima, minima, and tangents, paved the way to
the great inventions by Newton and Leibniz. Their calculuses can be seen
also as the starting point of a new phase of mathematical research.
Especially on the Continent, within a few years of Lebniz's first publication in 1684, several practitioners were making the theory of differential
equations the main area of advanced research of the time. In order to
appreciate the problems and subtleties of Leibniz's reaction to the
Principia and of his own theory of planetary motion it is helpful to gain
familiarity with some basic mathematical techniques of the late seventeenth century. As in the previous chapters, my aims here are very
selective and are largely dictated by the material presented in Part 2.[1]
This chapter also has an additional aim. I intend to show that the
mathematical formulation was not a 'neutral' tool and that the technique
cannot be easily separated from the physical or philosophical doctrine
because, far from being merely instrumental, the art of mathematical
representation interacted with the reflection on nature. On the basis of
the interpretation provided here, Chapter 4 will establish a correlation
between mathematics and mechanics, showing how fruitful it is to study
them in conjunction. This claim can be understood in several ways. The
cycloid, for example, was initially studied as a mathematical curiosity and
later found its applications in mechanics with Huygens's pendulum clock.
There is also a different sense in which my claim can be understood, and

---

[1] More general studies can be found in D. T. Whiteside, 'Patterns of mathematical
thought in the late seventeenth century', *AHES*, *1*, 1961, pp. 179–338; C. Boyer, *A History
of Mathematics* (New York, 1968).

this is the interpretation I am mainly concerned with. At least from the time of Galileo onwards, the transition from statics to the new science of motion was interwoven with the analysis of infinitely large and small quantities, as the debates on the force of percussion or on the 'infinite tardiness' of a body at the beginning of motion convincingly show. Thus the interplay between mathematics and mechanics involved results as well as the very conceptual bases of the disciplines, such as the notions of curve and differential, velocity and acceleration.

The challenging task of presenting an account of some central features of Leibnizian mathematics is made easier by the complementary works of Joseph Hofmann and Henk Bos. On the basis of previously unpublished manuscripts, Hofmann has reconstructed Leibniz's growth to mathematical maturity and the steps leading to the formulation of the differential calculus. Bos has studied the conceptual subtleties of his calculus, showing that this elegant and powerful construction must be studied in its own terms and in a seventeenth century context with regard to concepts, techniques, and notation. My account relies heavily on their works.

In the following section I introduce some early elementary results in infinitesimal geometry which are among the most typical in Leibniz's mathematics, namely the characteristic triangle and transmutation theorem. Section 3.3 discusses the relations between the notions of curve conceived as an infinitangular polygon, and of differential as a variable ranging over a sequence of values. I show the links between these geometrical and algebraic entities and outline the notion of curvature and its main properties. Lastly, Section 3.4 examines the Leibnizian calculus. I emphasize the lack of the notion of function, the problems related to the technique of summation or—as it was later called by Johann Bernoulli—integration, and the issues of order of infinity with respect to differentiation and integration. In the final portion of the last section I also present a brief characterization of Newton's fluxional calculus, trying to contrast some basic notions of Newtonian and Leibnizian mathematics. I focus in particular on the different role played by time in the two formulations, and on the method of first and last ratios employed by Newton in the *Principia*.

3.2 Early geometrical results

The story of Leibniz's mathematical career is an extraordinary one. On his arrival in Paris in 1672, 26 years old Leibniz was a novice in higher mathematics; by the time he left in 1676 he had become one of the very few leading mathematicians of his time. During those years he became

acquainted with Parisian mathematical circles, whilst in his correspondence with the Secretary of the Royal Society Henry Oldenburg and
during two visits to London in 1673 and 1676 he gained a deep understanding of the works being carried out across the Channel, especially by
James Gregory, John Wallis, and Isaac Newton. His mathematical
mentor in Paris was Christiaan Huygens, whose works—especially the
Horologium Oscillatorium—and advice influenced him profoundly.[2]

In the 1660s and 1670s advanced mathematics was based to a large
extent on a limited number of texts, partly coming from the Greek
tradition, partly composed in the seventeenth century. Among the latter
were the works by Bonaventura Cavalieri, Evangelista Torricelli, Blaise
Pascal, Pierre de Fermat—the last two included manuscripts circulating
within selected circles—and especially René Descartes, whose 1637
Géométrie in the 1649 and 1659–61 editions by the Dutch mathematician Frans van Schooten was extraordinarily influential. The common
sources for many of the problems, the difficulties and delays of publication, and the secretive attitude of mathematicians, resulted in an unprecedented and probably unsurpassed series of controversies over priority.
In their correspondences mathematicians would often reveal results
without giving details of the proofs and methods for finding the solutions, or even conceal the results with anagrams in order to secure
priority. The Dutch mathematician Hendrick van Heureat, in a priority
dispute with Huygens, appended to a letter a series of eight fake
anagrams—such as '4. Redeoque porci somnium'—mocking his
correspondent's style for establishing priority.[3] Significantly, although
both Newton and Leibniz had possessed the calculus for many years, a
major stimulus to organize their works for publication came from the
desire to secure priority. Although Newton had developed his method in
1665–6, he was led to compose the tract *De analysi* in 1669 as a
response to Nicolaus Mercator's *Logarithmotechnia* (London, 1668),
over which he wanted to prove his superiority.[4] In analogous fashion,
although Leibniz had formulated the differential calculus in 1675, he was
led to publish the *Nova Methodus* in the *Acta Eruditorum* for 1684 in

[2] Bos, 'Influence of Huygens'. A. R. Hall, 'Leibniz and the British Mathematicians:
1673–1676', *SL* Supplementa, *17*, 1978, pp. 131–52.

[3] *HOC*, 2, pp. 139–40, 24 Feb. 1658. Yoder, *Unrolling Time*, pp. 119–26. J. A. van
Maanen, 'Hendrick van Heuraet (1634–1660?): His life and mathematical work',
Centaurus, *27*, 1984, pp. 218–79; 'Die Mathematik in den Niederlanden im 17.
Jahrhundert und ihre Rolle in der Entwicklungsgeschichte der Infinitesimalrechnung', *SL*,
Sonderheft *14*, 1986, pp. 1–14.

[4] The hastily composed *De analysi*, however, remained unpublished at the time both
because of a depressed book trade, and because Newton was aware of some imperfections
in his work. See *NMW*, 2, p. 163f.

order to secure his priority over Ehrenfried Walter von Tschirnhaus, who had published an article on quadratures in 1683.[5]

Following these considerations, it is not surprising to read in Hofmann's monograph that Leibniz's 'first penetrating insight' into geometry was already known to ten other mathematicians including Isaac Barrow, who made it public.[6] It is worth outlining this result, called by Leibniz 'characteristic triangle', which is deployed by Leibniz very often in his later researches.

Given the curve $FABG$, its tangent AB and perpendicular AE in A, the infinitesimal triangle ABC is similar to the finite triangle AED. Setting $AB = s$, $AC = c$, $AE = a$, $AD = h$, we have $s:c::a:h$. This relatively straight-forward result allows infinitesimal lengths to be compared among themselves, namely the sides of the infinitesimal triangle ABC; moreover, the proportion allows us to consider the finite blow-up AED of the

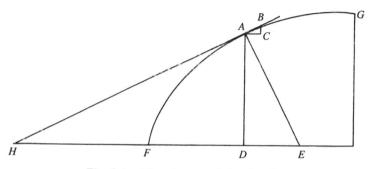

Fig. 3.1   The characteristic triangle.

infinitesimal triangle, such that ratios and especially tangents can be dealt with using finite magnitudes. It is worth pointing out that other finite triangles could be used, such as AHD, where H is the intersection of the tangent AB with the prolongation of FE. The proportion above had many applications in mathematics and mechanics, since the tangent to a curve notoriously represents velocity. We shall find in Part 2 and in Appendix 1 that Leibniz's analysis of orbital motion relied heavily on the characteristic triangle.

A mathematical development attained by Leibniz soon after the characteristic triangle is the so-called transmutation theorem. Usually an

[5] Hofmann, *Leibniz in Paris*, p. 64, n. 6, and p. 191, n. 27. See also H.-J. Hess, "Zur Vorgeschichte der 'Nova Methodus' (1676–1684)', *SL*, Sonderheft *14*, 1986, pp. 64–102, on p. 72.

[6] Hofmann, *Leibniz in Paris*, pp. 48 and 74–5. I. Barrow, *Lectiones Geometricae*, (London, 1670), lect. xi, prop. 1. See also M. S. Mahoney, 'Barrow's mathematics: between ancients and moderns', in M. Feingold, ed., *Before Newton*, (Cambridge, 1990), pp. 179–249.

area was approximated by narrow rectangles; his method was slightly more elaborate, but proved a useful tool for finding the quadrature of curves by transforming one closed curvilinear figure into another of equal area.

The basic idea consists in subdividing a curve *OAPQB* into infinitesimal triangles concurrent to the common point *O*. We shall see below how Leibniz tried to develop this idea while studying proposition 1 in

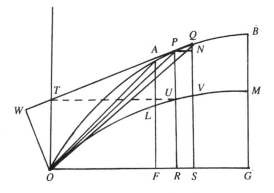

Fig. 3.2   The transmutation theorem.

the *Principia Mathematica*. The infinitesimal triangle *PQN* can be shown to be similar to the finite triangle *OTW*, constructed by taking the perpendicular from *O* to the tangent *QPW*; thus by a simple application of the characteristic triangle one has *TO*:*WO*::*PQ*:*PN*. Notice that the side *PQ* is treated by Leibniz as rectilinear; we shall come back to this problem below. From the proportion one has *WO·PQ* = *TO·PN*, which means that the area of the triangle *OPQ*, namely 1/2 *WO·PQ*, is equal to one half of the product *TO·PN*. Here *TO* = *UR*; since *PN* = *RS*, one has that the area of the triangle *OPQ* is one half of the area of the rectangle *UVSR*. Repeating an analogous procedure for other triangles such as *OAP*, whereby one finds point *L*, one finds that the area of the sector *OAB* is equal to one half of the area enclosed by the figure *FLMG*, where the curve *LUVM* is constructed by iterating the procedure just outlined.[7]

This result was the cornerstone of many of Leibniz's later works. A considerable portion of *De quadratura arithmetica circuli ellipseos et hyperbolae* is based on it. This essay was the most compendious treatise on infinitesimal geometry composed by Leibniz in Paris and, following Eberhard Knobloch, a summa of what was known in the field in 1676. Despite Leibniz's efforts, his treatise remained unpublished, but a small

---

[7] Leibniz, *Historia et origo calculi differentialis*, *LMG*, *5*, pp. 392–410, on p. 401; *LSB*, III, *1*, pp. 341–3, 347–8, 360–1. Hofmann, *Leibniz in Paris*, pp. 54–5.

portion of its results appeared in the other articles, such as Leibniz's first essay in the inaugural volume of the *Acta Eruditorum*, in 1682, *De vera proportione circuli*, and the 1691 *Quadratura arithmetica communis sectionum conicarum.*[8]

Having surveyed some preliminary geometrical results attained by Leibniz in 1673 and soon afterwards, during his stay in Paris, we move now to a deeper analysis of some of the most distinctive elements of his mathematics.

3.3 Curves and differentials

In the seventeenth century the notion of curve was transformed and the number of curves investigated by mathematicians increased considerably. From being purely geometrical entities, they became associated with algebra in the works of Viète, Descartes, and Fermat. As Michael Mahoney has emphasized, this crucial step was part of a more general process of algebraization of mathematics. The link between geometry and algebra, however, did not imply that curves were always defined by their equation. Indeed, within baroque mathematics we find a plurality of methods to define curves, such as the identification of some of their distinctive properties, or even of the operations required to draw them, as Henk Bos has shown in the case of the tractrix. Moreover, the boundary between 'exact' and 'non-exact' curves, or, to use the Cartesian distinction, 'geometrical' and 'mechanical', was often redefined. At the beginning of the century plane curves known to mathematicians had not changed much from antiquity, and included conic sections, conchoids cissoids, and Archimedean spiral. Towards the end of the century the range of curves commonly treated by mathematicians and considered to be exact had increased enormously; many of those curves emerged from the mathematical study of natural phenomena. Leibniz gave a considerable contribution to this process with the introduction of what he called transcendent magnitudes, which he considered to be perfectly legitimate mathematical objects. Transcendent lines were defined as those of no determinate degree, such as the logarithmic curve or the cycloid. Typical problems leading to transcendent magnitudes and curves were infinite series and quadratures. Often solutions to such problems consisted in

---

[8] E. Knobloch, 'Leibniz et son Manuscrit inédite sur la quadrature des sectiones coniques', in *The Leibniz Renaissance* (Florence, 1989), pp. 127–51. E. Knobloch is publishing a complete edition of this important work, which is known only through the partial edition by L. Scholtz, *Die exacte Grundlegung der Infinitesimalrechnung bei Leibniz*, Doktorarbeit (Marburg, 1934). G. W. Leibniz, 'De vera proportione circuli', *AE* Feb. 1682, pp. 41–6 = *LMG*, 5, pp. 118–22; 'Quadratura arithmetica communis sectionum conicarum', *AE* April 1691, pp. 178–82 = *LMG*, 5, pp. 128–132.

reducing a given quadrature to that of a transcendent curve to be used as standard. This procedure involved a redefinition of the notion of solution and led to a regrouping and reclassification of mathematical problems. The fact that curves which only a few decades earlier were either unknown, or were considered not to belong to the realm of proper mathematics, could be portrayed as adequate and rigorous solutions emphasizes how perceptions had changed during the century.[9]

In addition to such broad transformations of the ways curves were understood and represented, there are some specific aspects which require close attention. It is not uncommon to find seventeenth-century mathematicians following the Archimedean tradition based on the method of exhaustion of representing curves as polygons, and extending such representations to the cases when polygons have an infinite number of infinitesimal sides. Such infinitangular polygons are central to the understanding of Leibniz's mathematics and require a careful study. Very early in his career Leibniz referred to this notion, as in *Hypothesis Physica Nova* and in several texts of the Paris period.[10] In the 1684 tract *Additio ad Schedam de Dimensionibus Figuris Inveniendis,* Leibniz explained that '*a curvilinear figure must be considered to be equivalent to a polygon with infinitely many sides*'; in the *Nova Methodus pro Maximis et Minimis,* of the same year, he defined the tangent as the prolongation of the infinitesimal side of a polygon 'which for us is equivalent to the curve.'[11]

[9] M. Mahoney, *The mathematical career of Pierre de Fermat (1601–1665)* (Princeton, 1673); 'Infinitesimals and transcendent relations: The mathematics of motion in the late seventeenth century', in D. C. Lindberg and R. S. Westman, eds., *Reappraisals*, pp. 461–91. A. G. Molland, 'Shifting the foundations: Descartes' transformation of Ancient Geometry', *Historia Mathematica, 3,* 1976, pp. 21–49. H. J. M. Bos, 'On the Representation of Curves in Descartes' *Géométrie*', *AHES, 24,* 1981, pp. 295–338; "Arguments on Motivation in the Rise and Decline of a Mathematical theory; the 'Construction of Equations', 1637 –ca. 1750". *AHES, 30,* 1984, 331–380; 'Tractional Motion and the Legitimation of Transcendental Curves', *Centaurus, 31,* 1988, pp. 9–62, H. Breger, 'Leibniz's Einführung des Transzendenten', *SL*, Sonderheft *14,* 1986, pp. 119–32. Breger claims that the word 'transcendental' occurs first in a manuscript of 1675.

[10] Leibniz, *Hypothesis Physica Nova, LSB*, VI 2, p. 267; *LSB,* III, *1,* pp. 141–69, on p. 149, Leibniz to Huygens, Oct. 1674; pp. 336–55, on p. 341, Leibniz to La Roque, end of 1675; pp. 355–63, on p. 361, Leibniz to Gallois, end of 1675: 'Je suppose icy qu'un curviligne n'est qu'un polygone infinitangle, suivant la maniere de raisonner receue aujourdhuy; pour parler clairement et en peu de mots, puisque tant d'autres ont fait voir, qu'il est aisé de la reduire en cella des anciens per les inscrits et circonscrits, et on peut dire que cellecy est aussi rigoureuse que l'autre, puisque cette reduction a esté demonstrée generalment.' Hofmann, *Leibniz in Paris,* pp. 7–8.

[11] G. W. Leibniz, 'Additio ad Schedam de Dimensionibus Figuris Inveniendis', *AE* Dec. 1684, pp. 585–7 = *LMG, 5,* pp. 126–7: '... *figura curvilinea censenda sit aequipollere Polygono infinitorum laterum.*' 'Nova Methodus pro Maximis et Minimis', *AE* Oct. 1684, pp. 467–83 = *LMG, 5,* pp. 220–6, on p. 223; '... *tangentem* invenire esse rectam ducere, quae duo curvae puncta distantiam infinite parvam habentia jungat, seu latus productum polygoni infinitanguli, quod nobis *curvae* aequivalent.' H. J. M. Bos, 'Differentials, Higher-order Differentials and the Derivative in the Leibnizian Calculus', *AHES, 14,* 1974, pp. 1–90, on p. 14.

Given a curve, however, it is possible to represent it as an infinit-angular polygon in infinitely many ways. Following the standard conventions of naming x the abscissae, y the ordinates, and s the arclengths, it is possible to choose a polygon such that the x, the y, or the s are divided into equal parts, for example, or indeed according to any other rule. Following Leibniz, the difference between any pair of adjacent x, y, or s, is called dx, dy, ds, respectively; these entities are called 'differentials' and are indeterminate in two respects. First, from the freedom of choice of the polygon selected to represent the curve, and of the corresponding sequence of values of the x, y, s, it is possible to choose the dx, dy, ds, to be constant or variable according to the rule with which the sides of the polygon and the sequences of variables have been chosen. Second, the actual value of the differential is indeterminate regardless of the progression of the variable. Although Leibniz often referred to the differentials as well as to sides of the polygon as 'infinitesimal', strictly speaking he thought it was more correct to call them incomparably smaller than ordinary quantities. In the *Tentamen*, for example, Leibniz introduced infinitesimals as a grain of sand with respect to the sky, thus indicating that he conceived them as incomparably smaller, though not rigorously infinitesimal quantities. The differential can be chosen in such a way that by neglecting it—in the appropriate circumstances—the error in the result of the calculations is smaller than any given quantity. The differential of xy, for example, is $xdy + ydx + dx \cdot dy$, where the last term can be neglected because it is incomparably smaller than the other two. In a letter to Johan Bernoulli, Leibniz doubted the existence of infinitely small or large quantities; rather, he stressed the instrumental character of differentials, which he compared to imaginary roots in algebra, since in both cases 'fictitious' entities led to the correct result. This pragmatic image recurs often in his works.[12]

Leibniz was involved in several disputes on the justification and rigour of the calculus. Although he dealt with the philosophical aspect of the problem, his main concern seemed to be the spreading of the calculus as a successful mode of operation. Leibniz repeatedly stressed that there was no need to make the calculus dependent on metaphysical controversies.[13] In the dispute started in 1700 at the Paris Academy, Michel Rolle and Pierre Varignon argued about the existence of infinitely small

---

[12] *LMG*, 3, p. 524, Leibniz to Johann Bernoulli, 29 July 1698; in this important letter Leibniz discussed differentials in connection with the divisibility of matter and microscopic observations of *animalcula*.

[13] *LMG*, 4, p. 91, Leibniz to Varignon, 2 Febr. 1702; p. 98, 4 April 1702; p. 110, 20 June 1702; *LMG*, 5, p. 350. G. W. Leibniz, 'Cum prodiisset', in *Historia et Origo Calculi Differentialis*, ed. C. I. Gerhardt (Hannover, 1846), p. 43. This important essay is analysed by Bos, 'Differentials', p. 56f., in connection with the law of continuity.

and large quantities and the reliability of the calculus. The existence of infinitesimals had been defended by the Marquis de l'Hôpital in *Analyse des Infiniment Petits* (Paris, 1696), the first textbook on the calculus. Leibniz found himself in an awkward position because he had publicly stated that infinitesimals could be conceived as a grain of sand with respect to the earth, or the earth with respect to the distance of fixed stars. In a curious and typically Leibnizian letter to Pierre Dangicourt, he admitted that during the controversy he had to conceal his true opinons in order not to embarass his supporters in France, and especially l'Hôpital, who feared Leibniz was about to 'betray the cause'. In the same letter Leibniz frankly admitted that in his own views 'truly' infinite or infinitesimal quantities do not exist. Differentials were merely 'well founded fictions', namely useful entities which could be profitably employed for finding new results.[14]

The picture of Leibnizian mathematics presented thus far shows a correlation between the notions of curve conceived as an infinitangular polygon and of differential. More precisely, the infinitangular polygon is associated with a sequence of infinitely near values of a variable; the difference between two contiguous ordinates, or abscissae, or arclengths, is the corresponding differential:[15] 'The conception of the variables as ranging over infinite sequences of infinitely near values, and consequently the conception of the differentials and sums as new variables, is crucial to the understanding of the Leibnizian calculus. It marks, for instance, the contrast with Newton's fluxional calculus, which was based on a fundamentally different conception of the variable, namely as flowing along a continuum of values, rather than ranging over a sequence.' On the basis of this conception of variables, Leibniz grasped that tangents are related to the differences between the elements of a sequence associated with a variable, and quadratures are related to their sums. In a letter to John Wallis, Leibniz wrote: 'The consideration of differences and sums in number sequences had given me the first insight, when I realized that differences correspond to tangents and sums to quadratures.'[16] Before moving to the study of the calculus, which is the subject of the following section, it is convenient here to expand the

[14] G. W. Leibniz, *Opera Omnia*, ed. L. Dutens, 6 vols. (Geneva, 1768); *3*, pp. 500–501, Leibniz to Dangicourt, 11 Sept. 1716. M. Blay, 'Deux moments de la critique du calcul infinitesimal: Michel Rolle et George Berkeley', *RHS*, *39*, 1986, pp. 223–53. *JBB*, *2*, passim, esp. pp. 351–76. In order to defend the rigour of the differential calculus Varignon found nothing better than quoting section I of Newton's *Principia* (p. 352)! Bos, 'Differentials', pp. 55–6.

[15] H. J. M. Bos, 'Fundamental Concepts of the Leibnizian Calculus', *SL* Sonderheft *14* (1986), pp. 103–18, p. 108.

[16] *LMG*, *4*, p. 25, 28 May 1697, Leibniz to Wallis. Bos, 'Fundamental Concepts', pp. 104 and 106.

treatment of curves by discussing some important notions such as radius of curvature, osculating circle, evolute, and evolvent.

The notions of radius of curvature and evolute, which had been 'adumbrated' by Apollonius in the *Conics*, were developed in modern times by Christiaan Huygens in relation to his work on the pendulum clock, as we have seen in Section 2.3.[17] In order to make pendular oscillations isochronous, Huygens was prompted to constrain them within two arcs of a cycloid. The thread of the pendulum would thus unroll, constantly changing its length in such a way that the period of the oscillation is rigorously independent of its amplitude. The evolvent described by the bob of the pendulum is always perpendicular to the thread unrolling from the evolute. The whole third section of the *Horologium Oscillatorium* is devoted to the mathematical theory of these curves, which is developed far beyond the technical needs associated with horology.

In his 1686 *Meditatio Nova de Natura Anguli Contactus et Osculi* Leibniz focused on a related aspect. Instead of considering the relations between evolute and evolvent, he analysed the latter and its properties. A straight line is the most appropriate for determining the direction of a curve, since the straight line has a constant direction; similarly, the circle is the most appropriate for measuring curvature, because its curvature is constant. The straight line measuring the direction of a curve is called the tangent, and the circle measuring curvature is called the osculating circle. On taking into account the angle of contact, namely that magnitude between the arc of a circle and its tangent, Leibniz also defined the osculating circle as that forming the smallest angle of contact with the curve.[18] Each infinitesimal portion of a regular curve can be approximated by an arc of circle, namely of that circle osculating the relevant part of the curve. The advantage of considering a portion of a curve as an arc of a circle is immediately clear, since it is easier to deal with a circle than with any other curve. Particularly in the study of orbital trajectories, Leibniz and Newton adopted this mathematical technique. In Part 2 we shall see how Leibniz worked with osculating circles while seeking a generalization of Newton's law of centripetal force.[19]

---

[17] Boyer, *A History*, p. 414, quoted in Yoder, *Unrolling Time*, p. 98 and ch. 6.

[18] G. W. Leibniz, 'Meditatio Nova', *AE* June 1686, pp. 289–293 = *LMG*, *7*, pp. 326–9. M. Cantor, *Vorlesungen*, *3*, p. 189. On the angle of contact see Euclid, III, 16; T. Heath, *A history of Greek mathematics* (Oxford, 1921; reprinted New York, 1981), 2 vols., vol. 1, pp. 178 and 382. Hofmann, *Leibniz in Paris*, pp. 12–13. Leibniz's interest in the angle of contact dates at least from the 1669 *Doctrina Conditionum*, *LSB*, VI, *1*, p. 389, where he claimed that the angle of contact has no assignable ratio to a finite angle.

[19] On Newton and curvature see *NMW*, *1*, pp. 245ff. and 456; *3*, p. 151ff.; *6*, pp. 548–9, n. 25. B. Brackenridge, 'Newton's mature dynamics', *AS*, *45*, 1988, pp. 451–76; 'Newton's unpublished dynamical principles: a study in simplicity', *AS*, *47*, 1990, pp. 3–31.

3.4   The calculus

We have recently come to appreciate the subtle conceptual differences between Leibniz's and more modern formulations of the calculus. A notable feature which has emerged only implicitly in my account of Leibnizian mathematics thus far, is the lack of the notion of function. In the seventeenth century curves were not seen as a graph of a function $x \rightarrow y(x)$, where x is the independent variable, but as figures associated with a relation between x and y.[20] Thus differentiation and summation or integration do not act on functions of an independent variable, but on variables themselves. Consequently, the operations of the Leibnizian calculus do not involve the notion of derivative: Leibnizian differentiation associates to a variable x another variable dx infinitely, or better incomparably, small with respect to it, and conversely for integration. Differentiation and integration change the order of infinity; dimensions, though, are preserved. If x is a distance, for example, dx is also a distance, and likewise for higher order differentials such as ddx. When studying problems of motion it was necessary to consider time as well. Leibniz often represented the curve described by a body as a polygon selected in such a way that its sides are traversed in equal and constant elements of time dt. Since dt was constant, and $ddt = 0$, it was possible for Leibniz to neglect it in the relevant calculations. As we are going to see in more detail in the following chapter, this habit went hand in hand with the usage of proportions, in which variables are the focus of attention whilst constant factors occupy a marginal position. This is the reason why Leibniz often says that the differential of a length is as a velocity; what is meant is *dato tempore*, namely the differential of a length is proportional to a velocity when dt is constant.

It has often been emphasized that concepts and notation of the early formulation of the Leibnizian calculus are related to the work of the Italian mathematician Bonaventura Cavalieri. Cavalieri conceived areas as aggregates of lines and used the expression 'omnes lineae' to designate this aggregate.[21] In the fundamental 1675 tract *Analysi Tetragonistica* Leibniz started by denoting the sum of all lines l as *omn.l*, later replacing this by $\int l$, where the elongated $\int$ stands for 'summa'.[22] A further change in the notation mirrored important conceptual developments. By introducing the differential dx inside the sign of summation, Leibniz wanted to emphasize that for him quadratures are evaluated as sums of

    [20]   Bos, 'Differentials', p. 6.
    [21]   K. Andersen, 'Cavalieri's Method of Indivisibles', *AHES*, *31*, 1984, pp. 291–367; E. Giusti, *Bonaventura Cavalieri and the Theory of Indivisibles*, (Bologna, 1980).
    [22]   *LBG*, pp. 147–67, 25 Oct.–1 Nov. 1675.

area differentials rather than as aggregates of lines.[23] If the progression of the variable dx is selected such that dx is constant, Leibniz's procedure corresponds more closely to Cavalieri's. However, the possibility of choosing different progressions for the variable dx allows a considerable flexibility in the corresponding choice and even transformation of dx. Thus virtually from the start the Leibnizian calculus was geared to the technique of substitution of variable. Leibniz was perfectly aware of this considerable advantage of his technique and in his first publication on the integral calculus, *De geometria recondita*, he emphasized precisely this aspect:[24]

Before I finish, I add one warning, namely that one should not lightheartedly omit the dx in differential equations like the one discussed above $a = \int dx : \sqrt{1 - xx}$ because in the case in which the x are supposed to increase uniformly, the dx may be omitted. For this is the point where many have erred, and thus have closed for themselves the road to higher results, because they have not left to the indivisibles like the dx their universality (namely that the progression of the x can be assumed ad libitum) although from this alone innumerable transfigurations and equivalences of figures arise.

The reference to 'indivisibles' in this quotation follows the mid-seventeenth-century habit of conflating indivisibles and infinitesimals while dealing with quadratures, without worrying too much about philology. More rigorously one could say that indivisibles are constant entities with no magnitude, whilst infinitesimals are variables and have a magnitude—they are 'non quanta' and 'quanta' respectively.[25]

By 1675 Leibniz had developed the main principles of his calculus and in particular the reciprocal relation between differentiation or the search for tangents, and summation or the finding of areas. In the latter case the variations from more modern practices and concepts are even more marked than in the case of differentiation. Strictly speaking integration cannot be taken to be the reciprocal operation of differentiation because a variable has one differential, but the number of integrals are infinitely many depending on an arbitrary constant factor. In the seventeenth century there was no clear distinction corresponding to the modern notions of definite or indefinite integral; in certain cases this lack and the corresponding asymmetry between differentiation and integra-

---

[23] Bos, 'Differentials', p. 79.

[24] G. W. Leibniz, 'De geometria recondita et analysi indivisibilium atque infinitorum', *AE*, June 1686, pp. 292–300 = *LMG*, 5, pp. 226–33, p. 233, translated in Bos, 'Differentials', p. 79. *LMG*, 7, p. 387, Leibniz to Freiherr von Bodenhausen, late 1690s. See also G. W. Leibniz, 'Methodus Tangentium Inversa', dated 'July 1676', *LBG*, pp. 201–3, and the marginal annotations to Newton's *epistola posterior*, received by Leibniz in 1677, *LSB*, III, 2, pp. 93–4.

[25] K. Andersen, 'The Method of Indivisibles: Changing Understandings', *SL* Sonderheft *14*, 1986, pp. 14–25.

tion were not particularly significant, but in the study of differential equations a mistake regarding constant factors could produce devasting consequences in the result. Leibniz usually neglected the rule of adding constant factors, and even Newton did not follow this rule consistently. By contrast Johann Bernoulli, in his 1691–2 *Lectiones Mathematicae de Methodo Integralium* delivered in Paris to the Marquis de l'Hôpital, spelt out the need to add an arbitrary constant factor in the very first lecture.[26]

In this short characterization of some aspects of Leibnizian mathematics I have deliberately emphasized conceptual aspects and presented a partial and simplified picture; while stressing the importance of these conceptual observations, it is also worth recalling that often the calculus was used as a tool and that philosophical considerations about rigour could become less important than the attainment of a result. Bearing in mind these remarks, it is useful to characterize in a few sentences some basic notions of Leibniz's mathematics.

Leibniz

- Curves are conceived as infinitangular polygons consisting of incomparably many rectilinear segments. Tangents are the prolongations of such segments.
- Variables range over a discrete sequence of incomparably near values and differentials are the differences between contiguous pairs of such variables, such as ordinates, abscissae, or arclengths. Differentials are indeterminate because the sequences of the relevant variable or the associated polygon can be chosen in infinitely many ways. Moreover, differentials can be given arbitrarily small values in the calculations so that by neglecting them—in the appropriate circumstances—the error in the result is less than any given quantity.
- Differentiation and integration are operations on variables and change the order of infinity, not the dimension of a variable. The differential of a length is an incomparably small length, the integral of an incomparably small velocity is a finite velocity.

A similar characterization for Newton presents greater difficulties because, from his early mid-1660s tracts up to the mature treatises, he tried to adopt more and more rigorous formulations. In general, he thought of a variable as changing in time, as his predecessor on the Lucasian chair Isaac Barrow had done. Thus Newton focused on the finite speed of this change rather than on infinitesimal increments of a variable. This statement, though acceptable as a global characterization, requires a careful qualification, taking into account the context of

[26] *JBO*, *3*, pp. 285–558, on p. 287. On Newton see *NMW*, *3*, pp. 114–117.

Newton's work. In the 1669 *De Analysi*, for example, he wrote: 'Neither am I afraid to speak of Unity in points, or Lines infinitely small, since Geometers are wont not to consider Proportions even in such a case, when they make use of the Methods of Indivisibles.'[27] Later, however, and especially during the priority dispute with Leibniz, he emphasized the rigorous foundations of his method. In his 1713 response to Christian Wolff's review in the *Acta Eruditorum* of the revised tract *De Analysi*, Newton stressed the difference between his own 'totally vanishing' quantities, and Leibniz's infinitely small—though not vanishing—differentials:[28]

For in this method quantities are never considered as infinitely little nor are right lines ever put for arches neither are any lines or quantities put by approximation for any other lines or quantities to which they are not exactly equal, but the whole operation is performed exactly in finite quantities by Euclides Geometry until you come to an equation and then the equation is reduced by rejecting the terms which destroy one another and dividing the residue by the finite quantity *o* and making this quantity *o* not to become infinitely little but totally to vanish.

These observations clearly show the shift of emphasis between Newton's early and mature works, and sound as a warning against simplified approaches neglecting the changes in his formulations of the calculus.

In the scholium at the end of section 1 in the *Principia*, on first and last ratios, Newton wrote: 'Therefore if in what follows ... I should happen to mention least, evanescent, or last quantities, I do not understand them to be determinate, but always diminished without end.' In the 1691–2 tract *De quadratura curvarum* Newton first introduced the notation whereby fluxions are indicated with a dot on the corresponding variable. Fluxions express the rate of change of a continuous variable, usually with respect to time, and serve purposes similar to Leibnizian differentials, despite their conceptual differences. In a passage dating from the time of the priority dispute, Newton pointed to what he wished to present as a characteristic feature of his method of fluxions: 'This Method is derived immediately from Nature her self, that of indivisibles, Leibnizian differences or infinitely small quantities not so. For there are no *quantitates primae nascentes* or *ultimae evanescentes*, there are only

[27] *NMW*, *2*, p. 235. Quoted in P. Kitcher, 'Fluxions, Limits, and Infinite Littleness', *Isis*, *64*, 1973, pp. 33–49, p. 46. Cohen, *Newtonian Revolution*, sect. 3.1. Mahoney, 'Barrow's mathematics'; *NMW*, *3*, pp. 70–2, esp. nn. 80–4.

[28] *NMW*, *2*, pp. 263–73, on p. 264, and *Excerpts*, p. 484, n. 18. Among the documents on the priority dispute over the invention of the calculus there are several further remarks on this topic. Kitcher, 'Fluxions', pp. 46–9; F. De Gandt, 'Le style mathematique des *Principia* de Newton', *RHS*, *39*, 1986, pp. 195–222. The passage from *De Analysi* quoted above appeared unchanged in the 1711 edition: I. Newton, *Mathematical Works*, ed. D. T. Whiteside, 2 vols. (London and New York, 1964), *1*, p. 18. Compare also *Principia*, Book II, lemma 2, where the method is briefly explained.

rationes primae quantitatum nascentium or *ultimae evanescentium.*[29] In his article on 'Quadrature of curves' in John Harris, *Lexicon Technicum* (London, 1711), Newton wrote: 'I don't here consider Mathematical quantities as composed of Parts *extremely small*, but as *generated by a continuous motion.*' In an example taken from orbital motion, Newton wrote that the fluxion of a distance is a velocity, and the fluxion of a velocity is an acceleration or gravity. In the *Principia*, however, Newton employed the method of first and last ratios, whereby infinitesimals are avoided by considering not the finite speed with which a variable changes with respect to time, but the finite ratio between two vanishing variables, usually lengths. Further, in proposition 10, book II, the *o* refer to the distances on an axis and do not represent increments of time. Despite some important exceptions, however, on the whole kinematics for Newton appears to be not so much an application of the calculus, as something intrinsic to it and to its justification: time, in its most abstract and mathematical form, is a privileged variable. By contrast, time is extrinsic to the Leibnizian calculus, and kinematics is merely a special application. Moreover, as Enrico Giusti has observed, Leibniz's notation highlights that differentiation and integration are operations. The notation adopted by Newton, however, indicates the speed of change and hence an attribute of a variable. Although it is easy to rewrite Newtonian mathematics in a Leibnizian form and vice versa, their respective notations are indicative of important and deep conceptual differences.[30]

Before providing a schematic representation of those aspects of Newton's calculus which will occur again in the following, it is useful to outline some features of the mathematics of the *Principia*. Contrary to Newton's own claims that he had originally composed his masterpiece in the language of fluxions, Tom Whiteside has shown that there is a continuity between the preparatory manuscripts and the published version of the *Principia*. Its mathematical style has been aptly defined by François De Gandt as 'une géométrie de l'ultime'. Both the geometrical treatment and the method of first and last ratios were closely related to the objects of the investigation: the former, since Newton was dealing primarily with conic sections representing the trajectory of an orbiting body; the latter, since he needed to calculate velocities and accelerations

---

[29] Newton, *Principia* (first edition), p. 35 (my translation); Motte and Cajori, p. 39. *NMW*, 6, pp. 122–3, nn. 63–8 and p. 195. *NMW*, 3, pp. 17–18. The reciprocal of fluxions are called fluents, which correspond to Leibnizian summation and are related to quadratures.

[30] *NMW*, 7, p. 129. See also Hall, *Philosphers at war*, p. 269. The article in Harris, *Lexicon*, is reprinted in Newton, *Mathematical Works*, 1, pp. 141f. D. T. Whiteside, 'The Mathematical Principles Underlying Newton's *Principia Mathematica*,' *JHA*, 1, 1970,pp. 116–38; I owe the observation on proposition 10, book II of *Principia*, to D. T. Whiteside. E. Giusti, 'A tre secoli dal calcolo: la questione delle origini', *Bollettino della Unione Matematica Italiana*, ser. 6, vol. 3A, 1984, pp. 1–55, on pp. 39–40 and 53–54.

at specific points. As a result, the geometric figure was the protagonist in the mathematics of large portions of the *Principia*; in that mathematics points and segments move along curves, and the ratios between segments or between areas are computed. As Newton wrote at the end of section 1: 'And, to be sure, everything that is geometrical is legitimately employed in determining and demonstrating other geometrical things'. From his characterization of the mathematics of the *Principia* one should not jump to the conclusion that every proposition in it was written in exactly the same style. In proposition 10, book II, for example, one finds that series expansions are used in the investigation of projectile motion under constant gravity in a resisting medium. On the whole, however, this constituted the exception rather than the rule. A more representative example on which we shall come back later is lemma 9, which states:[31]

If the straight line *AE* and the curve *AC* given in position mutually intersect at the given angle *A* and *DB*, *EC* be applied as ordinates to that straight line at any given angle, meeting the curve in *B* and *C*, and if the points *B*, *C* then approach the point *A*: I assert that the areas of the triangles *ADB*, *AEC* will be ultimately to one another in the doubled ratio of the sides.

Lemma IX.

Si recta *A E* & Curva *A C* positione datæ se mutuo secent in angulo dato *A*, & ad rectam illam in alio dato angulo ordinatim applicentur *B D*, *E C*, curvæ occurrentes in *B*, *C*; dein puncta *B*, *C* accedant ad punctum *A*: dico quod areæ triangulorum *A D B*, *A E C* erunt ultimo ad invicem in duplicata ratione laterum.

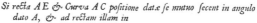

Etenim in *A D* producta capiantur *A d*, *A e* ipsis *A D*, *A E* proportionales, & erigantur ordinatæ *d b*, *e c* ordinatis *D B*, *E C* parallelæ & proportionales. Producatur *A C* ad *c*, ducatur curva A *b c* ipsi *A B C* similis, & recta *A g* tangatur curva utraq; in A; & secantur ordinatim applicatæ in *F*, *G*, *f*, *g*. Tum coeant puncta *B*, *C* cum puncto A, & angulo *c* A *g* evanescente, coincident areæ curvilineæ A *b d*, A *c c* cum rectilineis A *f d*, A *g e*, adeoq; per Lemma V, erunt in duplicata

Fig. 3.3 *Principia*, book I, lemma 9.

In the proof Newton employed triangles *Adb* and *Aec*, similar to triangles *ADB* and *AEC* respectively. *AFGfg* is the tangent to the curves *ABC* and

---

[31] Tom Whiteside has provided an extensive analysis of *Principia*, proposition 10, book II, and of the controversy between Newton and Johann Bernoulli in *NMW*, *8*, pp. 312–424. Concerning lemma 9, book I, see *NMW*, *6*, pp. 114–15, n. 49. The quotation from lemma 5 is on p. 113.

Abc in *A*. Now while points *B* and *C*, as well as *D* and *E*, coalesce with *A*, the angle *cÂg* becomes vanishing. Hence the curvilinear figures *Adb* and *Ace* will coincide with *Afd* and *Age*, respectively, and therefore will be as the squares of their sides *Ad* and *Ae*. This conclusion follows from lemma 5, stating that 'in similar figures all mutually corresponding sides, curvilinear as well as rectilinear, are proportional and their areas are in the doubled ratio of their sides.' In the present proof Newton omitted to clarify that *Ad* and *Ae* remain fixed and therefore finite while *B* and *C* move. Since triangles *ABD* and *ACE* are ever similar to *Abd* and *Ace*, also their areas will be ultimately as the squares of their sides *AD* and *AE*, respectively. Notice that Newton avoids vanishing areas and segments taken in isolation: he considers the finite ratio between vanishing quantities and in our case even the finite triangles homologous to the vanishing ones. Probably he did not wish to apply lemma 5 directly to the vanishing triangles *ADB* and *AEC*.

An algebraic, as opposed to a geometrical treatment, required a significant shift in the way mathematics and mechanics were practised. This transition from 'une géométrie de l'ultime' to the manipulation of differential equations was pioneered with mixed fortunes by Leibniz and later developed through several intermediate stages in the course of the first half of the eighteenth century. As we are going to see in Chapter 9, Pierre Varignon and Johann Bernoulli played an important role in this process.[32]

Newton

- Curves have continuous curvature. If they are treated as polygons, this procedure must be seen either as an approximation, or as a preliminary step in the calculations; ultimately the sides of the polygon become vanishingly small.
- Fluxions express the speed of change of a variable and are finite. They result from variables flowing continuously, almost always with respect to time. Hence kinematics is part of the foundations of the Newtonian calculus.
- Fluxions and fluents are attributes of a variable. They leave the order of infinity unchanged, that is, they remain finite. Dimensions,

---

[32] Whiteside, 'Mathematical Principles', p. 130, n. 53, and *NMW*, *6*; De Gandt, 'Style mathematique', p. 200; Mahoney, 'Diagrams and Dynamics: Mathematical Perspectives on Edgerton's Thesis', in J. W. Shirley and F. D. Hoeniger, eds., *Science and the Arts in the Renaissance* (Washington, 1985), pp. 198–220; and 'Algebraic vs. geometric techniques in Newton's determination of planetary orbits' (forthcoming). C. Truesdell, 'A program toward rediscovering the rational mechanics of the Age of Reason', *AHES*, *1*, 1960, pp. 3–36, esp. p. 9; P. Costabel, 'Newton's and Leibniz's Dynamics', in R. Palter, ed., *The 'Annus Mirabilis' of Sir Isaac Newton* (Cambridge, Mass., 1968), pp. 109–16, esp. p. 115.

however, vary according to the variable with respect to which they are calculated. The fluxion of a velocity with respect to time is an acceleration; the fluent of an area with respect to a length is a volume.

These brief and schematic observations show that although the Newtonian and Leibnizian formulations of the calculus could perform analogous operations, their equivalence presents considerable difficulties and cannot be accepted without major qualifications. The notion of equivalence becomes misleading unless similarities and differences with regard to their conceptual basis and notation are spelt out.

This attempt to recover the flavour of Leibnizian mathematics as opposed to Newtonian notions sets the scene for a deeper understanding of their respective notions in mechanics, which will be analysed in the following chapter. The correlation between mathematics and mechanics will emphasize that the differences in the formulations of the calculus cannot be treated simply as philological pedantry; on the contrary, mathematics interacted with mechanics and philosophy in a way that the definition and manipulation of mathematical entities cannot be isolated from physical and metaphysical views about the world.

4

MATHEMATICAL REPRESENTATIONS OF MOTION AND FORCE

4.1 Introduction

In the previous chapter we began to appreciate the links between mathematics and mechanics. One of the issues I have mentioned concerns the dimensions of physical quantities in relation to Leibnizian differentiation and integration, or to Newton's fluxions and fluents. The connections between pure mathematics and mechanics—significantly a branch of the mixed or applied mathematics—are certainly broader and deeper than this example may suggest, and are central to my understanding of Leibniz's theory and war with Newton. Already in Galileo with the notion of 'momento', or in Hobbes with the notion of 'conatus', the infinitely small was interwoven with the study of nature. By the end of the seventeenth century the analysis of curvilinear motion was gaining a prominent position in mechanics, and not surprisingly it interacted with the mathematical notion of curve.

In the following section I investigate this interaction between the way a curve was concceived, the principle of inertia, and accelerated motion. Leibniz had a clear predilection for rectilinear motion and impacts over accelerations, which he tried to dispense with as much as he could. Accelerations appear only as macroscopic effects explicable at the level of first-order infinitesimals in terms of impacts and rectilinear uniform motions. According to my interpretation, from the late 1680s onwards Leibniz perceived behind these representations a correlation with Newton's ideas of action at a distance in a void confronting his own physical explanations in terms of fluids. These reflections are indispensable in order to understand Leibniz's mechanical notions, such as solicitation, conatus, dead and living force. Section 4.3 deals with them and with their mutual relations as they result from the application of the rules of the Leibnizian calculus. In this section I pay particular attention to the problem of dimensional homogeneity, to conservation of force, to the links between dead and living force, and to the related issue of constant factors versus variables.

Since Leibniz sharpened his views on the links between physical and mathematical issues when confronting the *Principia Mathematica*, the problems I investigate are best covered by texts composed during his maturity, especially *Dynamica, Specimen Dynamicum, Illustratio Tentaminis,* and the correspondence with Pierre Varignon and Jakob Hermann. Many of the subtleties in the *Tentamen* and its preparatory manuscripts can be appreciated in connection with these later texts. Though different from Newtonian mechanics and from more modern formulations, Leibniz's science of motion and forces is a beautiful and effective theory which, like his mathematics, can be appreciated if it is studied in its own terms and in a seventeenth-century context with regard to concepts and techniques.

In this chapter I present only a partial account of Leibniz's mechanics and *dynamica*—a word he coined in Rome in 1689. A considerable portion of his efforts was devoted to controversies and debates on the conservation of force with the Cartesians as well as with Huygens and Johann Bernoulli. Moreover, Leibniz developed an elaborate array of notions relevant to a variety of problems and situations; although my presentation is restricted to the study of celestial motions, the idea of seeing his mechanics in relation to a philologically accurate examination of his mathematics can be extended, I believe, to other fields. Lastly, it is well known that Leibnizian dynamics had important metaphysical connotations; my focus, however, remains primarily on mathematical and physical issues.[1]

4.2 Mathematical representations of motion

We have seen in the previous chapter that Leibniz conceived a curve as a polygon with an infinite number of first-order infinitesimal sides. In the letter of 1676 to Claude Perrault this mathematical way of conceiving a curve was interestingly related to mechanics:[2]

And firstly I take as certain that everything moving along a curved line endeavours to escape along the tangent of this curve; the true cause of this is that curves are polygons with an infinite number of sides, and these sides are portions of the tangents. Then everything which moves circularly tends to escape along the tangent, and since the prolongation of this tangent goes away from the centre, it is for this reason, and in a certain manner accidental, that bodies moved circularly tend to escape from the centre.

---

[1] More philosophical accounts are in Gueroult, *Dynamique*; Buchdahl, *Metaphysics*, ch. 7. *SL*, Sonderheft *13*. See also D. Garber, 'Leibniz and the Foundations of Physics: The Middle years', in K. Okruhlik and J. R. Brown, eds., *Natural Philosophy*, pp. 27–130. A. Robinet, *Architectonique Disjonctive, Automates Systémiques et Idéalité Transcendantale dans l'Oeuvre de G. W. Leibniz* (Paris: Vrin, 1986).

[2] *LSB*, II, *1*, p. 264, quoted in Bertoloni Meli, 'Public Claims', p. 447, n. 56.

This interaction between the notion of curve and the law of inertia, or more generally between mathematics and mechanics, can be extended to other areas and is crucial for the understanding of mathematical representations of motion and force in Leibniz. Just as a curve is a polygon, curvilinear motion is composed of rectilinear uniform motions. In the lengthy *Dynamica*, largely composed in Rome in 1689, Leibniz wrote:[3]

All motions are composed of rectilinear uniform ones. In fact each motion in itself is uniform and rectilinear; but each action on bodies consists in motion. Thus rectilinear motion cannot be curved but by the impression of another motion by itself also rectilinear (albeit independent on the previous one), therefore the origin of curvilinear and non-uniform motion cannot be understood but by compositions of rectilinear uniform motions.

If in the letter to Perrault Leibniz had referred to mathematics, and in *Dynamica* he had focused on motion, in his last letter to Samuel Clarke in 1716 he expressed similar views with an emphasis on the primacy of impacts, which are the building blocks of his mechanics:[4]

A body is never moved naturally, except by another body which touches or pushes it; after that it continues until it is prevented by another body which touches it. Any other kind of operation on bodies is either miraculous or imaginary.

The first consequences, which I mention in passing, concern relative versus absolute motion and cohesion. Relativity of rectilinear motion was widely accepted in the second half of the seventeenth century; circular motion, however, was held to be absolute by Huygens, at least for some time, and by Newton. Leibniz believed that since the equivalence of hypotheses holds for rectilinear motions, and curvilinear motion is but a composition of rectilinear ones, the equivalence of hypotheses must hold for curvilinear motion too. By 'equivalence of hypotheses' he meant that—as far as phenomena are concerned—all representations of a system of bodies are equally valid regardless of the motion of the observer. In other words, each hypothesis about the motion or rest of a body of the system is acceptable.[5] In his account,

---

[3] *Dynamica. De potentia et Legibus Naturae Corporeae* (composed with the intention of publication), *LMG*, 6, pp. 281–514, on p. 502, quoted in Bertoloni Meli, 'Public Claims', p. 447, n. 57.

[4] *Leibniz-Clarke Correspondence*, Leibniz's fifth paper, par. 35. 'Antibarbarus Physicus', *LPG*, 7, pp. 336–44, on p. 338.

[5] On Huygens compare *Dynamica*, p. 508, referred to as 'viro praeclaro', and *LMG*, 2, pp. 184–5, Leibniz to Huygens, 22 June 1694, and the latter's reply, p. 192, 24 Aug. 1694. Later Huygens changed opinion on the matter: *HOC*, 16, pp. 189–200, 209–33. On Newton see *Principia*, scholium to definition 8. A more philosophical discussion is in L. Sklar, *Space, time and spacetime* (Berkeley and Los Angeles, 1974); J. Earman, *Worlds enough and space-time* (Cambridge, Mass., 1989).

cohesion and firmness arise from the tendency of rotating bodies to fly away along the tangent. Since in the universe there is no void, this tendency disturbs the motion of the surrounding fluid, which pushes the rotating bodies back towards the centre as if there were a magnetic attraction. In the 1695 *Specimen Dynamicum* Leibniz put the matter in the following terms:[6]

All motion is in straight lines, or compounded of straight lines. Hence it not only follows that *whatever moves in a curve strives always to proceed in a straight line tangent to it,* but there also arises here, the *true notion of firmness,* which one would hardly expect. For if we assume . . . some one of those bodies which we call firm . . . to rotate about its center, its parts will strive to fly off on a tangent; indeed, they really begin to fly off. But because this separation from each other disturbs the motion of the body surrounding them, they are thus repelled or crowded into each other again, as if there were a magnetic force in the center which attracts them, or as if there were a centripetal force in the parts themselves.

In a letter to Huygens of 1694 Leibniz recalled his doubts about relativity of motion:[7] 'When I told you one day in Paris that the true subject of motion is difficult to recognize, you replied that this can be done by means of circular motion. That stopped me for a while; and I remembered reading almost the same thing in Newton's book. But that was when I thought that I had already seen that circular motion has no advantage in this. And I see that you now agree with me.' It is conceivable that the polygonal representation outlined in the letter to Perrault was related to Leibniz's Parisian reflections on circular and relative motion. I recall that the letter to Perrault dates from the end of Leibniz's stay in Paris. Before exploring further the consequences of these views about curves and motion, I wish to raise an issue related partly to my presentation and partly to Leibniz's own texts. The previous chapter and this one, the order of the quotations above, and indeed Leibniz's own statement in the letter to Perrault might suggest a causal link from mathematics to mechanics; in other words one could argue that Leibniz believed in the existence only of rectilinear motions, with all its consequences, because of his mathematical notion of curve. In my opinion this reductionist view cannot be accepted. Similar issues concerning the identification of an alleged fundamental discipline are not

---

[6] *Specimen Dynamicum*, in Loemker, *Papers*, p. 449, and (Hamburg, 1982), pp. 54–9; Ariew and Garber, *Essays*, pp. 135–6. *Dynamica, LMG*, pp. 507–11; *Leibniz-Clarke Correspondence*, Leibniz's fifth paper, par. 53. See further *LPG, 4*, pp. 384–9, 'Animadversiones'; *Hypothesis Physica Nova, LSB*, VI, *2*, pp. 223 and 270; the 1677 letter to Fabri, *LSB*, III, *2*, p.136, proposition 13; the 'zweite Bearbeitung', *LMG, 6*, p. 162; 'De Causa Gravitatis', *AE* May 1690, pp. 228–39 = *LMG, 6*, pp. 192–203, on p. 198.

[7] *LMG, 2*, 14 Sept. 1694, p. 199, transl. Ariew and Garber, *Essays*, p. 308. Concerning the reference to Newton compare sect. 5.2 below.

uncommon in the literature on Leibniz. However, his system is based on an extraordinarily complex interplay of themes and disciplines with no fixed centre. Neither mathematics, nor logic, nor metaphysics, nor theology nor any other field, can be taken to be at the foundations of the whole system. The belief in relativity of motion, for example, is directly linked to mathematics and the polygonal representation, to physics and the primacy of impacts, to metaphysics and the theory of space and time, and none of these connections appears to me to be privileged over the others, even if at times Leibniz emphasized this or that aspect.[8]

A further issue related to Leibniz's views about curvilinear motion concerns the notion of acceleration. Since Galileo's studies of parabolic trajectories, Huygens's investigations of centrifugal force, and especially Newton's propositions in *Principia Mathematica*, we are accustomed to consider curvilinear trajectories as the resultant of rectilinear uniform motion composed with uniformly accelerated motion. At times, however, Newton's *Principia* appears as a hybrid in which different representations occur. In the second law Newton stated that 'the change of [the quantity of] motion is proportional to the motive force impressed', adding that the effect is the same 'whether that force is impressed altogether and at once, or gradually and successively'. Definition 4 states that 'impressed forces are of different origins, as from percussion, from pressure, from centripetal force.' Notice the coexistence of discrete and continuous representations. Is the reference to percussion relevant to finite impacts only, or to curvilinear motion as well? In proposition 1, proving that 'the areas which revolving bodies describe by radii drawn to an immovable centre of force do lie in the same immovable planes, and are proportional to the times in which they are described', the demonstration is based on the polygonal model: a series of instantaneous impulses deflects the orbiting body from its rectilinear path. Significantly, Newton added referring to triangles *SAB, SBC*, etc. (see Fig. 4.1): 'Now let the number of those triangles be augmented, and their breadth diminished *in infinitum*; and . . . their ultimate perimeter *ADF* will be a curved line: and therefore the centripetal force, by which the body is continually drawn back from the tangent of this curve, will act continually'. One may wonder at this point whether proposition 1 takes its form for mathematical reasons, or because Newton believed impacts to be ultimately responsible for curvilinear motion. Although before 1680 the

[8] An excellent multi-dimensional perspective is given by D. Mahnke, *Leibnizens Synthese von Universalmathematik und Individualmetaphysik* (Halle, 1925, reprinted Stuttgart, 1964), esp. p. 32. The controversy about relativity of motion erupted in the correspondence between Leibniz and Clarke. In addition to the brief discussion and references in sect. 5.2, the reader can see the following works selected from an immense bibliography: Sklar, *Space*, ch. 3; Earman, *Worlds enough*; H. Stein, 'Newtonian space-time', in C. Palter, ed., *Annus Mirabilis*, pp. 258–84.

[37]

SECT. II.

De Inventione Virium Centripetarum.

Prop. I. Theorema. I.

Areas quas corpora in gyros acta radiis ad immobile centrum virium ductis describunt, & in planis immobilibus consistere, & esse temporibus proportionales.

Dividatur tempus in partes æquales, & prima temporis parte describat corpus vi insita rectam *A*B. Idem secunda temporis parte, si nil impediret, recta pergeret ad *c*, (per Leg. 1) describens lineam B*c* æqualem ipsi *A*B, adeo ut radiis *A*S, B S, *c*S ad centrum actis, confectæ forent æquales areæ *A* SB, B S*c*. Verum ubi corpus venit ad B, agat vis centripeta impulsu unico sed magno, faciatq; corpus a recta B*c* deflectere & pergere in recta B*C*. Ipsi B S parallela agatur *c*C occurrens B*C* in

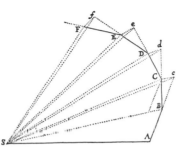

C, & completa secunda temporis parte, corpus (per Legum Corol. 1) reperietur in*C*, in eodem plano cum triangulo A S B. Junge S*C*, & triangulum S B*C*, ob parallelas S B, C *c*, æquale erit triangulo S B*c*, atq, adeo etiam triangulo S *A* B. Simili argumento si
vis

Fig. 4.1   Proposition 1 in the *Principia*.

latter may have appealed to Newton, I believe that by the time he was composing the *Principia* the polygonal representation was merely part of the scaffolding of the proof. As we have seen, Newton referred to a curved line resulting from a process *in infinitum* and to a centripetal force acting continually. The reader encounters finite impacts only at the beginning; ultimately the curve becomes smooth. Indeed, generally Newton favoured the continuous approach. Good examples of his method of representation can be found in lemma 10 and proposition 6. The former constitutes a generalization of Galileo's law of fall to the case when gravity is not constant. It states that 'the spaces which a body describes by any finite force urging it, are in the very beginning of the motion to each other as the squares of the times.' This lemma is a corollary of lemma 9; if in Fig. 3.3 the times are represented by *AD*, *AE*, and the velocities by *DB*, *EC*, the spaces will be as the areas *ABD*, *ACE*. Hence at the very beginning of motion the spaces are as the squares of the times. In a preliminary version Newton referred to Galileo's law of falling bodies. In the first edition proposition 6 is based on lemma 10, and establishes that if a body P moves along the curve APQ around a centre of force S, the centripetal force is proportional to the deviation RQ from the tangent RPZ and is inversely as the square of the time. Time is as the triangle SPQ, namely as $SP \cdot QT$, where QT is the altitude and SP the radius. Newton specifies that this theorem is valid in the limit when Q coincides with P.

In the proof Newton specifies that the figure $QRPT$ is indefinitely small; also in this circumstance PQ is taken to be an arc rather than a straight line, and motion along QR is accelerated rather than uniform

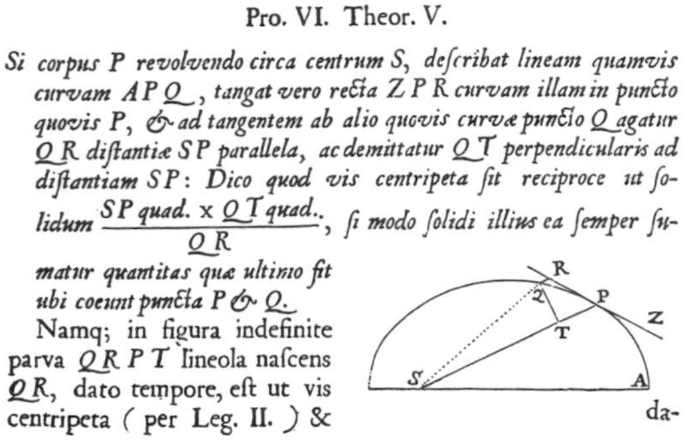

Pro. VI. Theor. V.

Si corpus P revolvendo circa centrum S, deſcribat lineam quamvis curvam A P Q , tangat vero reċta Z P R curvam illam in punċto quovis P, & ad tangentem ab alio quovis curvæ punċto Q agatur Q R diſtantiæ S P parallela, ac demittatur Q T perpendicularis ad diſtantiam S P : Dico quod vis centripeta ſit reciproce ut ſo-

lidum $\dfrac{S\,P\ quad.\ \times\ Q\,T\ quad.}{Q\,R}$, *ſi modo ſolidi illius ea ſemper ſu-*

matur quantitas quæ ultimo fit ubi coeunt punċta P & Q.

Namq; in figura indefinite parva $Q\,R\,P\,T$ lineola naſcens $Q\,R$, dato tempore, eſt ut vis centripeta (per Leg. II.) &

Fig. 4.2   Proposition 6 in the *Principia*.

da-

and rectilinear. Thus lemma 10 and proposition 6 are clearly related to accelerated motion, and the former can be reformulated in the following way: at the very beginning the motion produced by an arbitrary acceleration law can be conceived to be uniformly accelerated, namely centripetal force can be taken to be constant over a very small distance.[9]

However, it is possible to use different representations. Leibniz, for example, had a clear predilection for rectilinear uniform motions and dispensed with accelerations whenever he could; indeed, he sometimes did so even when he could not. If we consider a circumference centred in C and the infinitesimal arc $\widehat{AG}$, there are two ways of conceiving motion along it: either rectilinear uniform along the tangent AD and uniformly

[9] Quotations from the *Principia* are from the transl. by A. Motte and F. Cajori (Berkeley and Los Angeles, 1934), pp. 2, 13, 34, 40, 41. T. L. Hankins, 'The reception of Newton's second law of motion in the 18th century', *Arch. Int. d'Hist. des Sciences*, 1967, **20**, pp. 43–65, esp. pp. 55–6; E. J. Aiton, 'Polygons and Parabolas: Some Problems Concerning the Dynamics of Planetary Orbits', *Centaurus*, 1988, **31**, pp. 207–221. Westfall, *Force*, ch. 7. D. T. Whiteside has shown that taking accelerations into account, if we want to have rectilinear elements of the curve, they must be second-order infinitesimals; see 'Newtonian Dynamics' *History of Science*, 5, 1966, pp. 104–17; *NMW*, 6, pp. 34–9, n. 19, and ib., pp. 76–7, 97–8, 549–51, 563. I. B. Cohen, 'Newton's second law and the concept of force in the *Principia*', in R. Palter, ed., *The 'annus mirabilis' of Isaac Newton* (Cambridge, Mass., 1970), pp. 143–85. Cohen, *Newtonian Revolution*.

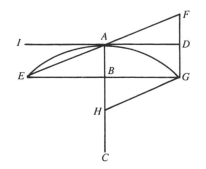

Fig. 4.3 Leibniz's continuous and discrete analyses of curvilinear motion in his 1706 *Excerptum ex epistola.*

accelerated along *DG*, as Newton did in proposition 6; or rectilinear uniform along *AF*, which is the prolongation of the chord *EA*, and along *FG* (*FG= AH*). Here *EA= AF*, and this means that the arcs $\widehat{EA}$ and $\widehat{AG}$ are traversed in equal times. Both alternatives lead to the same result for the central force or conatus, because the factor 1/2 involved in the choice of *DG* rather than *FG* cancels out the factor 1/2 resulting from motion along *DG* being uniformly accelerated. This seems to be the agreed interpretation concluding a debate between Leibniz and Pierre Varignon which lasted from December 1704 until 1706. This debate has been aptly defined as a 'comedy of errors' because Leibniz and Varignon exhausted virtually all possible representations, including a curve with continuous curvature without accelerated motion, giving half the correct value for centrifugal force, and a polygon with accelerated motion, giving twice the correct value. In the end Varignon preferred the solution which involves accelerations because he believed that centripetal or centrifugal force act continuously; in this respect he followed Newton's mathematical treatment. Leibniz, on the other hand, had no hesitations as to the interpretation he favoured: 'The simpler way is that in which acceleration does not occur in the elements, whenever this is not necessary. I have used it for more than thirty years.' And thirty years before takes us exactly to the letter to Claude Perrault quoted at the opening of this section. Notice that Leibniz prefers not to use accelerations in the elements, that is, in infinitesimal segments.[10]

In the *Tentamen* Leibniz tried to use similar considerations twice, for gravity and centrifugal force; in both cases he considered only uniform rectilinear motions. This led to a correct result for gravity, because he

[10] *LMG, 4*, pp. 150–51, 10 Oct. 1706: 'La voye est plus simple qui ne met pas l'acceleration dans les elemens, lorsqu'on n'en a point besoin. Je m'en suis servi depuis de 30 ans.' Aiton, 'The Celestial Mechanics of Leibniz', *AS, 16*, 1960, pp. 65–82, on pp. 77–82. Compare also the correspondence with Hermann in 1709, *LMG, 4*, pp. 344–53.

took the tangent as the prolongation of the chord. In the case of centrifugal force, however, he took the deviation from the rectilinear path to be 1/2 of the correct value without realizing the need for accelerations. This introduced the mistake by a factor 1/2 later corrected thanks to Varignon. Before Varignon's explanation, Leibniz neglected acceleration and believed that it was equivalent to consider motion along an infinitesimal arc or along its chord, because the difference between their lengths is incomparably small. The debate on these two different representations of curves and on their implications for mechanics lasted for several decades into the eighteenth century.[11]

It is worth explaining this issue with an explicit calculation (see Fig. 4.3). I follow a slightly modernized notation and pay special attention to constant factors. Take a body moving along the infinitesimal arc $\widehat{EAG}$ of a circumference with centre C. In the continuous representation, if the body is released in A, it will fly away along the tangent AD. The arc $\widehat{AG}$ and the segment AD are traversed in the element of time dt. DG is the deviation from the tangent to the circumference, which in the element of time dt can be taken to be parallel to AC; DG can be expressed as $DG = (AG)^2 : 2(AC)$. This is an elementary geometric relation based on the property that all triangles inscribed in a semicircle are right-angled; the diameter of the circle is $2(AC)$; further $BG = AD$. We move now to mechanics. DG is traversed with accelerated motion, since in the continuous representation the resultant $\widehat{AG}$ of the composition with the rectilinear uniform motion AD is curvilinear even for infinitesimal arcs. In the element of time dt the arc $\widehat{AG}$ of the circumference can be approximated by a parabolic arc, their difference being negligible. From the law of accelerated motion, $DG = (dt)^2 a : 2$, where a is the acceleration. From the two equations one has $(dt)^2 a : 2 = (AG)^2 : 2(AC)$, or $a = (AG : dt)^2 : AC$, namely acceleration is equal to the square of velocity over the radius. We consider now the polygonal representation with dt a constant, namely the sides of the polygon are equal, $EA = AG$. Now the body released in A moves along AF, and the deviation is FG. Since the body moves along a polygon, the resultant of AF and FG must be the rectilinear segment AG, hence motion along FG is rectilinear uniform. From simple geometry $FG = (AG)^2 : AC$. Introducing mechanics, we have $FG = bdt$, where b is a velocity. Notice, however, that if AC is finite, AG is infinitesimal of the first order and FG is infinitesimal of the second order. Hence b is an infinitesimal velocity $b = dv = ddr : dt$. From the previous equations one has $ddr = (AG)^2 : AC$. Since dt is constant, one can

---

[11] Leibniz, 'Excerptum ex Epistola', *LMG*, 6, pp. 276–80. *DG* is called *conatus centrifugus tangentialis*, and applies only at the beginning of motion; *FG* is called *conatus centrifugus arcualis* and applies to the body while rotating; in this essay Leibniz does not mention accelerations. On this problem see chapter 9 below.

say that the solicitation *ddr* is as the square of velocity over the radius. Notice that in this way accelerations are avoided in the first-order elements of the curve. For a finite arc, however, one can use accelerations regardless of the representation employed, even if the ingredients vary in the two cases. Further, in both representations velocity can be taken to be proportional to *BG*, because the difference between AG^2 and BG^2 is a fourth-order infinitesimal and therefore the measure of force or conatus is not affected. This approximation is often adopted by Leibniz.

In the cases which we have seen so far Leibniz tried to dispense with accelerations on mathematical grounds because of the way he conceived curves, and on physical grounds because of his belief that phenomena were ultimately explicable in terms of impacts. One may wonder at this point whether for Leibniz the polygonal representation corresponds to the real trajectory of the body, namely whether motion occurs along the segments *EA*, *AG*, etc. The answer to this question can only be a resounding no, and the reason is straightforward. The choice of the specific polygon entails a degree of arbitrariness depending on the progression of the variables. In our case *dt* is constant, hence the chords *EA*, *AG*, etc. are equal. However, different progressions of the variables could have been selected. In general, the vertices of the infinitangular polygon cannot be the actual places where impacts occur; Leibniz's mathematical representations of curvilinear motion are fictitious. The polygonal curve, however, corresponds in principle to physical actions in a way that the continuous curve does not. By correspondence 'in principle' I mean that mathematics mirrors the physical laws involved, rather than the infinitesimal details of the trajectory of the body. Moreover, as we have seen in Section 2.4 on elasticity, for Leibniz change takes place not instantaneously, but gradually and in accordance with the law of continuity. Thus each vertex ought to represent a smooth rather than a sudden transition.

In an essay of 1705, *Illustratio Tentaminis de Motuum Coelestium Causis*, Leibniz represented rectilinear uniformly accelerated motion by breaking up time into infinitesimal elements during which velocity remains constant.

QR represents time; *QS*, *SW*, *WX*, etc. are equal elements of time *dt*; *QTZw* is the curve of velocities showing that motion is uniformly accelerated. But at the level of first-order infinitesimals the curve is broken up into 'steps' representing uniform motion with velocity *ST* along *VT*, *WZ* along *YZ*, etc. and impulsive increases in velocity *QV* at the instant *Q*, *TY* at the instant *S*, etc. The area of the triangle *QRw* represents the space traversed in the time *QR*. The error introduced with this *motus scalariter acceleratus*, as Leibniz calls it, is proportional to the area of the rectangle *wRA* and is therefore negligible. Although Leibniz

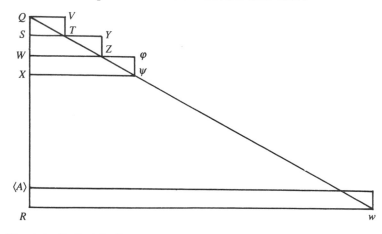

Fig. 4.4   Leibniz's discrete representation of rectilinear acceleration.

introduced this representation as a mathematical construction, a mechanical interpretation is straightforward: QV, TY, etc. are infinitesimal impulses or microscopic impacts, whilst accelerations appear merely as a macroscopic effect. We shall have to bear in mind these fundamental examples and diagrams in several instances below. Once again, the notion of equivalence between Newtonian and Leibnizian notions presents itself in a problematic fashion.[12]

4.3   Mechanical notions and their mutual relations

After these preliminary remarks, we are now equipped to examine some of Leibniz's most characteristic notions relevant to the texts discussed in Part 2. The task is rendered more difficult for the modern reader by the different treatment of the problem of dimensions. The introduction of factors with the appropriate dimensions, such as angular momentum per unit mass in the expression of Kepler's area law, occurs very rarely in early-modern mechanics. In the seventeenth century the most common way for avoiding the problem of dimensional inhomogeneity was to construct a suitable proportion. In order to express mathematically Kepler's area law, for example, one would write $A_1 : A_2 :: t_1 : t_2$, where A is the area and t time, and say that 'the areas are as the times'. Since such expressions often involved infinitesimals, one can say that the differential proportion was the primary tool in the mathematical investigation of

---

[12] 'Illustratio Tentaminis de Motuum Coelestium Causis', *LMG*, 6, pp. 254–76, is a longer preliminary draft of 'Excerptum ex epistola'; see esp. p. 259. See also Huygens, *Horologium*, part II, prop. 5.

nature. Strictly speaking, even the form in which the proportion above is written is slightly anachronistic, because the symbol of a variable with an appropriate index was not currently used. Mathematicians represented variables, including time, length, velocity, etc. by means of segments and areas of a suitable figure. Therefore proportions were written in a form closer to geometry, since variables were indicated by the corresponding letters in the figure, i.e. a segment as *AB* or *CD*, an area as *EFG* or *HILM*. Continental mathematicians often used algebraic symbols for such variables, for example *a* or *b* for constants, *x* or *y* for non-constant magnitudes, and *dx* or *dy* for their increments.[13] In differential proportions, however, constant factors were usually neglected or at least treated as secondary with respect to variables. This attitude led to a further problem related to dimensions in the mathematical representation of physical magnitudes. Leibniz in particular usually neglected constant factors, especially mass and time. Strictly speaking velocity, for example, is expressed by Leibniz as $x = dr:dt$, where dr and dt are the differentials of a length and of a time respectively. In the theory of planetary motion, and in general in all cases where the element of time dt is constant, Leibniz wrote velocity simply as dr, meaning *dato tempore*. It is worth recalling that the notation $dr:dt$ represents the ratio between two differentials and does not correspond to the notion of derivative of a function with respect to an independent variable. Quantity of motion and impetus, for example, are taken as synonyms meaning mass times velocity in *Dynamica* and *Specimen Dynamicum*, but in the *Tentamen* mass is neglected, velocity and impetus being taken as equivalent.[14] The term 'neglected' does not necessarily carry a negative connotation; it simply emphasizes that Leibniz focused on variables and often assumed that proportions were involved. In the study of differential equations, however, this attitude created several problems, as we are going to see. The modern reader ought not to be shocked if order of infinity and physical dimensions seem to be treated inadequately. Representations varied according to specific needs and circumstances. Proportions were often employed by Newton too, though with different preoccupations. As we have seen in Section 3.4, Newton generally avoided infinitesimals. Thus from his perspective proportions had the advantage of involving finite ratios rather than infinitesimal variables. An example of this usage can be seen in lemmas 9 and 10 of the *Principia*, book I.

Seventeenth-century studies of motion often involved reflections on infinitesimal velocity, or infinite tardiness. Howard Bernstein has claimed that Leibniz's reference to *conatus* or *solicitation* to indicate an

[13] Bos, 'Differentials', p. 47f. Cohen, *Newtonian Revolution*, sect. 1.3.

[14] *LMG*, 6, *Dynamica*, p. 398; *Specimen Dynamicum*, (Hamburg, 1982), pp. 10–11, Loemker, *Papers*, p. 437, Ariew and Garber, *Philosophical Essays*, p. 120.

infinitesimal velocity or a tendency to motion derived from Thomas Hobbes and had already appeared in the 1671 *Hypothesis Physica Nova*. Although it can be argued that Leibniz's usage of the same terms in analogous contexts derives from Hobbes, their exact meaning and mode of operation varied. Hobbesian 'conatus' was defined as 'motion made in less space and time than can be given'; despite this association with instantaneous velocity, Hobbes also stressed that conatus does 'not always appear to the senses as motion, yet it appears as action, or as the efficient cause of some mutation.'[15] Leibniz was able to provide a more precise definition closely linked to his mathematics, though some ambiguities remained. In the *Specimen Dynamicum* solicitation means an infinitesimal velocity, whilst conatus is used in the sense of velocity in one direction—a meaning close to the modern notion of vectorial velocity. In the *Tentamen*, however, solicitation and conatus—which can be rendered as 'endeavour'—are synonyms with regard to their mathematical expression and correspond to the differential of velocity dx, written as ddr because the element of time is constant. Possibly 'solicitation' conveys the idea of an external action, whereas 'conatus' suggests a tendency internal to the body. In spite of Leibniz's repeated statements that for him solicitations or endeavours are not accelerations, several commentators have overlooked this distinction.[16]

In a letter of 1673 to Edme Mariotte, Leibniz referred to pendular oscillations and mentioned a *force morte*, which is the weight of a body, and a *force violente ou animée*, which is the force of impact or percussion and is infinitely greater than dead force. This is his first known reference to these important notions. The most likely source for the notion of 'dead force' is *Saggi di Naturali Esperienze Fatte nell'Accademia del Cimento*, drafted by its secretary Lorenzo Magalotti, published in Florence in 1666 and read by Leibniz in 1673.[17] As in the case of

[15] Useful material on this topic regarding Galileo and his sources is in P. Galluzzi, *Momento. Studi Galileiani* (Roma, 1979). See also H. R. Bernstein, '*Conatus*, Hobbes and the Young Leibniz', *SHPS*, *11*, 1980, pp. 25–38. Hobbes, *De Corpore* (London, 1655), in Molesworth, *1*, pp. 206 and 342. The term 'conatus' occurs also in *Principia Philosophiae*, part III, prop 57. Leibniz read *De Corpore* shortly before his letter to Thomas Hobbes of 23 July 1670, *LSB*, II, *1*, pp. 56–9. Westfall, *Force*, pp. 109–14. D. Garber, 'Motion and Metaphysics in the Young Leibniz', in M. Hooker, ed., *Leibniz. Critical and Interpretive Essays*, (Minneapolis, 1982), pp. 160–84. On Newton see Herivel, *Background*, part I, ch. 3.

[16] See the correspondence with Jakob Hermann in the years 1712–13, *LMG*, *4*, pp. 368, 372, 378, 384 and especially 387, where Leibniz objects to Hermann's claim that solicitations are accelerations: 'Et primum observo mihi solicitationes et ipsa celeritatis incrementa esse idem.' 1 Febr. 1713.

[17] *LSB*, III, *1*, pp. 105–12, July 1673; pp. 107 and 109–10. Together with Huygens, Mariotte was Leibniz's mentor in mechanics. On this occasion he probably corrected a mistake by Leibniz concerning decomposition of motion, p. 109n. M. Fichant, "La 'Reforme' Leibnizienne de La Dynamique, d'apres des Textes Inédits", *SL* Supplementa *13*, 1974, pp. 195–214; 'Les Concepts Fondamentaux de la Mécanique selon Leibniz, en 1676', *SL* Supplementa, *17*, 1978, pp. 219–32. Magalotti, *Saggi*, pp. 141–2 and 202; Hess,

conatus with respect to Hobbes, however, it is by no means certain that at this early stage Leibniz considered violent force to be proportional to the square of velocity, as he did after the late 1670s. Dead and living force occur several times in Leibniz's works, but they are first mentioned in print in the *Tentamen*.[18] Dead force is mass times an infinitesimal velocity, while living force is mass times the square of velocity. Since mass is constant, Leibniz often said that solicitation is as dead force, namely as x or dr, and living force is as the square of velocity, namely as x^2 or dr^2. The following quotation from the *Specimen Dynamicum* provides examples of dead and living forces and explains their relations in a non-quantitative fashion:[19]

Hence *force* is also of two kinds: the one elementary, which I also call *dead* force, because motion does not yet exist in it, but only a solicitation to motion . . .; the other is ordinary force combined with actual motion, which I call *living* force. An example of dead force is centrifugal force, and likewise the force of gravity or centripetal force; also the force with which a stretched elastic body begins to restore itself. But in impact, whether this arises from a heavy body which has been falling for some time, or from a bow which has been restoring itself for some time, or from some similar cause, the force is living and arises from an infinite number of continuous impressions of dead force.

In this passage elasticity is interestingly treated as homogeneous to centripetal and centrifugal forces. This treatment is extended to living forces too, and this is certainly the area with the greatest philosophical significance.

The *Specimen Dynamicum* induced a debate on mechanics between Leibniz and Johann Bernoulli. In a letter to his Swiss friend dating from the same year as his essay on dynamics, Leibniz stated that the *ars aestimandi* consists in reducing phenomena *ad mensuram quandam congruam*, to a certain homogeneous measure which can be employed as the unity is for numbers. Leibniz identified in living force this unit

'Kurzcharacteristic', p. 199. Several commentators pointed out that the expression 'peso morto' had already been used by Galilei and mention Galileo's so-called sixth day of the *Discorsi* as Leibniz's source for the expression 'dead force'. In fact, the sixth day of the *Discorsi* was published for the first time in 1718 and cannot have been read by Leibniz.

[18] Compare 'De Arcanis Motus' in Hess, 'Kurzcharacteristik', p. 202; the letter to Antoine Arnauld in G. Le Roy, ed., *Arnauld*, pp. 143–50, on p. 149, 8 Dec. 1686. *Tentamen*, paragraph 10. This has been pointed out by A. Robinet, 'Dynamique et Fondements Métaphysiques', *SL* Sonderheft *13*, 1984, pp. 1–25, on p. 18. Both in the *Tentamen* and *Specimen Dynamicum* Leibniz refers to Galileo for an example of dead and living force. The example is the 'forza della percossa', which Galileo claimed to be 'interminata, per non dire infinita' with respect to static force; *Discorsi e Dimostrazioni Matematiche Intorno a Due Nuove Scienze* (Leida, 1638) = *GOF*, *8*, p. 313, fourth day.

[19] *Specimen Dynamicum*, (Hamburg, 1982), pp. 12–15. Transl. from Loemker, *Papers*, p. 438; Ariew and Garber, *Essays*, pp. 122–3. 'Essay de Dynamique', *LMG*, *6*, pp. 215–31, on p. 218.

measure common to all phenomena. The equivalence between cause and effect embodied in the conservation of living force allows a quantitative description of the world. If force were not conserved, its measure would be vague and inaccurate, depending on the different circumstances and points of view. Therefore the programme for a mathematization of nature and indeed Leibniz's own philosophy would collapse. His first publication on this issue is the famous *Brevis demonstratio erroris memorabilis Cartesii et aliorum* of 1686, in which he claimed that force is proportional to the square of velocity or the height of fall. In his celebrated essay Leibniz claimed that it is reasonable to suppose that living force—he says the sum of motive power—is conserved in nature, and recalled Descartes's view that force is equivalent to quantity of motion, that is, mass times velocity. On the basis of the two commonly accepted assumptions that a body falling from a certain height acquires a force sufficient to raise it back to its original height, and that the same force is required to raise a body of one pound to four feet or a body of four pounds to one foot, he showed that Descartes's measure of force leads to a contradiction, because a force could produce an effect greater than its cause. Leibniz's solution is that force is proportional to the height of fall or the square of velocity, not to simple velocity.[20]

The relation between dead and living force is further explained in a letter to the professor of mathematics and philosophy at Leiden, Burchard de Volder, to whom Leibniz wrote:[21]

And the impetus of living force is related to bare solicitation as the infinite to the finite, or as lines to their elements in my differential calculus . . . Consequently, in the case of a heavy body which receives an equal and infinitely small increment of velocity at each instant of its fall, dead force and living force can be calculated at the same time: to wit, velocity increases uniformly as time, but absolute force as distance or the square of time, that is, as the effect. Hence according to the analogy of geometry or of our analysis, solicitations are as dx, velocities are as x, forces as xx or as $\int x\,dx$.

In this passage and in the previous one accelerations seem to be treated as in *Illustratio Tentaminis*, namely as a *motus scalariter acceleratus*. Solicitations and velocities are linked by a single integration, $x = \int dx$, where time is not involved in the calculations. This is the obvious consequence of solicitations being as infinitesimal velocities and not accelerations. By contrast, the modern reader associates the transition from acceleration to velocity with an integration over time,

[20] *LMG, 3,* Leibniz to Johann Bernoulli, 29 July 1695, p. 208. Cassirer, *Erkenntnisproblem,* vol. 2, p. 165. *AE* March 1686, pp. 161–3 = *LMG, 6,* pp. 117–19; Loemker, *Papers* pp. 296–302.
[21] *LPG, 2,* pp. 153–63, on pp. 154 and 156; after Nov. 1698. Translation from Westfall, *Force,* p. 298. Compare also *Dynamica, LMG, 6,* p. 454.

$\int dt \cdot a = x(+ \text{constant})$, where a is the acceleration. As we have seen in Section 3.4, Leibnizian integration and differentiation change the size of the *integrandum* and *differentiandum* respectively to infinitely large or infinitely small. A distance is transformed into an infinite or infinitesimal distance, and a velocity into an infinite or infinitesimal velocity respectively. Their physical dimensions, however, remain unchanged.

There is a striking disagreement among historians on the interpretation of the second part of the quotation above and of other related passages relevant to the link between dead and living force. Some claim that living force was the integral of dead force times an infinitesimal distance. René Dugas in particular maintained that the previous quotation contains the theorem of living force and constitutes the great claim to glory of Leibniz's dynamics. Ernst Cassirer is the main advocate of the thesis that Leibniz discovered the concept of work, though his arguments are mainly philosophical rather than mathematical.[22] Others deny that Leibniz ever grasped the link between dead and living force. In their opinion this was a consequence of his metaphysical notion of substance or monad. According to these interpreters, since monads evolve in time, Leibniz would integrate dead force over time rather than over distance, and the integral of dead force over time would be proportional to simple velocity, not to velocity squared or to living force.[23] However, it is doubtful whether Leibnizian substances can be located in space and time, and any straightforward parallel between metaphysics and mechanics in this form must be treated with considerable scepticism. Other objections relate more closely to the present chapter; even if we neglect metaphysics, in a number of passages such as the quotation above from the *Specimen Dynamicum*, Leibniz seems to state that living force is the integral of dead force multiplied by the element of time. Since this integral would give simple velocity, he would have failed to establish the correct quantitative link between dead and

[22] Cassirer, *Leibniz' System*, ch. 6. Dugas, *Mécanique*, p. 490, renders 'Leibniz's' theorem as follows: $\int F \cdot dr = 1/2 \ mv^2$ (the integration constant is omitted); F is mass times vectorial acceleration and dr is also a vector. A similar position is held by J. O. Fleckenstein in L. Euler, *Opera Omnia*, II, 5, pp. xiv and xxxii.

[23] Westfall, *Force*, pp. 299–301, on pp. 299–300; 'What was readily accessible to his mathematics was obstructed by his philosophical considerations, and Leibniz never suggested that the generation of *vis viva* in free fall be seen as the integration of dead force over distance. No concept in his dynamics corresponds closely to the ideas of work and potential energy ... Indeed the whole tenor of Leibniz's philosophy sets itself against the possibility of a functional relation of force and distance.' And ib.: 'The law that governs a substance unrolls its course through time rather than space. Whereas the summation over time of a body's endeavours repeats mathematically the law of its being, a similar summation over space lacks all meaning.' K. Okruhlik, 'Ghosts in the World Machine: a Taxonomy of Leibnizian Forces', in J. C. Pitt, ed., *Change and Progress in Modern Science*, (Dordrecht, 1985), pp. 85–105, on pp. 97–9.

living force, which is as the square of velocity.[24] However, I find no convincing evidence that for Leibniz dead force is integrated in that way, namely $\int$(dead force)dt. Indeed, this belief seems to be linked to the erroneous representation of dead force or solicitation as proportional to acceleration. For Leibniz the integral of solicitation or dead force as such—not multiplied by the element of time—is proportional to simple velocity. Regarding Leibniz's alleged failure to realize that 'his' integration of dead force would produce simple velocity, I believe one ought not to translate too hastily his general statements into the rigorous language of the calculus. When he talks of a 'heavy body which has been falling for some time', he does not mean that the integral of dead force is multiplied by an element of time, but is simply providing a general description of the phenomenon. Further indication of Leibniz's failure to attain living force by integration would be the lack of the factor 1/2 in its expression. In other passages Leibniz rather enigmatically declares that the relation between dead and living force is analogous to that between a point and a line.[25] In my interpretation he wanted to stress that dead force is infinitesimal, whereas living force is finite. Chapters 5 and 7 below will show that Leibniz realized that dead and living force are related via a simple integration where dead force or conatus is multiplied by an infinitesimal distance. Moreover, I claim that the lack of the factor 1/2 in the expression of living force is a consequence of Leibniz's general habit of neglecting constant factors, and argue that his differential equations relevant to force can be plausibly interpreted in terms of the elasticity of the aether.

It may be useful to summarize some features of Leibniz's and Newton's mathematical representations of motion. Rather than trying to cover the entire spectrum of their researches, I focus on the relevant material from the *Tentamen* and *Principia*, starting from some general well known statements which will be briefly discussed in the following chapters.

Leibniz

● Empty space does not exist. The world is filled with a variety of fluids which are responsible for physical actions, including gravity.

[24] Together with Westfall and Okruhlik compare I. Szabó, *Geschichte der mechanischen Prinzipien* (Basel: Birkhaüser, 1979, second edition), pp. 70–1. In his opinion for the link between dead and living force we have to wait for Daniel Bernoulli, 'Examen Principiorum Mechanicae', *Commentarii Academiae Scientiarum Petropolitanae, 1,* 1726, pp. 126–42. However, compare Proposition 39 in Book 1 of Newton's *Principia Mathematica*, where the same result is proved in geometrical fashion; P. Varignon, 'Maniere Générale de Déterminer les Forces, les Vitesses, les Espaces, et les Temps', *MASP* 1700, pp. 22–7, on p. 27, and several other texts.

[25] P. Costabel, *Leibniz et la Dynamique*, (Paris, 1960), p. 104; compare also the quotation from the letter to de Volder above.

- Living force and its conservation are the fundamental notion and principle respectively, in the investigation of nature; however, they do not figure prominently in the study of planetary motion.
- Finite and infinitesimal variables are regularly employed in the study of motion and of other physical phenomena. Living force and velocity are finite; solicitation and conatus are infinitesimal.
- Accelerated motion, whether rectilinear or curvilinear, is represented as a series of infinitesimal uniform rectilinear motions interrupted by impulses. I call this 'polygonal representation'. Usually the polygon is chosen in such a way that each side is traversed in an equal element of time *dt*. In polygonal representations accelerations are reduced to a macroscopic phenomenon.
- Proportions are often used to safeguard dimensional homogeneity. Constant factors—such as numerical factors, mass, and the element of time—are usually ignored in the calculations.

Newton

- Celestial motions occur in spaces either empty or void of resistance. All bodies attract each other in proportion to their masses and inversely as the squares of their mutual distances.
- Force and acceleration are the key notions in the study of motion.
- Variables employed by Newton, such as velocities and accelerations, are rigorously finite.
- Accelerated motion, whether rectilinear or curvilinear, is represented by a continuous curve where force acts continually. I call this 'continuous representation'. Newton avoided infinitesimals by considering either the rate of change of a variable with respect to time, or the finite ratio between two variables. In general time plays a central role and is part of the foundations of Newtonian mechanics.
- Proportions are often used to safeguard dimensional homogeneity; further, they have the advantage of involving finite ratios between vanishing quantities rather than vanishing quantities taken in isolation.

It would be impossible to present a comprehensive study of the sources relevant to Leibniz's theory. Though not exhaustive, this preliminary account sets the scene for the study of the manuscripts and the dispute with Newton.

PART 2

THE TRANSFORMATION OF A WORLD SYSTEM: FROM THE *PRINCIPIA MATHEMATICA* TO THE *TENTAMEN*

5

THE PRIVATE ITINERARY

5.1 Introduction

This chapter outlines the itinerary leading from the *Notes*, namely Leibniz's working sheets containing his immediate thoughts on the *Principia*, to the *Tentamen*. All the manuscripts discussed here are unpublished: those marking the most significant stages in Leibniz's itinerary are reproduced in the Appendix, where I also provide a detailed textual analysis. In this chapter, however, I present a broader picture, including some manuscripts which are not reproduced later and the relevant *Marginalia*. Indeed, in some cases Leibniz's isolated annotations in his own copy of Newton's masterpiece can only be understood in relation to the texts presented here.

Leibniz's first impressions of the *Principia* can be reconstructed from a combined study of the *Notes*, *Marginalia*, and *Excerpts*. It is well known that Newton's masterpiece starts from the definitions and laws of motions, and contains two books on the motion of bodies through spaces void of resistance and in resisting media respectively, and a third book on the system of the world. Leibniz had a mixed reaction to the approximately 500 pages of the *Principia*: he criticized the lemmas on the first and last ratios in section 1, book 1, but was most impressed by the generalization of Kepler's area law in proposition 1. At the end of the *Notes* Leibniz was progressively abandoning any direct reference to the text and pursuing his own line of research; this was based on his transmutation theorem and on the attempt to determine the centripetal or centrifugal conatus for a body moving under the action of central forces. It will become clear in this chapter that it is impossible to isolate two separate phases of Leibniz's thought, namely that of the interpretation and that of the development of his own theory. Rather, I wish to stress that these two moments are present at the same time: Leibniz's *Notes* and essays on planetary motion form a continuum.

Leibniz's manuscript essays are very tentative, often unfinished or with later additions. The difficulties involved in dealing with the first steps of the new science of motion using the most advanced mathematical techniques available in his time largely account for Leibniz's line of thought being by no means straight. In spite of these problems, from a combined examination of several factors it is possible to reconstruct a

convincing intellectual development and strategy. The chronology which
emerges leads us, some time in the last quarter of 1688, from the
Principia Mathematica to the *Tentamen*.

Before embarking on the analysis of the manuscripts, I wish to outline
the criteria I have adopted for ordering and dating the texts. The first
criterion concerns papers and watermarks. All watermarked manuscripts
discussed in this chapter are of a type used by Leibniz in Vienna in 1688.
Although this element alone does not prove that the essays date from
1688, since Leibniz could have carried some Viennese paper with him to
Italy, it certainly establishes the possibility that they may have been
composed in Vienna, thus before the *Tentamen*. The second criterion
involves the conceptual development of Leibniz's theory and terminology employed, and provides a firmer basis for the dating. Considering
the care with which Leibniz chose his key expressions, such as *circulatio
harmonica* and *motus paracentricus*, it is reasonable to believe that the
occurrence of different terms for such crucial notions, or their occurrence with a different meaning, indicate an earlier date of composition. A
third related criterion concerns the sequence of the manuscripts, which
in many cases can be determined with great accuracy from several
factors: they include detailed cross references, a reasoning which is interrupted in an essay and then continued and developed in another text, the
perfect matching of the edges of two manuscripts, and the combination
of these factors. The relevant criteria are introduced and analysed at the
appropriate place below, so that each step is justified in the making.

5.2   The first reading of the *Principia*

Leibniz's style of working was often to comment on books or create
imaginary dialogues with other philosophers. Among the most famous
examples are the *Animadversiones in Partem Generalem Principiorum
Cartesianorum, Nouveaux Essais*, and *Essais de Théodicée*, which are
largely based on debates with René Descartes, John Locke, and Pierre
Bayle respectively. These works were either published or written with
the intention of publication. As will be immediately clear from the
transcriptions in the Appendix, this is not the case with the *Notes*.
Leibniz's manuscripts contain a private collection of thoughts and
calculations which he would not have wished to see in print. However,
the *Notes* shed new light on the development of Leibniz's thought and of
his relationships with Newton. Leibniz made several attempts both to
translate Newton's terminology and concepts into the scheme of his own
terms and ideas, and to attain similar mathematical results using his own
tools.

No attempt is made here to summarize the main results in the *Principia*, let alone to discuss the genesis of Newton's ideas; a partial effort in this direction can be found in Chapter 8. While focusing on Leibniz's own reading and understanding of the *Principia*, I have found it useful to take into account the relevant *Marginalia* and occasionally the *Excerpts* as well.

The *Notes* consist of two sheets folded in quarto; the first has no watermark, the second sheet has one of a type recurrent in the paper used by Leibniz in Vienna in 1688.[1] The two sheets appear to have been written in continuous succession and contain large portions of commentary. The text covers approximately the first 40 pages of the *Principia*, with occasional references to later material. I discuss a selection of the passages studied and commented upon by Leibniz, including the rotating vessel experiment, the second law of motion, and proposition 1.

The first pages of the *Principia* contain eight definitions with the famous scholium about absolute space and time (pages 1–11), and three laws of motion with their corollaries and a scholium (pages 12–25). Leibniz paid special attention to the choice and coinage of effective words and expressions: his philosophical system is often associated with such terms. Far from being merely a linguistic problem, the terminology used often implied the acceptance of a philosophy or the taking of sides in a controversy. Answering Johann Bernoulli's criticism, who had complained that Leibniz had given definitions rather than explanations, the latter replied: 'You say that I give definitions rather than explanations. But be this always the case! because definitions are explanations.' And the definitions are the only part of the *Principia* to be present in the *Notes* and in both sets of *Excerpts*. It is rare to find philosophers who paid attention to the definitions, George Berkeley being the most notable exception: neither the editor Edmond Halley, nor most of the readers of the *Principia* we know of, nor even the editor of the second edition Roger Cotes, showed any specific interest in them.[2] Here I examine definitions 1, 6, 7, and the final scholium.

In the first definition Newton states that the quantity of matter arises from its density and bulk conjointly. Leibniz's attention focused on the following explicative lines, where Newton claimed to 'have no regard in

[1] LH 35, 10, 7, f. 32–3 and 34–5.

[2] *LMG*, *3*, p. 551, Leibniz to Johann Bernoulli, 18 Nov. 1698: 'Ais me attulisse definitiones potius, quam explicationes. Sed utinam semper definitiones afferrentur! nam illis explicationes virtute continentur.' Compare *The Works of George Berkeley, Bishop of Cloyne*, ed. by A. A. Luce and T. E. Jessop, 9 vols. (London, 1948–1957); vol. 1, *Philosophical Commentaries*, pp. 1–140; vol. 4, *De Motu* (first published in London, 1721), pp. 1–51. W. A. Suchting, 'Berkely's Criticism of Newton on Space and Time', *Isis*, *58*, 1967, pp. 186–97. P. Casini, *L'Universo Macchina*, (Bari, 1969), chapter 8.

this place to a medium, if any such there is, that freely pervades the inter-
stices between the parts of bodies.' He could neglect this hypothetical
medium on the basis of the further claim—noted by Leibniz—that body
or mass 'is proportional to the weight, as I have found by experiments on
pendulums, very accurately made, which shall be shown hereafter.'[3] The
relevant passage announced by Newton is the lengthy general scholium
to proposition 40, book II (pages 339–54). Leibniz's interest in the
scholium can be inferred from the *Marginalia*: on page 346 Newton gave
some details of his experiment, explaining that it was important to be
extremely accurate because the demonstration of the existence of the
vacuum depended on it. The words 'demonstratio vacui' are underlined
by Leibniz. The 'crucial experiment' on pages 352–3 was described by
Newton from memory, since he had lost the original papers: it consists in
comparing the oscillations of an eleven foot pendulum whose bob, made
of a fir-wood box, is alternately empty or filled with a piece of metal. If
resistance to motion depended on the subtle aetherial fluid penetrating
the internal parts of bodies, the presence of the piece of metal in the box
ought to affect the oscillations. This would not happen, however, if
resistance depended only on the air, which would act exclusively on the
surface of the bob. The virtually identical behaviour of the pendulum
regardless of the material contained in the bob convinced Newton that
the resistance due to the internal parts is nil or imperceptible. In the
Excerpts from pages 352–3 Leibniz noticed exactly this point.[4] On pages
410–11, book III, one finds further references worth mentioning here. In
corollary 1 to proposition 6, Newton states that 'the weights of bodies do
not depend upon their forms and textures'. In a passage from the
Marginalia clearly related to definition 1 Leibniz noticed that the thick
pervading matter is not considered.[5] In corollary 3 to proposition 6,
Newton claimed that a vacuum is necessarily given, otherwise the
enormous density of the fluid filling the heavens would prevent bodies as
heavy as gold from falling onto the earth, since bodies do not fall in a
fluid whose specific density is greater than theirs. In the *Notes* Leibniz
copied the words 'vacuum necessario datur' and in the *Marginalia* he
added two comments. He objected that Newton's claim is not proved and
concerns only sensible experience, not the realm of what is possible
without contradicting the laws of nature. Leibniz subordinates to his
metaphysical principles the results of physical experiences. He objected

[3] *Principia*, first edition, p. 1; transl. by Motte and Cajori, p. 1. On the distinction
between mass and weight compare def. 1 in the Commentary to the *Notes*.

[4] *Excerpts*, p. 482: 'Nulla notabilis resistentia oritur ab internarum partium superficie-
bus in corporibus unde autor de liquido interfuso subtilissimo dubitat.' See also Cohen,
Newtonian Revolution, sect. 3.8.

[5] *Marginalia*, M 410: 'Scilicet in praesenti statu materiae crassae, qua interlabens non
computatur.' Motte and Cajori, p. 413.

further that Newton's reasoning does not apply to matter generating gravity—against the claim that gravity is universal—unless gravity arose from an inexplicable incorporeal cause.[6] These opening observations reveal the wide implications and ramifications of Newton's definition of mass, and show that Leibniz realized immediately the philosophical implications of the *Principia*.

Definitions 6 and 7 concern the absolute and accelerative quantities of centripetal force respectively. The former is considered by Newton in relation to its cause; in the *Marginalia* Leibniz provides the example of terrestrial versus lunar gravity. The latter definition refers to the effect, which is weaker at a greater distance from its cause, as gravity on the earth. In the *Marginalia* Leibniz called centripetal force *dc*: thus from the very beginning he did not consider that Newton's accelerative quantity of centripetal force is 'proportional to the velocity which it generates in a given time.'[7]

The most surprising feature of Leibniz's early response to the scholium on absolute space and time is the secondary position it occupies in his texts. The reader familiar with the sophisticated analyses in the correspondence with Samuel Clarke will notice that in the *Notes* Leibniz has nothing better to question than the accuracy with which the rotating vessel experiment has been performed, or Newton's account. Newton's reasoning is the following: a bucket attached to a long cord is turned around so that the cord is strongly twisted; then it is filled with water and released. At the beginning, when the water has a great relative motion with respect to the bucket, its surface is flat; later, when the water has acquired absolute rotation, its surface is concave even if its relative motion with respect to the sides of the bucket has decreased. Therefore, motion is not relative to other bodies, such as the sides of the bucket—incidentally, this would be Descartes's opinion. If motion is only relative, the forces to recede from the axis of rotation are nil; if motion is absolute, they depend on the quantity of motion. According to Leibniz, it is unlikely that at the beginning the motion of the bucket relative to the

---

[6] *Notes*, line 53. *Marginalia*, M 411 A: 'Hoc non est probatum; in possibilibus locum non habere, sed tantum de facto et in solis sensibilibus nec debet applicari ad ipsam materiam gravificam'; M 411 B: 'Ita sane si gravitas oriretur a causa incorporali inexplicabili, omnisque materia gravitaret pro portione suae quantitatis.' Compare also the variant readings in Koyré and Cohen, *Principia: Third Ediction*; E. McMullin, *Newton on matter and activity* (Notre Dame, 1978), esp. pp. 62–3; Koyré, *Newtonian Studies*, ch. 6; I. B. Cohen, 'Hypotheses in Newton's philosophy', *Physis*, 8, 1966, pp. 163–84; and the following articles by J. E. McGuire: 'Body and Void and Newton's *De Mundi Systemate*: Some New Sources', *AHES*, 3, 1966, pp. 206–48; 'Transmutation and Immutability: Newton's Doctrine of Physical Qualities', *Ambix*, 14, 1967, pp. 69–95; 'The Origin of Newton's Doctrine of Essential Qualities', *Centaurus*, 12, 1968, pp. 233–60; 'Atoms and the Analogy of Nature', *SHPS*, 1, 1970, pp. 3–58.

[7] *Marginalia*, M 3; *Notes*, lines 12–13; Motte and Cajori, p. 4.

water is considerable, because the bucket initially does not move much; moreover, it ought to be tested whether the water at the beginning does not follow the motion of the bucket.[8] It is extremely unusual for Leibniz to have recourse to experience rather than to laws of nature to counter an argument. But in one way or the other Newton's experiment has to be wrong, otherwise Leibniz's philosophical system would collapse. It is well known that Leibniz rejected absolute time, space and motion; as we have seen in Section 4.2, he claimed that since each curvilinear motion is composed of rectilinear ones, and since relativity of motion holds for rectilinear motions, it must hold for curvilinear ones as well.[9]

Leibniz's reaction to the laws of motion is relatively straightforward to grasp. The first law—often referred to as the law of inertia—was common knowledge by the late 1680s. We have seen in Section 1.2 that Leibniz was familiar with it before 1670, and that he had perceptively criticized Kepler and Galileo on this issue. Thus his lack of interest in Newton's first law is easy to account for.

The second law relies heavily on the exact meaning of the definitions. We have seen in Section 4.2 that Newton's law is a hybrid involving continuous and discrete notions. This interpretation is strengthened by the reference to impressed force in the statement of the law: 'The change of the quantity of motion is proportional to the impressed motive force.' I recall that from definition 4 impressed force changes the status of rest or rectilinear uniform motion of a body, and arises 'from impact, from pressure, from centripetal force.' In the *Notes* Leibniz transcribed a portion of the text, adding his commentary in brackets and asterisks:[10]

If any force generates a motion, a double or triple force will generate double or triple a motion. (+ I disagree +) (+ If the impression is only a conatus, namely an infinitely small velocity, I suppose that it can be admitted. +)

It seems that in his first survey of the *Principia* Leibniz was having some difficulties in getting accustomed to Newton's terminology: on reading the word 'vis' he probably thought immediately of his own living force. It is tempting to argue that Leibniz introduced the notions of living and

[8] *Notes*, lines 16–18 and 70–80. Further references to Newton's experiment are in *Dynamica*, LMG, 6, pp. 502 and 507; LMG, 2, p. 199, 14 Sept. 1694, Leibniz to Huygens; *Specimen Dynamicum*, (Hamburg, 1982), pp. 58 and 74–5, and Loemker, *Papers*, pp. 449–50; Ariew and Garber, *Essays*, pp. 136–7; Leibniz's fifth letter to Clarke, sect. 53.

[9] The bibliography on this topic is enormous; see for example A. Koyré and I. B. Cohen. 'The Case of the Missing *Tanquam*: Leibniz, Newton and Clarke', *Isis*, 52, 1961, pp. 555–66; 'Newton and the Leibniz-Clarke Correspondence', *Archives Internationales d'Histoire des Sciences*, 15, 1962, pp. 63–126; M. Fox, 'Leibniz's Metaphysics of Space and Time', *SL*, 2, 1970, pp. 29–55; J. E. McGuire, 'Existence, Actuality and Necessity: Newton on Space and Time', *AS*, 35, 1978, pp. 463–508.

[10] *Notes*, lines 19–22. For the previous translations see Motte and Cajori, pp. 2 and 13.

dead force for the first time in print in paragraph 10 of the *Tentamen* specifically to contrast Newton's definitions.

The third law, stating the equality of action and reaction, is discussed by Newton in the lengthy scholium to the laws of motion both for impacts and attraction. Leibniz found the third law very appealing: he mentioned it again in the *Specimen Dynamicum*, among the systematic rules of motion, and in the *Theodicy*.[11]

The first section of book I, on the method of first and last ratios, contains 11 lemmas and a scholium. Newton claimed that the propositions in the rest of the book are demonstrated on the basis of those lemmas; however, they appear to be more a retrospective justification than a heuristic device. In the concluding scholium he implied that rigour and elegance played a part in his choice:[12]

I set these lemmas in introduction to avoid the monotony of adducing complicated proofs by *reductio ad absurdum* in the manner of the ancient geometers. For, of course, proofs are rendered more compact by the method of indivisibles. Yet, because the hypothesis of indivisibles is a rather harsh one, and for this reason that method is reckoned less geometrical, I have preferred to reduce proofs of following matters to the last sums and ratios of vanishing quantities and the first ones of nascent quantities.

We have seen in Section 3.4 that the justification of the Newtonian calculus was based on kinematics. In the *Principia*, however, motion was the object of the investigation. Thus, by adopting the finite ratio between two vanishingly small space variables, Newton was possibly trying to avoid a circular argument while dispensing with indivisibles.

Despite their relatively elementary character, Leibniz commented on the lemmas with increasing disappointment: I discuss here those numbered 9–11. We have seen in Chapters 3 and 4 that lemmas 9 and 10 are closely related, and that lemma 10 allowed Newton to generalize Galileo's law of uniformly accelerated motion to the cases with a non-constant regular force, though only at the very beginning of motion. Leibniz embarked on lengthy calculations on lemma 9, leading to the astonishing conclusion that 'here then the subtlety of the very clever Newton suffered a failure.'[13] Lemma 11 states that 'the vanishing subtense of the angle of contact is ultimately in the doubled ratio of the subtense of the bounding arc.'

---

[11] *Notes*, lines 23–5, 29–51, and 81–9. *Specimen Dynamicum*, (Hamburg, 1982), pp. 22–3; Loemker, *Papers*, 441. Ariew and Garber, *Essays*, p. 125. *Theodicy*, par. 346. See also R. W. Home, 'The third law in Newton's mechanics', *BJHS*, 4, 1968, pp. 39–51; I. B. Cohen, 'Newton's third law and universal gravity', *JHI*, 48, 1987, pp. 571–93.

[12] *Principia*, p. 35, transl. in *NMW*, 6, p. 121; see also ib., pp. 107–9.

[13] *Notes*, lines 206–65, esp. 262–3. More sober judgements soon prevailed, since in the *Excerpts*, p. 481, Leibniz seems to accept Newton's lemma.

Lemma XI.

Subtenſa evaneſcens anguli contaƈtus eſt ultimo in ratione duplicata ſubtenſæ arcus contermini.

Caſ. 1. Sit arcus ille A B, tangens ejus A *D*, ſubtenſa anguli contaƈtus ad tangentem perpendicularis *B* D, ſubtenſa arcus A B. Huic ſubtenſæ A B & tangenti A D perpendiculares erigantur A *G*, B G, concurrentes in G; dein accedant punƈta D, B, G, ad punƈta *d*, *b*, *g*, ſitq; *J* interſeƈtio linearum B G, A G ultimo faƈta ubi punƈta D, B accedunt uſq; ad A. Manifeſtum eſt quod diſtan- tia

[33]

tia G *J* minor eſſe poteſt quam aſſignatan quævis. Eſt autem (ex natura circulorum per punƈta *A B G*, *A b g* tranſeuntium) *A* B quad. æquale *A G* x *B* D & *A b* quad. æ- quale A *g* x *b d*, adeoq; ratio A B *quad.* ad A *b* quad. componitur ex rationibus A G ad A *g* & B D ad *b d*. Sed quoniam *J* G aſſu- mi poteſt minor longitudine quavis aſſigna- ta, fieri poteſt ut ratio A G ad A *g* minus differat a ratione æqualitatis quam pro differentia quavis aſſignata, adeoq; ut ratio *A* B quad. ad *A b* quad. minus differat a ra- tione B D ad *b d* quam pro differentia quavis aſſignata. Eſt ergo, per Lemma I, ratio ultima *A B quad.* ad *A b quad.* æqualis rationi ultimæ B D ad *b d*. *Q. E. D.*

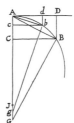

Caſ. 2. Inclinetur jam B *D* ad *A D* in angulo quovis dato, & eadem ſemper erit ratio ultima B D ad *b d* quæ prius, adeoq; ea- dem ac *A B quad* ad *A b* quad. *Q. E. D.*

Fig. 5.1   Subtense of the angle of contact.

The angle of contact is the magnitude between a curve *AbB* and its tangent *AdD*; the subtense of the angle of contact is *BD* or *bd*, whilst the subtense of the bounding arc is the chord *AB* or *Ab*. Notice in the diagram the small letters *b*, *c*, and *d*, employed by Newton in order to prove the statement by means of proportions involving finite ratios between vanishing quantities, namely that the last ratio between AB^2 and Ab^2 is as *BD* to *bd*. This lemma is implicitly based on the properties of the osculating circumference and tells us that any vanishing or infinitesimal arc of a curve with finite curvature can be approximated by the arc of a suitable circle. Leibniz's cursory reading led him to believe that lemma 11 is valid only when the curve *AbB* is a circle. Probably referring to his correspondence with Newton in 1676–7, he concluded his annotations on section 1 with the extraordinary words:[14]

Hence it seems to me that he has somewhat lost his excessive subtlety. At last, having considered everything, I suspect that all this subtlety is void and that at the very beginning it is possible to indicate improvements in infinitely many ways. It remains to be seen where these things are applied to a real problem.

[14]   *Notes*, lines 290–4. The correspondence with Newton involved mainly infinite series.

One can speculate that Newton's method of avoiding infinitesimals, based on the finite ratio between two vanishing variables, created some initial problems to Leibniz, whose calculus was centred on incomparably small differentials. The demonstration of lemmas 9–11 is indeed based on such ratios between vanishing variables; in his comments on lemma 9 in the *Marginalia*, Leibniz focused precisely on this aspect.[15] The quotation above seems to suggest that at that point Leibniz had read only section 1 and was therefore yet unaware of the number of 'real problems' tackled by Newton.

His attitude changed considerably moving to section 2, on the determination of centripetal forces. One can surmise that Leibniz's belief in the enormous difference in quality betwen sections 1 and 2 stimulated him to produce a reply based on his own mathematics. Proposition 1 states that the areas described by the radii to a body orbiting around a centre of force are proportional to the times. The importance and originality of this proposition, opening a bridge between mechanics and astronomy, can scarcely be overestimated. Although in the statement Newton referred to a centre of force, in the demonstration and figure he took the force to be centripetal. This led Leibniz to comment: 'I am also surprised that Newton did not consider that his theorem is not reciprocal, but is valid also for the centrifugal force'. However, Newton was perfectly aware that if the areas are as the times, force could be centripetal or centrifugal. Leibniz's reformulation reads:[16]

If any body already set in motion is driven in no other way than by the conatus of gravity or levity, it describes a curve, whose areas swept out by the radii are proportional to the times; if it is not already set in motion it describes a straight line; and the curved line differs from Galileo's parabolas in no other way than because for Galileo the centre is conceived to be infinitely distant, and gravity or levity acts everywhere uniformly. Further Newton, considering the matter in the most general fashion, found the property known to everyone, which Kepler proved.

The reference to Kepler's areas law as 'omnibus communis' echoes a similar pronouncement by Newton in book III, hypothesis 8, where he claimed that Kepler's law is 'Astronomis notissima'.[17]

The last three sides of the *Notes* are largely devoted to an attempt

---

[15] *Marginalia*, M 31 A. See also *NMW*, 6, pp. 114–17.

[16] Notes, lines 329–31 and 367–75. In proposition 12, book I, Newton proved that if a body moves along a hyperbola, centripetal force towards the focus is inversely proportional to the squared distance, adding that if the force were centrifugal, the body would move along the conjugate hyperbola; in corollary 3 to proposition 41 he investigated the curves described by a body acted upon by a centripetal or centrifugal force inversely proportional to the third power of the distance.

[17] *Principia*, first edition, p. 404; Motte and Cajori, p. 405, phenomenon 5; see Section 1.3 above.

based on the transmutation theorem to solve some problems inspired by proposition 1. The transmutation theorem was an obvious choice because curvilinear figures are divided into triangles concurrent to one point, as we have seen in Section 3.2; a similar partition into triangles occurs in proposition 1 and in general in the study of orbital motion with central forces. However, this type of investigation led to no significant result. At the end of the text, though, in a passage later crossed out and expanded in other manuscripts, Leibniz abandoned his attempts based on the transmutation theorem and tried instead to determine the centripetal or centrifugal conatus for a body acted upon by central forces. This time he used the evolvent of the curve described by the body. As we have seen in Section 2.3, a similar approach was adopted by Huygens for the circumference: it is possible that Leibniz vaguely remembered some conversations with his mentor on this issue. If the curve is not a circumference, however, it is easy to realize that the evolvent takes no role in the analysis of orbital trajectories, because motion along the tangent is uniform while motion along the curve is not, therefore Huygens's reasoning discussed in Section 2.3 (especially Fig. 2.2) cannot be applied.

As we have seen in the general introduction, Leibniz claimed that he had seen the *Principia* in Rome in 1689 for the first time. However, several elements point to a date of composition of the *Notes* preceding that of the *Tentamen*. Although in his commentary Leibniz had planetary motion in mind, he referred neither to the *circulatio harmonica*, nor to the *motus paracentricus*, terms and concepts which he regularly used in and after the *Tentamen*. In the *Notes* he defined the curve described by a body acted upon by central forces 'linea projectitia', an expression occurring again in the manuscript essays immediately following the *Notes*. The same terminology was also employed in other essays, which are discussed in the following sections, where Leibniz was more resolutely developing his own line of research rather than interpreting Newton. Considering the care with which Leibniz chose his terminology, it is reasonable to argue that usage of different key expressions with respect to the *Tentamen* indicates an earlier date of composition: from other documents which can be independently dated, we know that after 1688 Leibniz consistently adopted the terminology of the *Tentamen*, as we shall see in Chapter 7. The conceptual analysis of central problems, such as curvilinear motion, is at least as important as the terminology employed. At the end of the *Notes* he took centrifugal conatus of a body moving along a curve under the action of central forces to be equal and opposite to centripetal conatus; since this approach is developed by Leibniz in a later manuscript discussed in the following section, for convenience my analysis is deferred to that point.

5.3 The early developments: elaborating on Newton's results

The essay following immediately after the very end of the *Notes* is *De Conatu Centripeto vel Centrifugo (vel Generaliter Paracentrico) Mobilis in Aliqua Curva Incedentis*.[18] This manuscript contains the continuation of the attempt to determine centripetal or centrifugal conatus: Leibniz crossed out his preliminary attempt in the *Notes* and started again in *De Conatu*. This succession can be inferred especially from the similarity between the respective figures—15 and 16—and the proportions which Leibniz establishes. The link in contents is complemented by a further factor: the edge of folii 29–30, containing *De Conatu*, matches perfectly that of the second manuscript sheet containing the *Notes*, folii 34–5. They originally formed one folio which was divided in two and used in continuous succession. It is also worth recording the occurrence in the title of the word 'centripetal', a distinctly Newtonian expression.

The opening paragraph outlines a theory of curvilinear motion radically different from that found in the *Tentamen*:[19]

Let *CCC* be a curved line whatsoever and given a point *P* whatsoever, it is to be determined how great is the centrifugal conatus of a moving body carried along the curve, namely with what velocity while moving it tends to recede from the point *P* taken as the centre; or, what turns out to be the same, with what force of the conatus tending to the centre *P* it is needful to retain the body in its orbit, so that it does not fly away along the tangent. Whence from the nature of the curve the centrifugal conatus of the moving body is the same as the centripetal conatus retaining it in the orbit. Moreover, the conatus of a body to recede along the tangent from a certain centre placed outside the concave part of the curve, is the conatus of the same body to approach along the tangent a certain centre placed in the concave part. And this centrifugal or centripetal conatus differs greatly from the conatus along a ruler moved around *P*, which composed with the other motion can describe the curve *CC*, for centrifugal conatus is infinitely smaller than the conatus along the ruler *PC* composing the motion along the curve; besides, the motion along the ruler composing the motion along the curve can be conceived to be varied, according to one composing motion, another is taken at the same time with it; but centrifugal conatus is always the same.

Leibniz states that centrifugal and centripetal endeavours are equal and opposite, and that they are infinitesimal with respect to the conatus along a rotating ruler: both observations are worth examining in detail. First, in paragraphs 10 and 11 of the *Tentamen* there is a clear distinction between outward conatus ('conatus excussorius'), which relates to an arbitrary curve, and centrifugal conatus ('conatus centrifugus'), which is a special case of the outward tendency and relates exclusively to the

---

[18] LH 35, 10, 7, f. 29–30. This text is reproduced below.
[19] *De Conatu*, lines 3–20.

circumference. Centrifugal conatus is also called 'conatus excussorius curculationis'. As we shall see in Section 7.4, this distinction is drawn even more clearly in later texts, but is lacking in the *Notes* and *De Conatu*. Second, in the *Tentamen* orbital motion is based on two opposite tendencies, one attractive and the other repulsive, which are generally different. Even after Leibniz corrected the mistake by a factor of two for centrifugal conatus in 1706, he still believed in the imbalance between centripetal and centrifugal endeavours. As we have seen, in the *Tentamen* Leibniz conceived them, as it were, along a rotating ruler. The difference between gravity and twice centrifugal conatus gives rise to non-circular orbits and in particular to motion in ellipses. In the *Notes* and *De Conatu*, centripetal and centrifugal endeavours are always equal and opposite. Moreover, although in *De Conatu* Leibniz introduced the term 'paracentric', he meant a one-term expression, either centripetal or centrifugal: in the *Tentamen* the same term referred to a two-term expression, namely the combination of centripetal and centrifugal endeavours. The unlikely proposal that in 1689 Leibniz was developing a theory different from that which he had already published in the *Tentamen* can be dismissed by considering his analysis of the orders of infinitesimals. In the passage just quoted Leibniz compared centrifugal or centripetal conatus measured from the tangent to the curve, to the conatus along a rotating ruler, claiming that the former is infinitely smaller than the latter. This statement contradicts one of the basic results of the *Tentamen* and shows convincingly that *De Conatu* dates from 1688, when Leibniz had not yet developed his theory.

After preliminary remarks about taking constant elements of time by dividing the curve into infinitesimal triangles of equal areas, and about the curve being a polygon with infinitesimal sides, Leibniz used the evolvent of the curve described by the body in order to determine paracentric conatus. An analogus attempt can be found in the *Notes*. While writing *De Conatu*, however, Leibniz realized that the construction of the evolvent was unnecessary for his purposes: 'For the same could have been already derived from the previous considerations and therefore there would have been no need for the evolution of the line.'[20] Immediately afterwards he attained the following general result: paracentric conatus, for a body moving along a curve under the action of central forces, is directly as the square of orbital velocity and as the secant of the angle between the paracentric radius and the curve, and inversely as the radius of the circumference osculating the curve.[21]

This result helps us to understand a comment in the *Marginalia*. In book I of the *Principia*, corollary 7 to proposition 4 is related to the

---

[20] *De Conatu*, lines 59–60.

[21] In modern notation we have: paracentric conatus $= v^2/\rho \sin\phi$, where v is orbital velocity, ρ the osculating radius, and ϕ the angle.

present discussion. Proposition 4 states that bodies uniformly rotating along circular lines are attracted by centripetal forces directed towards the centre, proportional to the square of the arcs described in equal times, and inversely proportional to the radii. Thus centripetal forces are as v^2/r, where v represents velocity and r the radius. Corollary 7 states:[22] 'Identical assertions regarding the times, speeds and forces in, at and by which bodies describe similar portions of any similar figures having their centres similarly placed ensue in every case from applying thereto the proof of those preceding.' This corollary was marked by Leibniz in his own copy of the *Principia* with the *distillatur*, an alchemical sign meaning that the matter deserves further investigation. Immediately below he wrote a comment which he subsequently modified; this is an important element in dating the *Marginalia*. The first stage reads: 'Since I do not admit the generality of Lemma XI, I also doubt the generality of this Corollary 7'. Probably soon after he added: 'On the contrary this is true, because the considerations on the secant of the angle between the radius from the centre and the curve, and on the radius of the circle osculating the curve, vanish on account of the similitude between the figures.'[23] Hence in the second part of the marginal note Leibniz accepted Newton's corollary. In fact, since proposition 4 is stated in the form of a proportion between the homologous elements of two similar figures, their similarity cancels out the dependence of paracentric conatus on the secant of the angle and on the osculating radius: the former is equal in the two figures; concerning the latter, the ratio between the osculating radii is identical with the ratio between the paracentric radii. With the same notations employed above, naming further ρ the osculating radius and θ the angle between the paracentric radius and the curve, we have for two similar figures called 1 and 2:

$$\frac{v_1^2}{r_1} \;:\; \frac{v_2^2}{r_2} \;::\; \frac{v_1^2}{\rho_1 \sin \theta} \;:\; \frac{v_2^2}{\rho_2 \sin \theta} \;;$$

the link between the second part of the marginal note and *De Conatu* is straightforward. This correlation, while clarifying an otherwise obscure

---

[22] *Principia*, first edition, p. 42; transl. in *NMW*, 6, p. 130–1, n. 86.

[23] *Marginalia*, M 42 A: 'Quia lemma XI generale nondum admitto, etiam de generali isto Corollario 7 dubito.' The second part reads: 'Imo verum est, quia considerationes secantis anguli quem facit radius ex centro ad curvam, et radii circuli curvam osculantis, ob similitudinem figurarum evanescunt.' I have amended Fellmann's transcription: after 'considerationes' he has 'angulorum', which was crossed out by Leibniz. Furthermore, in his opinion this note was written in continuous succession. In *Marginalia*, M 41 B, as in the *Notes*, Leibniz claimed that lemma 11 is valid only in the case of the circumference: 'Suspectum hoc Lemma generale. In circulo tamen res vera speciali ratione, quia abscissae in circulo sunt ut quadrata chordarum.' For this reason I translate 'Lemma generale' and 'Corollarium generale' as 'the generality of the Lemma' or 'of the Corollary' respectively. This generality is juxtaposed to the particular case of the circumference.

passage in the *Marginalia*, suggests that Leibniz started writing the *Marginalia* in Vienna in 1688, thus while he was composing the *Notes*.

At the end of *De Conatu* Leibniz's interests shifted towards physical causes. His attempts to formulate a vortex theory in a mathematical fashion emerge again from a different context below. Later, however, he was to concentrate on the inverse-square law and Kepler's laws, while here his two preoccupations were relative motion and vertical descent. Concerning the former, he studied trajectories observed from within or without the rotating vortex. The latter was a vexing problem in vortex theories of gravity, because it was difficult to explain how a body descends vertically if the cause of its descent is a vortex rotating perpendicularly, or very nearly so, to the line of descent. Here he proposed that the vortex moves more swiftly the further away it is from the centre, following a law which is the exact opposite of the harmonic circulation and which cannot be easily extended to planetary motion.

A manuscript which undoubtedly follows in close succession is *Inventum a me est et alia scheda explicatum*. In the first paragraph Leibniz recalls word by word the results on paracentric conatus attained in *De Conatu*, which is the *alia scheda* referred to in the opening line. The text reads:[24]

It has been found and explained by me in another paper that the endeavours of a body moving along a curved orbit with respect to a certain given point as a centre, or paracentric, are as (1) the velocities of the body in the orbit, (2) the angles of deflection, and (3) the secants of the angles made by the radii drawn from the centre with the curve; or are directly as the squares of the velocities and as the secants of the same angles which the paracentric radii make with the curve, and inversely as the radii of the osculating circles.

Below Leibniz reconsidered the decomposition of motion involving the rotating ruler which he had mentioned in *De Conatu*, and established the basis of a fundamental result. He states that orbital motion can be seen either as the resultant of rectilinear inertial motion and gravity, or as the composition of circular motion on a rotating ruler and of rectilinear motion along the ruler:[25]

The same line can be conceived to be described by a motion composed of the circular motion of an indefinite ruler $P\urcorner$ around the centre P, and rectilinear of the body C on the ruler, tending towards the centre, although in such a projectile line it can never reach it. From $_3C$ let the perpendicular $_3C_2L$ fall on P_2C, it is apparent that in the inassignable angle $_2CP_3C$ the perpendicular can be taken as the arc, and it is just as if the body were transported from $_2C$ to $_3C$ by a motion

---

[24] LH 35, 10, 7, f. 36–7: f. 36r., lines 1–8. This is reproduced below, and is on a type of paper used by Leibniz in Vienna in 1688. With regard to the paper see the following section.

[25] *Inventum a me est*, lines 25–32.

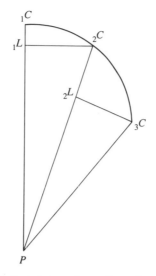

Fig. 5.2 Leibniz's transformation of proposition 1.

composed of $_2L_3C$ and $_2C_2L$. Therefore the very *circular progressions* $_2I_3C$ are as *the angles of circular progression and as the radii combined.*

Notice the lack of any clear statement of the imbalance between opposing tendencies for the motion along the rotating ruler. On the basis of this alternative decomposition of motion Leibniz was able to formulate the following fundamental proposition:[26]

Thus generally the increments of time are as the paracentric radii and as the circular progressions combined. Therefore taking equal increments of time in the motion of projectiles, the paracentric radii PC will be inversely as the circular progressions $L(C)$ of the moving body.

By $L(C)$ Leibniz means $_nL_{n+1}C$, where n is an arbitrary index; we shall find this notation again below. In this passage Leibniz introduced the notion of harmonic circulation as a corollary from proposition 1, book I. However, this denomination is not yet used, since Leibniz refers to a 'progressus circularis', and, most importantly, his discourse remains on a mathematical level. It will not take long to him to realize that his equivalent mathematical representation went beyond a mere change of notation and could serve other purposes.

The conclusion of *Inventum a me est* is also purely mathematical and contains the equation for the element of descent or ascent in a projectile line with an explicit expression for the osculating radius. Despite the preliminary investigations regarding the rotating radius, paracentric conatus is still measured as a deviation from the tangent to the curve.

[26] *Inventum a me est,* lines 49–52.

Thus *Inventum a me est* contains a hybrid approach between *De Conatu* and *Tentamen*. Leibniz's equation is demonstrated on two separate sheets, the draft *Inquisitio in Semidiametrum Circuli Osculantis si pro Ordinatis Convergentes Adhibeatur Ope Calculi*, and the clean version *Investigatio Semidiametri Circuli Curvam in Proposito Puncto Osculantis*.[27] This general result is one of the best achievements of Leibniz's purely mathematical investigations. The resources developed in these early attempts were soon to be reshaped and deployed in other works.

5.4   On the wrong track: 'pseudo-Galilean' motion

The manuscript *De Motu Gravis in Linea Projectitia* illustrates very effectively the difficulties encountered by Leibniz and the non-linearity of his itinerary. The first three sides contain an attempt to determine the increments of paracentric or radial descent, which are proportional to paracentric velocity and to time. For reasons which will become clear presently, it is important to stress that such increments are measured by the deviation from the tangent to the curve. On the internal sides of the manuscript Leibniz applied his reasoning to conic sections. The occurrence of expressions such as 'linea projectitia', 'progressus circularis', and 'paracentricus', suggests a link in contents with *Inventum a me est*. This link is strengthened by the fact that the margin of *De Motu Gravis* matches perfectly that of *Inventum a me est*: hence it is reasonable to suppose that they date from approximately the same time.[28]

The last side of the manuscript appears to be unrelated to the preceding ones. Its most interesting feature is the following: curvilinear motion is decomposed not along the tangent *CT* and the radius *CP*, but along a line *CΣ* parallel to the tangent to the initial point and the radius, as in the following diagram reproducing the relevant portion from the manuscript.

I call this approach 'pseudo-Galilean' motion, because Galileo adopted a similar decomposition in his study of the composition of rectilinear inertia with a constant gravity acting along parallel lines. The actions of gravity could be easily added together, and Galileo found parabolic trajectories.[29] Leibniz, however, added the impulsions of

---

[27] LH 35, 10, 7, f. 41v. and LH 35, 10, 7, f. 31 respectively.

[28] LH 35, 10, 7, f. 16–17; on the basis of the matching edges it can be inferred that the present manuscript is on a type of paper used by Leibniz in Vienna in 1688, watermark 510 in the catalogue at the NLB, letters 'M R' and posthorn.

[29] G. Galilei, *Discorsi e dimostrazioni matematiche intorno a due nuove scienze* (Leida, 1638) = *GOF, 8*, pp. 272–3; transl. by S. Drake (Madison, 1974), pp. 221–2. Leibniz studied it in Paris: Hess, 'Kurzcharacteristik', p. 199.

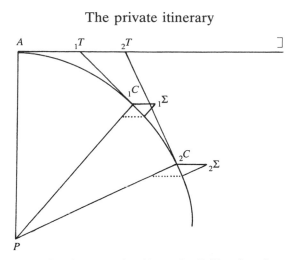

Fig. 5.3   An example of 'pseudo-Galilean' motion.

gravity towards a centre without considering their different orientations, and found a wrong result. The word 'wrong' is not the historian's retrospective judgement, since some time later Leibniz added the word 'Erronon' in correspondence with the relevant passage. His approach is very surprising for two reasons. First, he adopted it after having decomposed motion correctly at the beginning of the manuscript. Indeed, even on the present side he explained that the tangents CT represent direction and velocity in the orbit.[30] Second, at the end of the manuscript side he referred to Kepler's discovery of the area law, claiming that Newton rendered this in a general fashion: 'But the very times, because of Kepler's most wonderful discovery (which Newton rendered in a general fashion), are proportional to the areas.'[31] The attribution to Newton of the generalization of Kepler's area law in a text where Leibniz's own theory looks by his own standards hopelessly inadequate shows that before seeing at least the review of the *Principia* Leibniz had not developed the notion of harmonic circulation and in general a coherent theory. Notice also that Leibniz's clear attribution to Newton and emphasis on the generality of the demonstration echoes his own comment in the *Notes*. The type of paper used, the conceptual analysis of curvilinear motion, the lack of any reference to the harmonic circulation, and the terminology, clearly point to a date of composition preceding the *Tentamen*.

---

[30] See f. 17v.: 'Ducatur recta $_1C_2C$, quae exprimet directionem curvae, ea producta usque in A]erit curvae tangens, nempe $_1C_1T$, $_2C_2T$.'

[31] Ib.: 'Sed eadem tempora pulcherrimo Kepleri invento, (quod Neutonus generale reddidit) sunt areis proportionales.' Leibniz is fully aware that proposition 1 is a general statement about motion of bodies acted upon by central forces entailing Kepler's area law

Four manuscripts similar to the last side of the essay which we have just seen contain a series of unsuccessful attempts to study the curve described by a body under the action of central forces. The most interesting among them are *Galilaeus* and *Repraesentatio Aliqua*.[32] The common feature of these texts is the mistake in the decomposition of motion: instead of taking one component along the tangent to the curve and the other along the radius, Leibniz took one along the radius and the other along a straight line parallel to the tangent at the initial point. In all other points of the curve this component is not along the tangent, but along a sheaf of straight lines parallel to the first tangent.

It is not immediately clear why Leibniz adopted this approach, which from his point of view must have looked disappointing also for the lack of physical explanations. In *Galilaeus*, for example, Leibniz seemed to believe that his procedure had been followed by Newton. The title of another essay—*Nova Methodus*—suggests that he was looking for a new method of representing orbital motion alternative to Newton's. Thus although the analysis of motion is unsatisfactory, in spirit these manuscripts are close to the *Tentamen* because they show how Leibniz explored a range of mathematical representations of motion and forces. We would now expect to find a coherent theory different from Newton's.

In the opening paragraph of *Galilaeus* Leibniz clearly outlined the inverse problem of central forces as a generalization of Galilean parabolas. He took the centre of force at a finite distance, such that the endeavours are not parallel among themselves as for Galileo, and considered a conatus varying according to the distance from the centre, beginning with the supposedly simplest case of a constant conatus.[33] The

---

only as a special case. In Pfautz's review proposition 1 is omitted and in general section 2, propositions 1 to 17, are treated very superficially without any reference to the fundamental results there contained. The link between central forces, the area law and Kepler's third law is referred to in the section on book III without any detail: C. Pfautz, *AE*, 1688, p. 311.

[32] *Sit MM linea in qua gravia projecta motu aequaliter accelerato tendunt ad centrum terrae*, LH 35, 10, 7, f. 8 (octavo sheet). At the bottom there is a note on motion in a resisting medium. *Nova Methodus Tractandi Lineas Corporum Gravium vel Levium Projectorum*, LH 35, 10, 7, f. 40. This manuscript is on a type of paper used by Leibniz in Vienna in 1688, watermark number 564, 'cross with cavalier'. *Galilaeus tractare incipit de linea gravium projectorum*, LH 35, 10, 7, f. 18v.-19r. This essay is on a type of paper used by Leibniz in Vienna in 1688, watermark number 695, letters 'M R'. With regard to the paper compare the following section. The text, which is crossed out, is discussed in the commentary to *Repraesentatio Aliqua* in Appendix 1. *Repraesentatio Aliqua Curvae Quae Describitur a Gravi Projecto*, LH 35, 10, 7, f. 12. This manuscript is reproduced in Appendix 1 and is on a type of paper used by Leibniz in Vienna in 1688.

[33] A similar problem was discussed by Robert Hooke and Newton a few years before: *NC*, 2, Newton to Hooke, 13 Dec. 1679, p. 307; Hooke replied on 6 Jan., p. 309, explaining that he was considering an attractive force inversely proportional to the square of the distance. See chapter 8 below and *NMW*, 6, pp. 148–53.

opening paragraph in *Galilaeus* sets the agenda for further investigations which we are going to examine in the following section.

5.5 Mathematics, mechanics, and physical causes: Leibniz's theory takes shape

In *Si mobile aliquod ita moveatur* we find the concept—not the name though—of harmonic circulation in connection with the motion of a vortex. Since the edge of the manuscript matches perfectly that of *Galilaeus*, it is reasonable to suppose that they date from the same time. The introduction of vortices in connection with the mathematical formalism represents a new and crucial stage, and will remain a characteristic feature of Leibniz's attempts. In *Si mobile aliquod ita moveatur* he took constant elements of time by dividing the curve described by the body into infinitesimal triangles of equal area, as in *De Conatu* and *Inventum a me est*. He also conceived orbital motion as the resultant of a circular motion due to a vortex rotating with a velocity inversely proportional to the centre, composed with a rectilinear radial motion due to gravity.[34] On the third side Leibniz sought the particular law of descent such that the resulting curve is an ellipse. In the margin there is a note corresponding to a passage where Leibniz states that neglecting the circulation and being left only with inertial motion and gravity as in the 'motus projectorum', attractive conatus is as the second-order differential of the radius.[35] The marginal note begins with some differential equations which are reproduced adopting the following conventions: the numbering is mine; 'a' is the transcription from Leibniz's text, 'b' is a modern form which has been introduced in order to highlight the difference from Leibnizian mathematics, and has to be read in connection with Section 3.4. Some observations on the differential equations introduced here can be found in Section 7.3. In Leibniz's

---

[34] LII 35, 10, 7, f. 38–9. The last side contains calculations on ellipses which are not immediately related to the rest of the text. On f. 39r.: 'Intelligi potest mobile duplici motu, uno circulationis velocitatem habentis vicinitatibus ad centrum proportionalem, altero rectilineo quocunque pro lineae natura vario, ferri, quorum ille a fluidi deferentis rotatione, hic ab aliqua vi gravitati analoga oriatur.'

[35] Ib.: 'Itaque investigandum jam erit, quae nam sit lex conatuum descendendi, in variis ab umbilico distantiis, ut orbita fiat elliptica.' And: ib.: 'Porro descensus in hac motus compositione sunt ut incrementa radiorum paracentricorum. Itaque si fingeremus eandem lineam describi sine circulatione, sola compositione motus alicuius aequabilis, seu impetus semel concepti cum impulsu gravitatis quemadmodum in motu projectorum, erunt descensus ut impetus a gravitate impressi, itaque si linea eadem motu projectionis describeretur, forent conatus novi continue a gravitate impressi, seu incrementa impetus versus centrum, ut differentiae differentiarum a radiis, seu ut differentiae secundae radiorum.'

notation a horizontal bar over an expression means that the terms below are enclosed in brackets; this method is not employed consistently, however, and Leibniz often made it clear how an equation ought to be read by spacing out its terms. At times he also used a comma as a separation symbol, as we are going to see below.

$$\overline{ddp}/w = w/p^2, \tag{1a}$$

$$d^2p/dt^2 = -K/p^2, \tag{1b}$$

p is a distance and w a constant possibly representing both an infinitesimal element of time and an arbitrary factor. In equation (1b) K is a constant and t time. Leibniz's equation means that the conatus ddp is inversely proportional to the square of the distance. Multiplying by dp we have

$$dp \cdot \overline{ddp} = w^2 dp/p^2, \tag{2a}$$

$$dp \cdot d^2p/dt^2 = -dp\, K/p^2, \tag{2b}$$

and then calculating the integral, with E constant, one finds

$$\tfrac{1}{2}\,\overline{dp^2} = w^2/2p \ + w; \tag{3a}$$

$$\tfrac{1}{2}\,(dp/dt)^2 = (K/p) + E. \tag{3b}$$

Leibniz carried out the integral without changing sign in the right member, where he erroneously inserted the factor $\tfrac{1}{2}$; moreover, he wrote the integral constant with the same symbol w he had already used, thus restricting the generality of the procedure and mixing two conceptually different factors. In spite of these shortcomings, these equations clearly show how according to Leibniz the integral of conatus times an infinitesimal distance is equal to one half the square of velocity; their relevance to the debate on dead and living force referred to in Section 4.3 is straightforward. They are also among the first uncertain steps towards an algebraic science of motion; Leibniz's uncertainties and difficulties testify to the complexity of the task. The marginal note continues with an attempt doomed from the start to calculate the integral. Leibniz multiplies both members by 2 and takes the square root:

$$dp = \sqrt{\frac{w^2 + 2wp}{p}}\,, \tag{4a}$$

$$dp = dt\sqrt{(2Ep + 2K)/p}\,, \tag{4b}$$

multiplying both members by p Leibniz found

$$pdp = \sqrt{w^2p + 2wpp}\,, \tag{5a}$$

$$pdp = \sqrt{2Ep^2 + 2Kp}\ dt;\qquad\qquad(5b)$$

here one sees immediately how problematic it is to integrate equation (5a). In fact, Leibniz is led to write:

$$\tfrac{1}{2}pp = \int dp\sqrt{w^2p + 2wpp},\qquad\qquad(6a)$$

$$\int dp \cdot p / \sqrt{2Ep^2 + 2Kp} = t.\qquad\qquad(6b)$$

Leibniz could not separate the variables because time is not made explicit and is represented with the same symbol as other constants. In equation (6a) he arbitrarily inserted at second member the differential dp, which is not even constant. The marginal note continues from a modified reading of equation (4a), where he neglects the integration constant. Leibniz says that dp is inversely proportional to $\sqrt{p}$, thus $dp\sqrt{p}$ is constant and is proportional to the element of time. Therefore $\int dp\sqrt{p}$, or $p\sqrt{p}$, is proportional to time and the third power of the radius is as the square of the period of revolution.[36] Leibniz's calculations are aimed at a relation between time and distance entailing Kepler's third law; his procedure, however, is unsuitable to this task because in order to attain the third law it is necessary to impose the condition that the orbit is an ellipse. We shall find him dealing with this problem several times in the following essays.[37]

Contents and terminology of *Si mobile aliquod ita moveatur* are directly related to the most important manuscripts in the formation of Leibniz's theory, namely the two folded folii of *De Motu Gravium vel Levium Projectorum*. The text begins with a statement about the inverse problem of central forces resembling the opening paragraph of *Galilaeus*:[38]

---

[36] Ib.: 'dp reciproce ut $\sqrt{p}$. Ergo ut dp$\sqrt{p}$ est constans. Ergo proportionalis temporis elementis. Ergo $\int$dp$\sqrt{p}$ seu p$\sqrt{p}$ proportionalis temporibus. Ergo cubi distantiarum proportionales arearum seu temporum quadratis.'

[37] Newton found a modified version of the third law in proposition 60, book I, where he showed that the proportionality constant between the square of the times and the third power of the major axes depends on the sum of the masses of the planets and Sun. This proposition is referred to in the *Excerpts*, p. 492, but I am not aware of any commentary by Leibniz on it. Newton referred to Kepler's third law also in the scholium to proposition 4, book I, and in corollary 6. He showed that Kepler's third law is equivalent to the inverse square law when the orbit is a circumference. In the *Marginalia* Leibniz criticized this claim with the words: 'Sed non quadrat satis, quia figurae motuum non sunt similes, nec circuli' (M 42 B). Newton generalized his statement in proposition 15, which is transcribed in the *Excerpts*, p. 490, without commentary.

[38] LH 35, 10, 7, f. 1–25 and 2–3: f. 1r. (lines 4–11). Both manuscripts are on a type of paper used by Leibniz in Vienna in 1688, watermark 564 in the catalogue at the NLB, cross with cavalier. The texts are reproduced in Appendix 1. The structure of the texts is as follows: f. 1 and 25r. (referred to as 'essay') have been written in continuous succession; they end with an attempt of solving the inverse problem of central forces; f. 2 contains a continuation of the same calculations, where Leibniz refers to proposition 11, book 1 of

If the conatus of gravity or levity is along straight parallel lines where the centre is conceived to be infinitely or incomparably removed, and it is always constant, which is the simplest case, the line of projection will be a common parabola, as Galileo showed. The next case is with respect to a certain centre at a finite distance, and the conatus with respect to the centre, or paracentric, is also everywhere constant; although I do not know yet whether the line is determined, nevertheless first of all we expose these generalities about centroparabolic lines, for so I like to call them.

The idea that the case of a central constant conatus is the next simplest case after Galilean parabolas is questioned at the end of the quotation. 'Centroparabolic lines' are just another denomination of 'projectile lines'—the word 'projectaria' occurs in line 18. On the basis of calculations analogous to those in *Inventum a me est* and *Si mobile aliquod ita moveatur*, Leibniz is able to state the following fundamental result:[39]

Therefore it is discovered that centroparabolic lines, with which I am now dealing and are described in like manner by the projection of bodies with gravity or levity, are the same as vortical lines of heavy bodies, of dinobaryc, which are described by a body with gravity or levity carried in a vortex with a speed which is smaller, in proportion, to the greater distance from the centre of the vortex; and at the same time tending to the centre of the vortex because of gravity, or receding from it because of levity.

The surprising identity between curves described with rectilinear uniform motion and the action of gravity or levity, and curves described by a body pushed by a vortex rotating with a velocity inversely proportional to the distance from the centre, allows Leibniz to make the crucial transition from Newton's central forces as an explanation for the area law, to vortical motion. This metamorphosis allowed Leibniz to raise the profile of his interpretation by presenting it not as a modification of Newton's theory, but as the result of an independent discovery dating from several years earlier. Needless to say, the move from central attraction to vortical motion is carefully concealed in the *Tentamen*; Leibniz's claim to originality is based on his demonstrations of the area law and inverse-square law being different from Newton's. *De Motu Gravium* predates the *Tentamen* not only because it is on a type of paper used by

---

the *Principia* (addition 2). In the demonstration Newton employed lemma 12 ('all parallelograms circumscribed to a given ellipse have equal area'), which Leibniz tried to prove on f. 3r. (addition 3); f. 25v. contains an attempt to solve the direct problem of central forces, starting from ellipses (addition 1); f. 3v. contains an independent essay on planetary motion (addition 4).

[39] LH 35, 10, 7, f. 1r., lines 61–7. The adverbs 'centroparabolice' and 'dinobaryce' occur in *Si mobile aliquod ita moveatur*, f. 39r. They do not occur in known texts dated after the *Tentamen*.

Leibniz in Vienna in 1688, or because the harmonic circulation is named differently, but also for the following reason. In the texts seen thus far Leibniz understands by 'paracentric conatus' a one-term expression, either centripetal or centrifugal. Here he started as in those texts, but later he tried to calculate the conatus along the rotating radius. His attempts differ from Newton's for the following reason: Newton calculated the deviation from the tangent to the curve; Leibniz calculated the variation of the distance from the centre, comparing the distances at different times by a rotation of the radius. At the beginning he thought that paracentric conatus was still a one-term expression:

Whence it is necessary that $ddr = m$, namely m or $_2MG$, which represents the new impression of gravity, is the difference between the differences between two nearby radii, or between two descents, $_1G_1L$, $_2G_2L$, or, what turns out to be the same, between two nearby whole impetuses to descend, which could have also been foreseen.

Namely, the difference between two radii is dr, and the difference between two such differences, which can be called dr_1 and dr_2 respectively, is $dr_2 - dr_1 = ddr$. Later in the essay, however, he realized that he had committed a mistake: 'Therefore at last we have corrected our calculations, and together with the conatus to descend from gravity, the centrifugal force from the circulation has to be conjoined.' Therefore paracentric conatus along a rotating radius becomes a two-term expression:[40]

Hence it is clear that if all endeavours from gravity m are added together, and from this all centrifugal endeavours k are detracted, one will have the impetus of descent $_1G_1L$, namely $dr = \int m - \int k$, that is, $ddr = m - k$.

This imbalance between opposite tendencies will remain one of the most characteristic features of Leibniz's theory of planetary motion, and constitutes a new result of his investigations. If his starting point can be clearly located in the *Principia*, in *De Motu Gravium* Leibniz attained truly original findings.[41] If we reconsider the manuscripts of the previous

[40] *De Motum Gravium*, lines 79–83, 141–3, and 152–5 and figure 26.
[41] In the *Tentamen* endeavours are taken with opposite signs, and the expression $a^2\theta^2:r^3$ represents twice centrifugal conatus. From 1688 to 1705 (*LMG*, *4*, Leibniz to Varignon, 27 July 1705, p. 128) Leibniz interpreted his equation in the same way. In the 'zweite Bearbeitung' of the *Tentamen*, for example (see Section 7.4), he admits that his result is surprising. *LMG*, *6*, p. 184: 'Unde etiam analysis conatus paracentrici geometrica in solicitationem gravitatis et duplum conatum centrifugum, *id est duplum eius qui esse debere vederetur*, ...' (my emphasis). In a letter to Johann Bernoulli of 28 Jan. 1696 he probably wanted to make this point clear: 'Verissimum est, quod ais, et a me quoque comprobatum in *Tentamine de Motuum Coelestium Causis*, vires centrifugas in ratione composita esse ex duplicata directa celeritatum et reciproca simplice radiorum; neque id contemnendum est in rem nostram, etsi enim hae vires, vel potius sollicitationes, differant a viribus ipsius per se circulantis; sufficit quod illis sunt proportionales. Interim revera nihil aliud sunt quam celeritates elementares.' *LMG*, *3*, p. 241. See also the discussion below.

section, we see that finding a new decomposition of motion was a main concern for Leibniz. He began investigating this problem very soon after reading the *Principia*.

Having established a new equation, Leibniz tackled the inverse problem of central forces with his differential calculus. He worked with several differential equations which, while showing a considerable improvement on those seen above, still failed him in his task of finding ellipses and Kepler's third law. Leibniz adopted different strategies for finding the integral of his equation: he could not find the integral when gravity is constant; by setting gravity composed of two terms, the second of which cancels out centrifugal conatus, he attained a solution, but even apart from the artificial character of the procedure, he did not find ellipses. In a renewed series of attempts in the second folded folio of *De Motu Gravium* (addition 2), he set gravity proportional to an arbitrary power of the distance, namely proportional to r^n, trying to determine the exponent a posteriori in order to find ellipses. On failing again to find a result, he set first $n = -2$ and then $n + 1 = -2$, but the result he found had no relation to ellipses. After some further calculations based on common geometry, in a passage starting with the words 'Ad Neut. p. 50', he excerpted proposition 11 from the *Principia*, where it is proved to Leibniz's satisfaction that if a body moves along an ellipse, the force towards the focus is inversely proportional to the square of the distance. Indeed Newton proved more than this, and established that centripetal force is equal to the square of the altitude of the triangle swept out by the radius in a vanishing element of time over the latus rectum of the ellipse. Since all such triangles have equal areas, for each of them the altitude is inversely as the radius. Leibniz transcribed this equation as $Lm = k^2$, where $L = 2a$ is the latus rectum, m centripetal force, and $k = \theta a : r$ the altitude of the triangle; since θa is the area of the triangle and r the radius, Leibniz introduces here a trivial mistake by a factor of two. Following his notation derived from proposition 11 of the *Principia*, one has that centripetal force is equal to $a^2\theta^2 : 2ar^2$, or simply $a\theta^2 : 2r^2$. These excerpts resemble closely the corresponding annotations in the *Marginalia*. On the basis of this result he was able to start a new series of calculations, which are here reproduced:[42]

$$ddr = a\theta^2 : 2r^2 - a^2\theta^2 : r^3; \tag{7a}$$

$$d^2r : dt^2 = (-h^2 : r^2a) + (h^2 : r^3). \tag{7b}$$

It is extraordinary to notice how Leibniz included the result of proposition 11 in the different context of his own attempts. Here *ddr* is

---

[42] LH 35, 10, 7, f. 2. Newton, *Principia*, first edition, pp. 50–51; *Marginalia*, M. 50 D; *De Motu Gravium*, Addition 2, lines 46–60.

the second-order differential of the radius; $a\theta^2:2r^2$ represents gravity, where the factor 2 ought to be at the numerator, because of the mistake mentioned above; θa is the infinitesimal area swept out by the radius in the time θ ($\theta = dt$); a is also half the latus rectum, which typically is indicated with a letter already employed, possibly in the attempt of simplifying the final equation; $\theta a:r$ is as the velocity of rotation (a factor 2 is missing) and its square over the radius, namely $a^2\theta^2:r^3$, is taken to be simple centrifugal conatus. In equation (7b) a is the semi-latum rectum and h is the angular momentum.[43] Leibniz's equation clearly expresses the second-order differential of the radius as the difference between two terms with opposite signs. Multiplying by dr we have:

$$\int dr \cdot \overline{ddr} = \theta^2 a, \int dr:2r^2 - a\int dr:r^3; \tag{8a}$$

$$\int dr \cdot d^2r:dt^2 = \int dr[(-h^2:r^2a)+(h^2:r^3)]; \tag{8b}$$

the comma in equation (8a) means that $\theta^2 a$ multiplies the entire expression to its right. Calculating the integral one finds:

$$1/2\,\overline{dr^2} = \theta^2 a,\ 1:2r - a:r^2\ :2; \tag{9a}$$

$$1/2\,(dr:dt)^2 = (h^2:ra) - (h^2:2r^2) + E; \tag{9b}$$

where the factors $1/2$ are cancelled on both sides of (9a), and in (9b) E is the integration constant. Leibniz interpreted the first and second term at second member as integrals of centrifugal force and gravity times an infinitesimal distance; again, their relevance to the discussions in Section 4.3 is straightforward.[44] Unlike equation (4a), here Leibniz is able to separate the variables:

$$\theta = dr \cdot r:\sqrt{ar - uu}, \tag{10a}$$

$$dt = dr \cdot r:\sqrt{2Er^2 + 2r(h^2:a) - h^2}; \tag{10b}$$

and operates the substitution $\sqrt{ar - aa} = v$ in order to carry out the integration. Leibniz shows no uncertainty as to the substitution of variable. He sets $r = v^2 + a^2,:a$[45]

$$\text{time} = \int 2v\,dv:v + \int a^2 dv:v \quad \text{or} \quad \text{time} = 2v + \int a^2 dv:v. \tag{11a}$$

[43] Equation (7b) can be easily obtained by taking twice the derivative of the polar equation of an ellipse with respect to time. It is worth recalling that $h = r^2 da:dt$, where da is the differential of the angle of circulation. The polar equation is $r = c:(1 + \varepsilon \cos \alpha)$, where c is the parameter of the conic, equal to the semi-latum rectum, and ε its eccentricity; if ε is nil the curve is a circle; if $\varepsilon = 1$ a parabola; if ε is positive but smaller than 1 an ellipse; lastly, if ε is greater than 1 a hyperbola.

[44] Similar calculations can also be found in a later essay *Si mobile feratur motu composito*, LH 35, 9, 9, f. 9–10; f. 10v. Compare also the calculations in Gerland, *Schriften*, pp. 32–3.

[45] Because of a trivial slip Leibniz wrote $dr = 2vdv + a^2,:a$; later, however, he did not take the term a^2 into account.

As in *Si mobile aliquod ita moveatur*, his final goal is a relation between time and distance satisfying Kepler's laws. From equation (10a) we should have instead $t = 2v + 2\int v^2 \, dv : a^2$. Leibniz also tried to take into account that the radius does not start from being infinitesimal, but ranges between the distances at aphelion and perihelion. In other words, he realized that he had to calculate what we call a definite integral; he set $r = z - h$, where z is a new variable and h a constant, and failed again to find the result he was seeking. Although these and similar calculations failed him in his task of finding the third law of planetary motion, they represent an extraordinary document of the early usage of differential equations, and of techniques such as substitution and separation of variables.

In these calculations one sees at the same time the power and the difficulties of the new algebraic representations in mechanics. The reader has certainly realized the similarity between Leibniz's analysis of orbital motion and modern vectorial representations of motion under central forces. This similarity is somewhat artificially strengthened by the modern version marked with the letter 'b'. Some aspects concerning mathematics and the problem of dimensional homogeneity have been outlined in Chapters 3 and 4: I recall the difference between Leibniz's differential and the derivative of a function, or between his infinitesimal velocity and acceleration. Moreover, constant factors such as angular momentum h and energy E play a central role in modern representations, whilst Leibniz treated the former as a relatively unimportant constant, and did not take the latter into consideration. Further, I wish to mention the problems of interpretation of the differential equations with regard to real versus fictitious entities. In one respect both Leibniz's and the modern representations are fictitious: Leibniz conceived radial motion to take place along a rotating ruler which, with regard to planetary motion, was certainly imaginary; likewise, the rotating radius-vector of modern mechanics is a purely ideal entity. With respect to other aspects, however, the similarity vanishes. For Leibniz centrifugal conatus was a real tendency due to a circular motion induced by a material rotating fluid. By contrast, in the modern account centrifugal force appears as a fictitious entity due to the choice of the representation along the rotating radius. The similarity between equations 'a' and 'b' hides a conceptual gulf in mathematics and in the physical interpretation as well.

The last side of the second folded folio of *De Motu Gravium* (addition 4) contains an essay on planetary motion where Leibniz tried to explain the causes of the physical actions. He claimed that gravity, for example, results from the difference between the centrifugal forces of the vortex and of the body. Since they rotate with the same velocity, the difference between their centrifugal forces must depend on their respective densities. In a passage partially crossed out Leibniz hinted at a connection between his theories of planetary motion and of motion in resisting medium. He seemed to suggest that the circulation of the body depends on its surface, whereas gravity or levity depend on its solidity. This passage resembles Descartes's theory in *Principia Philosophiae*, especially for the reference to 'soliditas', and is based on notions corresponding to 'absolute' and 'respective' resistance, the two kinds of friction described by Leibniz in his *Schediasma de Resistentia Medii*.[46] This is his only known reference to the theory of motion in a resisting medium in the context of planetary motion.

5.6   The application to ellipses and the direct problem

On the last side of the first folded folio of *De Motu Gravium* (addition 1), Leibniz tried to apply some of the results on the circulation and descent of a body to the specific case of motion along an ellipse. In particular, he wanted to determine the ratio between the differential dr of the radius and the velocity of rotation $a\theta{:}r$ in terms of the parameters of the ellipse by means of the known properties of conic sections. This project leads directly to paragraph 18 of the *Tentamen*. His attempt represents a retreat from the inverse to the direct problem, and from the integral calculus to common geometry and some elementary differentiation rules. Similar calculations can be found in other manuscript essays, where Leibniz was able to attain his aim by means of the characteristic triangle and of simple geometry.[47] In one of them, *Calculus Motus Elliptici*, one reads that θa is twice the area swept out by the radius in the element of time; thus the error mentioned above is eliminated. Leibniz's proportion is:

$$\frac{\theta a}{r}{:}dr{::}b{:}\sqrt{e^2 - p^2};$$

where b and q are the minor and major axes of the ellipse respectively, e its eccentricity, a the latus rectum, no longer its half, and $p = 2r - q$; further, it is worth recalling that $e^2 = q^2 - b^2$ and $b^2 = aq$. Differentiating

---

[46] LH 35, 10, 7, f. 3v., lines 21–5. See Section 2.2 above.

[47] LH 35, 10, 7, f. 4v-5: *Inveniendus Est Calculus Differentialium Ellipseos per Solas Eductas ex Umbilicis*; LH 35, 10, 1, f. 12 3 (not written in continuous succession): *Calculus de Elementis Radiorum Ellipseos ex Umbilico Eductorum per Radios et Velocitatem Circulationis Inveniendis, et Speciatim de Casu Quo Circulationes Sunt Radiis Reciproce Proportionales*; LH 35, 9, 9, f. 5–6 (not written in continuous succession): *Calculus Motus Elliptici Aequales Areas Aequalibus Temporibus ex Foco Abscindentis*.

the equation obtained from this proportion and eliminating dr by substitution one finds the equation of paracentric motion:

$$ddr = \theta^2 a^2 : r^3 - \theta^2 a2 : r^2;$$

In *Calculus de Elementis Radiorum Ellipseos*, as in *De Motu Gravium*, this equation is interpreted as centrifugal force minus gravity, whereas in *Calculus Motus Elliptici* the first term is identified as twice centrifugal conatus, as in the *Tentamen*.[48] Leibniz's equation of paracentric motion is attained by means of a mixed technique, first geometric and then algebraic. The final result does not contain an interpretation of its terms, which has to be given separately.

In all three manuscripts under discussion Leibniz gave numerical examples together with literal expressions, as if he wanted to reassure himself of their soundness.[49] In two manuscripts, for example, he fixed the radius $r = M\Theta = 5$; the major axis AV of the ellipse $q = 9$, and the latus rectum $a = bb{:}q = 8$, where $bb = 72$ is the square of the minor axis; the eccentricity $e = 3$. On the basis of the proportion above we have that the differential dr of the radius, or $LD = \theta8{:}15$. At this point Leibniz fixed the value of $\theta = 0.000015$, representing the infinitesimal time dt; thus $LD = 0.000008$; the circulation $a\theta{:}r = 0.000024$. Here Leibniz took his differentials as very small numbers. While Newton dealt with finite forces and accelerations, the main ingredients in Leibniz's analysis were infinitesimal displacements. In another manuscript essay Leibniz took the radius equal to 1 000 000 000 000, namely an incomparably large number, and checked again his results with numerical examples.[50]

The last side of *Calculus Motus Elliptici* contains some general remarks on finding the orbit given the two relations between paracentric impetus and time, and between angle of circulation, radius and time.[51] From the proportion $dr{:}dt{::}r^e{:}c^e$, where e is an arbitrary exponent and c is a constant introduced in order to preserve dimensional homogeneity ('ut servetur lex homogeneorum'), taking the integral we have a relation between time and radius. The second relation, namely $da{:}dt{::}cc{:}rr$, where da is the differential of the angle, can be used in different ways. If the circulation is harmonic, and the differential of time constant, the infinitesimal angles are inversely proportional to the square of the radii. With the aid of the preceding

---

[48] LH 35, 10, 1, f. 13v. and LH 35, 9, 9, f. 6v.

[49] In *Calculus Motus Elliptici*, f. 5v., Leibniz even applied his *ars characteristica* by associating numbers to concepts.

[50] LH 35, 9, 9, f. 5r. and LH 35, 10, 1, f. 12v. See also LH 35, 10, 7, f. 23–4, on f. 24v. Leibniz often used this procedure of numerical control. For example, at the end of an essay on motion in a resisting medium of 1688, LH 35, 9, 5, f. 27v., he wrote: 'Deprehendetur res adhibitis ubique quantum licet numeris.' This manuscript has been studied by Aiton, 'Resisting Medium'.

[51] LH 35, 9, 9, f. 6v.

proportion one can have an equation dependent either on the angle and time, or on the angle and the radius, namely the equation of the orbit. The relations used are too simple to be applied to planetary motion. They seem to be a second attempt, after the failure to solve the inverse problem of central forces in *De Motu Gravium*. Although these calculations are relevant to the inverse problem, here Leibniz starts from velocities, or more precisely from *dr*, rather than from forces. These results will be mentioned in paragraph 13 in the *Tentamen*.

In the manuscript *Ad Relationem Actorum Junii pag. 303 seqq.*, which is a preliminary draft of *De lineis opticis*, there is a passage in which Leibniz mentions two hypotheses explaining planetary motion. The former states that the aetherial matter and the planets are repelled from the Sun proportionally to their weight or mass, and stems from the assumption that centrifugal force is proportional to the mass of a body; the latter states that the powers or living forces of aether rings and of planets are equal among themselves. This is a primitive attempt to ensure equilibrium in the vortex.[52] Although the latter hypothesis will be adopted again by Leibniz (see Section 7.5), neither is mentioned in the *Tentamen*, nor does he ever explain how elliptical orbits arise. The equality of force between the vortex rings and the planets, together with the harmonic circulation, determines a relation between mass and distance. If mass times the square of velocity is constant, and velocity is inversely proportional to the distance from the centre, mass must be directly proportional to the square of the distance. Indeed, there is a manuscript in which Leibniz sets the linear density of each ring proportional to the radius, and since the length of the ring is proportional to the radius, mass is directly proportional to the square of the radius.[53]

Ad Relationem Actorum is followed in continuous succession by *Tentamen de Legibus Naturae Mundi*, an essay intended for publication in the *Acta* together with the three papers of January and February 1689.[54] The text is composed of 20 numbered paragraphs on dead and living force, composition of motion, the law of continuity, cohesion, the impact laws, elasticity, the non-existence of the vacuum and of atoms. Referring to a falling body Leibniz wrote:[55]

[52] LH 35, 10, 4, f. 1. The text covers the first side of a folded folio and the beginning of the second (other drafts of the same essay are in LH 35, 10, 1, f. 14v., and LH 35, 15, 2, f. 1). On the distinction between weight and mass see definition 1 in the commentary to the *Notes*.

[53] LH 35, 9, 9, f. 3–4, *Incrementa angulorum circulationis harmonicae sunt in ratione duplicata reciproca radiorum*. The relevant passage is on f. 3v.-4r.

[54] LH 35, 10, 4, f. 1v.-2. At the begining of f. 1v. Leibniz mentioned four essays on the laws of nature, the system of the world, optical lines and motion in a resisting medium. On f. 1v., referring to conservation of force, he wrote: 'Eadem *alibi in his Actis* demonstravi', and this shows that the essay was intended for publication in the *Acta*.

[55] Ib., partially in a marginal note on f. 1v.: 'Vim corporis in motu positi recte aestimo ex altitudine ad quam se potest elevare, similiter ex producta intensione alicuius elastri vis

I measure the force of a body set in motion by the height to which it can raise itself; similarly, the force can be measured by the stress produced in a spring.

Leibniz established an equivalence between *vis viva*, the height to which a body can raise itself, and the force of a compressed spring. The analogy between gravity and elasticity also extends to the case of fall and to the realm of living forces.

Lastly, a few words on the final stage of our journey, the *Tentamen de Systemate Universi*.[56] This essay is composed of numbered paragraphs which resemble the *Tentamen* in structure and contents. The first paragraph begins with the words: '*Circulationem* voco *harmonicam*', and corresponds closely to paragraph 3 of the *Tentamen*. This and the great number of corrections in the text reveal the manuscript as a preliminary draft. Leibniz later added an introduction and two opening paragraphs, shifting the others consequently. A notable exception with respect to the *Tentamen* is that here Leibniz makes clear that the harmonic circulation is equivalent to a motion composed of rectilinear inertia with central solicitation.[57] As he had discovered in *De Motu Gravium* with respect to the *lineae projectitiae* and *vorticales*, these two different representations of orbital motion are equivalent in accounting for the area law.

The manuscript essays referred to so far contain the main stages in the development of Leibniz's theory. They can be summarized as follows: a purely mathematical elaboration of Newton's results, involving a general expression for centripetal or centrifugal conatus; the discovery that if the areas are proportional to the times, circular progression is inversely proportional to the radius; a series of attempts to find an alternative decomposition of motion based on what I have called 'pseudo-Galilean' motion; the introduction of a vortex and the recognition that *lineae centro-parabolicae* are the same as the *lineae vorticales*; the discovery that paracentric conatus along the rotating radius is a two-term expression, namely the difference between centrifugal conatus and gravity; the failure to solve the inverse problem and the success with the direct problem, namely to find paracentric conatus starting from ellipses. Although proposition 1 in the *Principia* played a central role in the development of Leibniz's theory, many

---

aestimari potest.' Compare also *Dynamica*, *LMG*, 6, p. 452: 'Vis etiam in Elastro sustinendi aliquod est *Mortua*, et similiter vis in pondere coercendi Elastrum ... Sed Vis ponderis quam habet ad Elastrum comprimendum v. gr. aerem ex statu ordinario redigendum intra aliquod spatium arctius, *Viva* est; opus enim est descensu ex aliqua altitudine, seu impetu concepto, nec solum mortuum pondus sufficit.'

[56] LH 35, 9, 9, f. 1–2; related drafts are in LH 35, 9, 9, f. 7–8, *Si mobile ex centro semel emittatur aut repellatur*; and LH 35, 9, 9, f. 9–10, *Si mobile feratur motu composito*.

[57] LH 35, 9, 9, f. 1v.: 'Coincidunt inter se circulatio harmonica, et motus compositus ex impetu priore semel concepto et solicitatione gravitatis vel levitatis nova.'

of the results discussed in this chapter represent original discoveries which did not follow automatically from Newton's masterpiece. The representation of orbital motion along the rotating radius cannot be defined to be equivalent to the *Principia* without major qualifications. After having seen the private development of Leibniz's theory and the surprising results he attained, we possess all the elements to study the text which Leibniz chose to make public.

6

PUBLICATION

Introduction

The *Tentamen* appeared in the *Acta Eruditorum* of February 1689. At the same time Leibniz left Vienna for Venice, where he arrived on 4 March. Therefore, the essay was certainly composed in the Imperial capital. We have the manuscript draft, the final manuscript version having probably been sent to the editors of the *Acta* in Leipzig.[1] The text would have been sufficiently challenging even without several misprints. Moreover, the figure was lost and had to be redrawn by Christoph Pfautz, the mathematician collaborating with the editor of the *Acta* Otto Mencke. As a result, the segment $_3MG$ and the letters G at the bottom of the figure were omitted. On the basis of a manuscript diagram, I have restored the figure inserting the missing letters in angle brackets. In the footnotes I indicate Leibniz's later emendations and explain some technical points.[2]

While trying to present a text which can be followed by a modern reader, my translation of the *Tentamen* corresponds fairly closely to the original. Capitalization follows modern conventions, while punctuation has been altered occasionally for clarity, and mathematical symbols have been italicized. In my work I have been helped by two partial translations. In the chapter on astronomy of his *Treatise of Fluxions* (London, 1704), Charles Hayes closely paraphrased large portions of the *Tentamen*: this work is particularly interesting because it is contemporary to Leibniz. Further, in his articles in *Annals of Science* Eric Aiton translated some crucial propositions.

An essay on the causes of celestial motions

It is well known that the ancients, especially those who followed the beliefs of Aristotle and Ptolemy, did not yet understand the splendour of

---

[1] LH 35, 9, 2, f. 56–9, consisting of an entire folded sheet and two separate half sheets. The last side and about three quarters of the last but one side contain some complementary remarks. In the margin Leibniz added some passages leading to the 'zweite Bearbeitung' of the *Tentamen*; this later version, first published by Gerhardt, was not available to Leibniz's contemporaries.

[2] LH 35, 9, 2, f. 67r. and v.; *LSB*, I, *5*, p. 607, Mencke to Leibniz, 9 July 1690.

nature, which has at last shone forth in our century and in the preceding one; this ever since Copernicus showed that the most beautiful hypothesis of the Pythagoreans, which they seem perhaps to have proffered tentatively rather than correctly determined, recalled from obscurity, satisfied the phenomena with utmost simplicity. Moreover *Tycho*, having followed Copernicus in the principal points of the system (apart from transposing Sun and Earth), cast his mind to observations more accurate than usual, and removed from the heavens the wholly unseemly apparatus of solid orbs. And yet he did not gather sufficient results from his Herculean labours, partly because he was hindered by certain prejudices, partly because death forestalled him. Divine Providence, though, arranged that his observations and efforts passed into the hands of the incomparable *Johann Kepler*, whom the fates had destined to be the first mortal man to make public[3]

the laws of the heavens, the order of nature, and the precepts of the Gods.

He then discovered that any planet describes an elliptical orbit in which the Sun occupies one of the foci, following that law of motion whereby the areas swept out by the radii from the Sun to the planet are always proportional to the times. The same man found that the periodic times of the several planets of the same system are in the sesquialterate ratio of their mean distances from the Sun; and he would certainly have been about to triumph wonderfully had he known (as Cassini eminently noticed) that the satellites of Jupiter and Saturn also observe the same laws with respect to their planets as these do with respect to the Sun.[4] But he could not yet determine the causes of so many and so unvarying truths, both because his mind was hindered by thoughts of Intelligences or unexplained radiations or sympathies, and because at that time the more profound mathematics and the science of motion were not yet as far advanced as they are now. However, he also opened the way to investigation of the causes. For to him we owe the first indication of the true cause of gravity and of this law of nature on which gravity depends, that rotating bodies endeavour to recede from the centre along the tangent; thus if stalks or straws are afloat on water moving in a vortex by the rotation of a vessel, the water being denser than the stalks and therefore being driven out from the centre more strongly than they are, will push them towards the centre. Kepler himself clearly explained this in two or more places in *Epitome Astronomiae*, though he was still

[3] For this quotation from Claudian referring to Archimedes see Section 1.2.

[4] That is, the square T^2 of the period of revolution is proportional to the third power q^3 of the mean distances, or of the major axis. With regard to Cassini's observations see Section 1.3.

somewhat in doubt and ignorant of his own means, and insufficiently aware of how many things would follow therefrom in physics and especially in astronomy. But later Descartes made brilliant use of these reasonings, though in his usual manner he concealed their author. Further, I often marvel that Descartes did not even try to provide reasons for the celestial laws discovered by Kepler, as far as we know, either because he could not reconcile them sufficiently with his own opinions, or because he remained ignorant of the fruitfulness of the discovery and did not consider it to be so accurately followed by nature.

Further, since it seems not at all the province of physics, and indeed unworthy of the admirable workmanship of God, to assign to the stars individual *Intelligences* directing their course, as if He lacked the means for accomplishing the same by laws governing bodies; and to be sure *solid orbs* have some while now been rejected, while *sympathies*, magnetisms and other abstruse qualities of that kind are either not understood, or, when they are, they are judged to be effects consequent on corporeal impressions—I myself judge there is no alternative left but that the cause of celestial motions should originate in the motions of the aether, or, using astronomical terms, in *orbs* which are *deferent*, yet *fluid*. This opinion, though very ancient, has been neglected: Leucippus in fact expressed it even before Epicurus to the extent that, in fashioning his system, he employed the very name δίνηζ (*vortex*), and we have learnt how Kepler foreshadowed gravity in the motion of water driven round in a vortex. And from Monconys' book on voyages[5] we learn that Torricelli (and I suspect also Galileo, whose pupil he was) was already of the opinion that the entire aether with the planets is driven round by the motion of the Sun about its centre, just as water is if a stick is rotated about its axis in the middle of a vessel at rest; and like straws or stalks floating on the water, so too heavenly bodies closer to the centre revolve faster.

But these more general considerations come to mind without difficulty. We, however, intend to explain more distinctly the laws of motion themselves, which will prove to be a matter needing a far deeper investigation. And since some light has dawned on us in this matter, and our research seems to proceed extremely favourably and naturally, I am raised to hope that we have come close to the true causes of celestial motions.[6]

(1) To tackle the matter itself, then, it can first of all be demonstrated that according to the laws of nature *all bodies which describe a curved line in a fluid are driven by the motion of the fluid*. For all bodies describ-

[5] De Monconys, *Journal des Voyages*, first part, pp. 130–1.

[6] Some general observations on this introduction and the following two paragraphs can be found in Sections 1.2, 7.2, and 7.4.

ing a curve endeavour to recede from it along the tangent (because of the nature of motion), it is therefore necessary that something should constrain them. There is, however, nothing contiguous except for the fluid (by hypothesis), and no conatus is constrained except by something contiguous in motion (because of the nature of the body), therefore it is necessary that the fluid itself be in motion.

(2) Hence it follows that *planets are moved by their aether*, namely they have fluid orbs which are deferent or moving. In fact by universal agreement they describe curved lines, and it is not possible to explain phenomena by supposing rectilinear motions alone. Therefore (by the preceding paragraph) they are moved by an ambient fluid. The same can otherwise be demonstrated from the fact that the motion of a planet is not uniform, or describing equal spaces in equal times. Whence, also, it is necessary that a planet be driven by the motion of the ambient fluid.

(3) I call a *circulation* a *harmonic* one if the velocities of circulation in some body are inversely proportional to the radii or distances from the centre of circulation, or (what is the same) if the velocities of circulation round the centre decrease proportionally as the distances from the

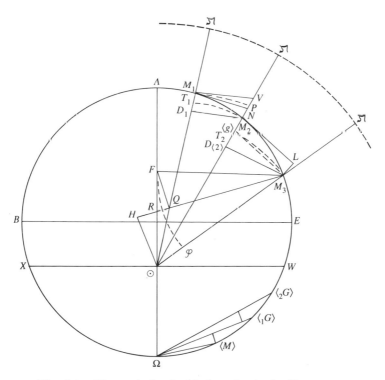

Fig. 6.1   The analysis of orbital motion in the *Tentamen*.

centre increase; or, most briefly, if the velocities of circulation increase proportionally to the closeness. Thus if, for instance, the radii, namely the distances, increase uniformly or arithmetically, the velocities will decrease in harmonic progression. Accordingly, harmonic circulation may occur not only in the circle arc, but also in any other curve whatsoever to be described. Let us suppose that the moving body M is carried along in an arbitrary curve $_3M_2M_1M$ (or $_1M_2M_3M$) and that it describes the elements of the curve $_3M_2M$, $_2M_1M$ in equal times; its motion can be conceived to be composed of a circular one around some such centre as $\odot$ (as for instance $_3M_2T$, $_2M_1T$), and a rectilinear one such as $_2T_2M$, $_1T_1M$ (where $\odot_2T$ is taken equal to $\odot_3M$ and $\odot_1T$ to $\odot_2M$). This motion can also be conceived to be such that, while a ruler or an indefinitely long and rigid straight line $\odot\overline{\mathcal{A}}$ moves around the centre $\odot$, at the same time the body M moves along the line $\odot\mathcal{A}$. Further, it does not matter what is the rectilinear motion of approach to or recess from the centre (which I call *paracentric motion*), provided that a circulation of the moving body M, such as $_3M_2T$, is to another circulation of it, $_2M_1T$, as $\odot_1M$ to $\odot_2M$, namely if the circulations completed in equal elements of time are inversely as the radii. For since these arcs of elementary circulations are as the times and speeds combined, and the elements of time are taken to be equal, the circulations will be as the velocities, and consequently the velocities inversely as the radii, and therefore the circulation will be called harmonic.

(4) *If a moving body is carried with a harmonic circulation* (whatever its paracentric motion may be) *the areas swept out by the radii drawn from the centre of the circulation to the body will be proportional to the times required, and vice versa.* For since the elementary circular arcs, such as $_1T_2M$, $_2T_3M$, are incomparably smaller than the radii $\odot_2M$, $\odot_3M$, the differences between the arcs and their sines (such as between $_1T_2M$, and $_1D_2M$) will be incomparable with the arcs and sines, and therefore (*for our analysis of infinites*)[7] can be taken as non-existent, and the arcs and sines as coincident. Therefore $_1D_2M$ is to $_2D_3M$ as $\odot_2M$ to $\odot_1M$, namely $\odot_1M$ times $_1D_2M$ is equal to $\odot_2M$ times $_2D_3M$, therefore also their halves, namely triangles $_1M_2M\odot$ and $_2M_3M\odot$, are equal. Since these triangles are elements of the area $A\odot MA$, and we have assumed by hypothesis equal elements of time, also the elements of the area are equal, and vice versa, and therefore the areas $A\odot MA$ are proportional to the times with which the arcs AM are traversed.

(5) In the demonstrations I have employed *incomparably small quantities,* such as the difference between two finite quantities, incomparable with the quantities themselves. Such matters, if I am not

---

[7] Leibniz, 'De Geometria recondita et Analysi Indivisibilium atque Infinitorum', *AE* 1686. See more specifically the following paragraph.

mistaken, can be exposed most lucidly as follows. Thus if someone does not want to employ *infinitely small* quantities, one can take them to be as small as one judges sufficient as to be incomparable, so that they produce an error of no importance and even smaller than allowed. In the same way as the Earth is like a point, or the diameter of the Earth as an infintely small line with respect to the sky, so it can be demonstrated that if the sides of an angle have a basis incomparably smaller than them, the angle they enclose will be incomparably smaller than the right angle, and the difference between the sides will be incomparable with the sides themselves; and the diffcrence between the whole sine, the sine of the complement, and the secant, will be incompable with the terms of the difference, as the difference between the sine, the chord, the arc, and the tangent.[8] Therefore, since these quantities are infinitely small, the differences will be *infinitely many times infinitely small,* and also the versed sine will be infinitely many times infinitely small, thus incomparable with the sine.[9] Further, there are infinitely many orders both of infinite and infinitely small quantities. Moreover, it is possible to use finite triangles similar to the *inassignable* ones, which are most useful for finding tangents, maxima, minima, and for unfolding the curvature of lines; likewise, almost in every application of geometry to nature; for, if motion is represented by a finite line, which is traversed by a body in a given time, the impetus or velocity is expressed by an infinitely small line, and the element of velocity, which is solicitation of gravity or centrifugal conatus, by a line infinitely many times infinitely small. I reckoned these matters were to be noted down here as *lemmas* for *our Method of incomparable quantities and analysis of infinites,* just as *elements* of this new doctrine.[10]

(6) From this now it follows that *planets move with a harmonic circulation* round the Sun as a centre, satellites round their planet. For with the radii drawn from the centre of the circulation they sweep out areas proportional to the times (by observations). Therefore taking equal elements of time, the triangle $_1M_2M\odot$ is equal to the triangle $_2M_3M\odot$, thus $\odot_1M$ is to $\odot_2M$ as $_2D_3M$ to $_1D_2M$, namely the circulation is harmonic.

[8] Here the *sinus* of an angle α is understood to be $R\sin\alpha$, where R is some convenient power of 10 chosen in order to avoid decimals. The whole sine is R; the sine of the complement $R\cos\alpha$; the secant must be taken to be $R/\cos\alpha$. The clarifying comma between 'the sine, the chord (sinus, chordae)' was inserted by Leibniz in 1706.

[9] The versed sine is the difference between the whole sine and the sine of the complement, that is, $R - R\cos\alpha$. By 'infinitely many times infinitely small' Leibniz means second-order infinitesimals. Thus, if α is a first-order infinitesimal, $\sin\alpha$ is also a first-order infinitesimal, whilst the versed sine is infinitesimal of the second order.

[10] Leibniz will refer to this important paragraph as *Lemmata Incomparabilium* or *Lemmes des Incomparables*; *LMG*, *3*, p. 524, Leibniz to Johann Bernoulli, 29 July 1698; *LMG*, *4*, p. 92, Leibniz to Pierre Varignon, 2 Febr. 1702.

(7) It is also reasonable that the aether or *the fluid orb of each planet moves with a harmonic circulation*. In fact it has been shown above that a body in a fluid does not move spontaneously in a curved line, therefore there will be a circulation also in the aether; further, it is reasonable to believe it to be in harmony with the circulation of the planet, so that the circulation of the aether of each planet is also harmonic; that is, if the fluid orb of the planet is mentally divided into innumerable concentric circular orbs of hardly any thickness, each orb will have its own circulation which gets proportionally faster the nearer to the Sun. But a more accurate account of this motion in the aether will be given elsewhere.

(8) Thus we suppose that *a planet moves with a double motion composed of the harmonic circulation of its fluid deferent orb, and the paracentric motion*, as if it had a certain gravity or attraction, namely an impulsion *towards the Sun or*—if it is a satellite—*the planet*. Further, the circulation of the aether makes the planet circulate harmonically not, as it were, by its own motion, but almost as if swimming smoothly in the deferent fluid, whose motion it follows. Thus the planet does not retain the faster impetus of circulation it had in the inferior or closer orb, but this impetus weakens while crossing the superior orbs (resisting a velocity greater than their own), decreases continuously, and adapts imperceptibly to the approaching orbs. Conversely, while the planet tends from the superior to the inferior orbs, it acquires their impetus. And this happens all the more easily because when its motion agrees once with the motion of its actual orb, then it hardly differs from the motion of the orbs nearby.

(9) Having explained the harmonic circulation, we must come to the *paracentric motion of the planets, born of the outward impression of the circulation and solar attraction* combined. Moreover, it may be permitted to call it an attraction, although in reality it is an impulse, inasmuch as the Sun in a certain sense can be conceived to be a magnet; the magnetic actions themselves, however, are derived doubtless from the impulsions of fluids. Whence we shall also call it *the solicitation of gravity*, conceiving a planet to be a heavy body tending towards a centre, namely the Sun. The type of orbit, however, depends on the particular law of attraction. Let us see then which law of attraction produces an elliptical path. In order to attain this, it is necessary to enter a while the adytum of geometry.

(10) Since every moving body which describes a curved path endeavours to recede along the tangent, one may call this *the outward conatus*, as in the motion of the sling, for which there is required an equal force constraining the moving body, lest it flies away. *This conatus can be measured by the perpendicular from the following point to the tangent at the inassignably distant preceding point*. When the curve is a circle, the renowned Huygens, who was the first to investigate it mathematically,

called this *force* produced by repeated endeavours *centrifugal.*[11] Moreover, every outward conatus is infinitely smaller than the velocity or impetus acquired from the conatus repeated for some time, in just the same way as the solicitation of gravity, whose nature is homogeneous to it. Whence also it is confirmed that they have the same cause. It is consequently not surprising, as Galileo thought, that percussion is infinitely greater in comparison to simple gravity, or, in my terms, to simple conatus, whose *force* I usually call *dead*; while acting and receiving the impetus from repeated impressions, dead force is at last rendered living.

(11) *Centrifugal conatus, namely the outward conatus of circulation, can be expressed by PN, the versed sine of the angle of circulation* $_1M\odot N$ (or by $_1D_1T$, which turns out to be the same because the difference between the radii is inassignable); for the versed sine is equal to the perpendicular drawn from one end-point of the arc of a circle to the tangent from the other end-point, whereby we expressed the outward conatus in the preceding paragraph. (Centrifugal conatus can be also expressed by PV, the difference between the radius and the secant of the same angle; the distance between their difference and the versed sine is *infinitely infinitely many times infinitely small,* and so wholly insignificant with respect to the radius.)[12] Hence, furthermore, since the versed sine is as the square of the chord, it follows that *centrifugal endeavours of bodies describing equal circles with uniform motion are as the squares of the velocities, and those of bodies describing circles of different size as the squares of the velocities and inversely as the radii.*

(12) *Centrifugal endeavours of bodies circulating harmonically are inversely as the cubes of the radii.* For (by the preceding paragraph) they are inversely as the radii, and directly as the squares of the velocities, that is (since the velocities of the harmonic circulation are inversely as the radii), inversely as the squares of the radii; and combining the simple inverse and the square inverse, the ratio becomes the inverse cube.[13] For

[11] In the 'zweite Bearbeitung' and 'Excerptum ex epistola' Leibniz crossed out the misleading words 'produced by repeated endeavours'; indeed, the impetus, not the force, is produced by repeated endeavours.

[12] The versed sine PN is equal to $R - R\cos\alpha$; PV is equal to $(R/\cos\alpha) - R$; their difference $PV - PN$ is equal to $R(1 - \cos\alpha)^2/\cos\alpha$; thus, if α is infinitesimal of the first order, $PV - PN$ is a fourth-order infinitesimal. In addition, notice that $\odot V$ depends on the cosecant of the angle of circulation.

[13] In the following calculation Leibniz conflates two alternative representations of motion. Since he is using a continuous curve, he ought to consider that PN is traversed with uniformly accelerated motion. Alternatively, if he wanted to neglect accelerations, he ought to consider the tangent as the prolongation of the chord. The latter is the solution he chose in the list of errata published in the 1706 'Excerptum ex Epistola', where he stated that in paragraphs 11, 12, 15, 21, 27, and 30 one should read just the centrifugal conatus instead of its double, and half of it instead of the whole. See also Section 4.2 above.

the sake of calculation let θa be a fixed area always equal to twice the elementary triangle $_2M_3M\Theta$, or to the rectangle $_2D_3M$ times the radius Θ_2M or r; then $_2D_3M$ is $\theta a{:}r$ or θa divided by r; now the centrifugal conatus $_2D_2T$ is equal to the square of $_2D_3M$ divided by twice Θ_3M and is therefore equal to $\theta\theta aa{:}2r^3$.

(13) *If the paracentric motion* (receding from the centre Ω or approaching it) *is uniform*, and the circulation harmonic, *the trajectory ΩMG will be a spiral* starting from the centre Ω, with the property that the *segments $\Omega GM\Omega$ are proportional to the radii*, namely in this case to the *chords ΩG* drawn from the centre: in fact, both the areas, that is, the segments, and (because of the uniform motion of recession) the radii are proportional to the times. There are many other notable properties of this spiral, and the construction is not difficult. Indeed, in the harmonic circulation *a general method is given* of constructing the curves, supposing at least their quadratures, if from the radii the times are given, or the velocities of paracentric motion, or at least the elements of the impetuses or the solicitations of gravity.[14]

(14) *Paracentric solicitation, whether of gravity or levity, is expressed by* the straight line $_3ML$ drawn from the point $_3M$ of the curve to the tangent $_2ML$ (produced to L) of the preceding, inassignably distant point $_2M$ parallel to the preceding radius Θ_2M (drawn from the centre to the preceding point $_2M$).

(15) *In every harmonic circulation the element of paracentric impetus* (namely the increment or decrement of the velocity of descent towards the centre or of ascent from the centre) *is the difference or sum of the paracentric solicitation* (namely the impression due to gravity or levity or a similar cause) *and of twice the centrifugal conatus* (arising from the harmonic circulation itself); the sum indeed if levity is present, the difference if gravity. When the solicitation of gravity prevails, the velocity of descent increases or the velocity of ascent decreases, but the contrary happens when twice centrifugal conatus prevails. From $_1M$ and $_3M$ let $_1MN$ and $_3M_2D$ be normal to Θ_2M; then, because of the harmonic circulation the triangles $_1M_2M\Theta$ and $_2M_3M\Theta$ have been shown to be equal, their altitudes $_1MN$ and $_3M_2D$ will also be equal (because of the common base Θ_2M). Now, taking $_2MG$ equal to L_3M, let $_3MG$ be drawn parallel to $_2ML$ itself;[15] the triangles $_1MN_2M$ and $_3M_2DG$ will then be

---

[14] The word 'segment' refers to the plane portion of a spiral figure. The case of a uniform paracentric motion implies an attraction inversely proportional to the cube of the distance and was studied in the second addition to *De Motu Gravium*, lines 18–20. Here the lower part of diagram 6.1 is tacitly considered as a separate figure unrelated to the planetary ellipse.

[15] The tangent $_2ML$ is the prolongation of the chord $_1M_2M$. Leibniz's mathematical reasoning in this paragraph is correct; the mistake lies in paragraph 12, since PN is only half the centrifugal conatus (see Section 4.2). The reader may find it easier to follow the argument with the help of the following diagram, which is rigorously based on Leibniz's text.

congruent, and $_1M_2M$ equal to G_3M, and N_2M equal to G_2D. Further, in the straight line $\odot_2M$ (produced if necessary, as I always understand) take $\odot P$ equal to $\odot_1M$, and $\odot_2T$ equal to $\odot_3M$; P_2M will then be the difference between the radii $\odot_2M$ and $\odot_3M$. Now P_2M is equal to (N_2M or) $G_2D + NP$, and $_2T_2M$ equal to $_2MG + G_2D - _2D_2T$, therefore $P_2M - _2T_2M$ (the difference of the differences) will be $NP + _2D_2T - _2MG$, that is (since the versed sines NP and $_2D_2T$ of two angles and radii whose differences are incomparable, coincide) twice $_2D_2T - _2MG$. Now the difference between the radii expresses the paracentric velocity, the difference of the differences expresses the element of paracentric velocity. Further $_2D_2T$ or NP is the centrifugal conatus of the circulation, being namely the versed sine (by 11), and $_2MG$ or $_3ML$ is the solicitation of gravity (by the preceding paragraph). Thus the element of paracentric velocity is equal to the difference between twice the centrifugal conatus NP or $_2D_2T$, and the simple solicitation of gravity G_2M, or (as it follows in the same way) the sum of twice the centrifugal conatus and the simple solicitation of levity.

(16) Given the increments or decrements of the velocity of ascent or

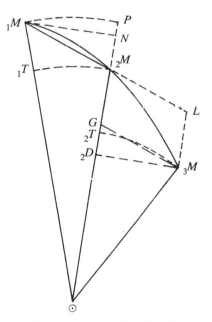

Diagram from paragraph 15 of the *Tentamen*

$P_2M = N_2M + PN$; $_2M_2T = _2MG + G_2D - _2D_2T$; Leibniz was able to find the difference $P_2M - _2M_2T$, namely $PN + PN - _2MG$, by setting $NP = _2D_2T$ and $N_2M = G_2D$; these two equalities were later criticized by Newton and are discussed in Section 8.4.

descent, the solicitation of gravity or levity is given, and vice versa. This is clear from the preceding paragraph, for I hold centrifugal conatus to be always given, since it is inversely as the cube of the radii (by 12).

(17) In equal elements of time the increments of the angles of harmonic circulation are inversely as the squares of the radii. For the circulations are as the angles and the radii combined, and since the elementary circulations are harmonic, they are inversly as the radii, therefore the elementary angles are inversely as the squares of the radii. Such are virtually the apparent diurnal motions as seen from the Sun (for in this instance days are sufficiently short periods of time, especially for the outer planets), which are approximately inversely as the squares of the distances, so that at double the distance only the fourth part of the angle is covered in the same element of time, at triple the distance only the ninth.[16]

(18) *If an ellipse is described by a body moving with a harmonic circulation round a focus as a centre of the circulation, one has these following three related magnitudes: the circulation $_2T_3M$ or $_2D_3M$* (these do not differ by a comparable quantity), *the paracentric velocity $_2D_2M$, and the velocity of the same body* (composed of them) *in the elliptical orbit itself,* namely $_2M_3M$. *They are as these three respectively*: the minor axis BE; the mean proportional between the difference and the sum of the distance $F\Theta$ of the foci one from the other, and the difference $\Theta\phi$ between the distances of the point $_3M$ of the orbit from the foci; and, lastly, twice the mean proportional between Θ_3M and F_3M, or the distances of the same point from the two foci.[17] *The same* is true for the *hyperbola* in its own way. *In the parabola* with quantities which are infinite, and elsewhere are vanishing, the circulation, paracentric velocity, and the velocity composed by them, which is in the same orbit, will be respectively as the latus rectum; the mean proportional between the latus rectum and the excess of the radius over the shortest radius (which is the fourth part of the latus rectum), and lastly, twice the mean proportional between the radius and the latus rectum. The correctness of these statements can be derived from the common elements of the conic sections if one supposes that the straight line $_3MR$[18] perpendicular to the curve (or to its tangent)

[16] Since Leibniz is talking of elementary or infinitesimal circulations, he thinks that one day is a sufficiently short interval of time with respect to the entire period of an orbit, to be represented by a differential (compare his observations in paragraph 5 above). The results of this paragraph were attained in *Inventum a me est*. Notice, however, that for Mars a day corresponds approximately to half a degree.

[17] This remarkably cumbersome passage can be clarified in the following way: circulation $_2T_3M$ is as BE; paracentric velocity $_2D_2M$ as $\sqrt{(F\Theta+\Theta\phi)\cdot(F\Theta-\Theta\phi)}$; and orbital velocity $_2M_3M$ as $2\sqrt{\Theta_2M\cdot F_3M}$. Notice that, strictly speaking, paracentric velocity ought to be $_2T_2M$; however, the difference from the expression given by Leibniz is a second-order infinitesimal.

[18] The letter 'R' was added by Leibniz in 1706.

in $_3M$ meets the axis $A\Omega$ in R, and that the perpendiculars FQ, ΘH from the foci are drawn normal to it; it is clear that $_3MH$, $H\Theta$,[19] Θ_3M, are proportional to $_2M_2D$, $_2D_3M$, $_3M_2M$, that is, the paracentric velocity, the circulation, and the velocity in the orbit. Therefore, it is sufficient to show that the sides of the triangle $_3MH\Theta$ are to one another as we have stated. This can the more easily be done by observing that the triangles $_3MQF$, and $_3MH\Theta$ are similar, and further that F_3M is to Θ_3M as FR to ΘR, whence the statement will be demonstrated by common analysis.[20] Hence it follows that, even if the foci are interchanged, so that one instead of the other becomes the centre of the harmonic circulation, in any point the ratio between circulation and paracentric velocity remains the same as before.

(19) *If a moving body having gravity*, or which is drawn to some centre, such as we suppose a planet is with respect to the Sun, *is carried in an ellipse* (or another conic section) *with a harmonic circulation, and the centre both of attraction and of circulation is at the focus of the ellipse, then the attractions or solicitations of gravity will be directly as the squares of the circulations, or inversely as the squares of the radii or distances from the focus.* We find this as follows by a not inelegant specimen of our differential calculus or analysis of infinites. Let $A\Omega$ be q; BF[21] be e; BE be b (that is $\sqrt{qq - ee}$); Θ_2M the radius r; $\Theta\phi$ (or $\Theta_2M - F_2M$) $2r - q$, or for brevity p; and the latus rectum WX be a, equal to $bb{:}q$. Let twice the element of the area, or twice the triangle $_1M_2M\Theta$, which is constant, be θa, supposing a to be the latus rectum, and representing the constant element of time by θ; and the circulation $_2D_3M$ will be $\theta a{:}r$ (see 12 above). Moreover, call the difference $_2D_2M$ of the radii dr,[22] and the difference of the differences ddr. Further, from the preceding paragraph dr (or $_2D_2M$) is to $\theta a{:}r$ (or $_2D_3M$) as $\sqrt{ee - pp}$ to b. Therefore $brdr = \theta a\sqrt{ee - pp}$, which is the *differential equation*. But this *differential equation differentiated (following the Laws of the calculus explained by us elsewhere in these Acta)* is $bdrdr + brddr = 2pa\theta dr{:}\sqrt{ee - pp}$, and eliminating dr from these two equations, so that only ddr remains, it becomes $ddr = bbaa\theta\theta - 2aaqr\theta\theta{:}bbr^3$,[23] from which the proposition follows. In fact, the element of paracentric velocity ddr is the difference between $bbaa\theta\theta{:}bbr^3$ or $aa\theta\theta{:}r^3$, which is twice the centrifugal conatus (by 12

---

[19] Read 'ΘH, H_3M', so corrected by Leibniz in 1706.

[20] The reader interested in this calculation can consult the 'zweite Bearbeitung' of the *Tentamen*, *LMG*, 6, pp. 173–6.

[21] Read 'ΘF', so corrected by Leibniz in 1706.

[22] If we were to take second-order infinitesimals into account this difference would be $_2T_2M$. Leibniz's reasoning, however, is perfectly adequate because in this instance his choice does not affect the result.

[23] The clarifying comma in the equation was inserted by Leibniz in 1706; thus, we should read $ddr = (bbaa\theta\theta - 2aaqr\theta\theta){:}bbr^3$.

above), and $2aaqr\theta\theta{:}bbr^2$, that is, (since $bb{:}q = a$) $2a\theta\theta{:}rr$; it is therefore necessary (by 15) that $2a\theta\theta{:}rr$ be the solicitation of gravity, which multiplied by the constant $a{:}2$ gives the square of the circulation $aa\theta\theta{:}rr$. Therefore the solicitations of gravity are directly as the squares of the circulations, and hence inversely as the squares of the radii. The same conclusion follows both *in the hyperbola and parabola,* and especially in the *circle,* which is the simplest ellipse. The reason, however, for the difference between these conic sections, and why it should be that circles and ellipses are generated in preference to the other conics, will appear below.

(20) *The same planet is attracted by the Sun in different ways, namely as the square of its closeness,* in such a way that it is continually solicited to descend towards the Sun by a certain new impression four times stronger if twice as near, nine times stronger if three times as near. This is clear from the preceding paragraph, supposing that the planet describes an ellipse, circulates harmonically, and in addition is continuously impelled towards the Sun. I see that this proposition was already known also to the renowned Isaac Newton, as it appears from the review in the *Acta,* although from the review I cannot determine how he attained it.

(21) Moreover, it is clear that *the solicitation of gravity on a planet is to the centrifugal conatus of the planet* (or the outward conatus derived from the harmonic circulation snatching it in its orb and thus trying to drive it away) *as its present distance from the Sun to the fourth part of the latus rectum of the planetary ellipse,* namely as r to $a{:}4$; and therefore the ratios of gravity to centrifugal conatus are proportional to the distances of the planet from the Sun.

(22) *The velocity of the planet round the Sun is everywhere greater than the paracentric velocity, that is, of approach to or recess from the Sun.* For since circulation is to paracentric velocity as b to $\sqrt{ee-pp}$ (by 18, and add the calculation in 19), the former will be greater than the latter if $bb + pp$ is greater than ee; this is certainly the case, because bb is greater than ee, namely the minor axis b is greater than the distance e between the foci. Certainly this always happens in planetary ellipses known to us, which do not differ much from circles.

(23) In *Aphelion A and Perihelion Ω* there is only the *circulation* without approach or recess, *greatest* in Perihelion, *smallest* in Aphelion. Further, *at the mean distance* of the planet from the Sun (which is at the end-points B and E of the minor axis) the velocity of approach and recess is to the circulation as the distance between the foci to the minor axis, namely as e to b. There, in fact, p vanishes.

(24) *The greatest velocity of the planet of approach to the Sun, or recess from it,* is when the distance $W\odot$ or $X\odot$ of the planet from the Sun is equal to the semi-latus rectum of the ellipse; then in fact (by 19 and 21)

$ddr = 0$, since $r = a{:}2$. Thus if from the Sun as a centre a circle is described with a radius equal to the semi-latus rectum $\odot W$, this circle will intersect the planetary ellipse in two points W and X where the greatest paracentric velocity occurs, which in one point such as W will be of approach, in the other X of recess. The *smallest* paracentric velocity is nil and is in Aphelion or Perihelion, namely in both vertices A and Ω of the ellipse.

(25) In the ellipse and therefore *in the planet the centrifugal conatus* of recess from the Sun, or the outward conatus of the *harmonic circulation, is always smaller* than the solicitation of gravity or the central *attraction of the Sun.* For (by 21) attraction is to centrifugal conatus as the distance from the Sun, or the focus, to the fourth part of the latus rectum, and in the ellipse the distance from the focus is always greater than the fourth part of the latus rectum.

(26) *The impetuses which a planet acquires in its path by the continued attraction of the Sun are as the angles of circulation,* that is, the angles enclosed between the radii drawn from the Sun to the initial and final points of the path, or as the apparent motion or path seen from the Sun. Thus the impetus impressed during the path $A_1 M$ is to the impetus impressed during the path $A_3 M$ as the angle $A \odot_1 M$ to the angle $A \odot_3 M$. In fact the increments of the angles are as the impressions of gravity (by 17 and 19), therefore their respective sums will also be proportional, namely the completed angles of circulation are proportional to the sums of the impressions or to the impetuses acquired from the beginning. Hence in the point W, where an ordinate from the Sun intersects the ellipse, the impetus acquired from the Aphelion A is half the impetus acquired from Aphelion to Perihelion; for in W the distance OW from the Sun is the semi-latus rectum. Further, the impetus acquired in any path is to that acquired in a semirevolution as the angle of circulation to two right angles. Now, however, I am considering the impetuses impressed by gravity or attraction by themselves, without subtracting or computing the contrary impetuses impressed by the outward conatus.

(27) Now, however, from the assigned causes it is worth *explaining the whole revolution of a planet and the degree of approach to and recess from the Sun.* A planet then placed at the greatest distance A from the Sun, namely in Aphelion, experiences both a weaker centrifugal conatus of the circulation driving it away and attractive conatus of the solicitation of gravity than if it were nearer to the Sun. At that distance, however, namely at the vertex further away from the Sun, gravity is greater than twice centrifugal conatus (by 21), because the distance $\odot A$ of the Aphelion or vertex further away from the Sun, or the focus, is greater than the semi-latus rectum $\odot W$. Therefore the planet descends towards the Sun along the path $AMEW\Omega$, and the impetus of descent increases

continuously, as in heavy accelerated bodies, as long as the new solicitation of gravity remains greater than twice the new centrifugal conatus; as
long as this happens, the impression of approach increases over the
impression of recess, therefore velocity of approach increases absolutely,
until it reaches the point where those two new contrary impressions are
equal, that is, the point W, where the distance $\odot W$ from the Sun is equal
to the semi-latus rectum. There accordingly the velocity of approach is
greatest, and stops to increase (by 24). Thence, however, although the
planet continues to move towards the Sun as far as Ω, yet the velocity of
approach decreases, since twice the centrifugal conatus prevails over the
impression of gravity; and this continues until the sum of the centrifugal
impressions from the beginning A thus far, destroy exactly the sum of
the impressions of gravity also from the beginning thus far, namely when
the entire impetus of recess (acquired only from the centrifugal
impressions taken together) equals at last the entire impetus of approach
(acquired from the continuously repeated impressions of gravity), where
any approach stops; and this very place is the Perihelion Ω, in which the
planet is closest to the Sun. Thereafter, however, the planet continues its
motion, and while thus far it was approaching, now it begins to recede
and tends from Ω through X towards A. In fact twice centrifugal
conatus, which had begun to prevail over gravity from W to Ω, still
continues to prevail from Ω to X; and therefore since the planet from Ω
starts almost to move anew, because the previous contrary impetuses
cancel each other out, recess prevails also from Ω, and the velocity of
recess increases continuously as far as X; but the increment thereof or
the new impression decreases until this new impression of recess, namely
twice centrifugal conatus, equals again, obviously in X, the new impression of approach[24] or gravity. Therefore in X the velocity of recess is
greatest. Thence, further, gravity or the new impression of approach
prevails, although the entire impetus of recess, namely the sum of all the
impressions of recess acquired from Ω, prevails still for some time over
the whole impetus of approach impressed from Ω. But since after X the
latter increases more than the former, at last they become equal in A,
where they mutually destroy each other, and recess stops, that is, the
planet has returned to the Aphelion A. Thus, then, all previous equal and
opposite impressions cancel each other out, and the situation returns to
its original state; and everything takes place anew in perpetual games
until one distant day—time having completed its course—will bring a
notable change in the ordering of things.

(28) Hence we have *in the elliptical motion of a planet six especially
noteworthy points*; four, to be sure, are obvious: Aphelion A and

---

[24] The text says 'of recess'.

Perihelion Ω, and likewise the mean distances E and B (for $\odot B$ or $\odot E$ is half the major axis $A\Omega$, and consequently the arithmetic mean between the greatest digression $\odot A$ and the least one $\odot\Omega$). Two we have added: the end-points W and X of the latus rectum WX or the ordinate of the axis at the focus $\odot$, which are the points of greatest velocity, in W of recess and in X of approach (by 24). Here, also, (by 26) the impetus acquired by the continuous impression of gravity from A to W is exactly half of that acquired in the entire descent from A to Ω; similarly, the impetus acquired from Ω to X is half of that acquired from Ω to A; and, indeed, the impetuses acquired by gravity through AW, $W\Omega$, ΩX, XA, are equal.

(29) It is now time to recount *the causes which determine the type of planetary ellipse*. Given the focus $\odot$ of the ellipse, which is the Sun's position, and given also now the place A where the Sun begins to attract the planet, for example at the planet's greatest distance, the more distant vertex of the ellipse from this focus is also given. Further, given the ratio of the gravity or virtue with which the Sun begins to attract the planet, to the centrifugal conatus by which the circulation in the same point drives the planet away and strives to repel it from the Sun, also the ellipse's principal latus rectum WX is given, namely the ordinate at the focus $\odot$. For the given $\odot A$ is to the semi-latus rectum $\odot W$ in the given ratio of the solar attraction to twice the centrifugal conatus. Thus if now the fourth part of the latus rectum is subtracted from the greatest given digression $A\odot$, the remainder will be to $A\odot$ as $A\odot$ to $A\Omega$: therefore the ellipse's major axis $A\odot$ or its minor axis is given. Hence given the points $\odot$, A, W or X, also Ω is given, and consequently also the centre C of the ellipse, its other focus F, and the minor axis BE, and thereby the ellipse itself. No less are all given if at the beginning Ω had been given instead of A.

(30) From these considerations it is clear at the same time *how an ellipse, or a circle*, which is an instance of it, *is described by the planets, and not another conic section*. The *circle*, to be sure, results when the attraction of gravity and twice the centrifugal force arising from the circulation are at the beginning of attraction equal; for they will remain thus equal, since there is no cause for approach or recess; but when at the beginning (or in the state in which the previous contrary impetuses of recess or approach, which is equivalent to the beginning, cancel each other out, namely in Aphelion or Perihelion) attraction and twice centrifugal conatus are different, provided (by 25) that the simple centrifugal conatus is less than the attraction, an *ellipse* is described; and if the attraction prevails the start is the Aphelion, but if twice centrifugal conatus prevails it is Perihelion. If the simple centrifugal conatus is equal to the attraction, a *parabola* is described; if it is greater, a hyperbola, whose internal focus is the Sun. If, however, the planet were endowed

with levity and not gravity, and were not attracted but repelled from the Sun, an *opposite hyperbola* would arise, whose external focus would obviously be the Sun.[25]

Two particularly outstanding points now remain to be accounted for in this argument: one, to explain what motion of the aether makes the planets heavy, namely drives them towards the Sun, and this as the squares of their closeness; next, what may be the cause of the relationship of the motions of the different planets of the same system, that the periodic times are in the sesquialterate proportion of the mean distances, or what is the same, the major axes of the ellipses; that is, one ought to explain more clearly the motion of the solar vortex, namely the aether, constituting each individual system. But since these matters have to be re-examined more deeply, they cannot be included within the brevity of this essay. What seems fitting to us will be explained separately in a more appropriate fashion.

[25] This classification of the orbits is discussed in Section 7.2.

PART 3
THE FORTUNES OF LEIBNIZ'S
RESPONSE TO NEWTON

7

REFLECTIONS ON LEIBNIZ'S THEORY AND ITS DEVELOPMENT

7.1   Introduction

The previous two chapters have charted the formation and development of Leibniz's theory from his reading of the *Principia*, through a series of manuscript essays and calculations, to the published *Tentamen*. A comparative study of the material presented so far and of further private manuscripts, letters, and public texts discussed in this chapter allows us to grasp the problems met by Leibniz. Taking into account the contrast between these different genres, and the style of writing in the *Tentamen*, it is possible to appreciate the development of his theory and strategy.

The following section examines the structure and style of the *Tentamen*, emphasizing the difference between the genesis of Leibniz's ideas and the order of presentation. I also select two examples relevant to the relationships between mathematical theory and physical interpretation, namely the generalization of the expression for centripetal force attained in *De Conatu*, and the law of the harmonic circulation. My contention is that in the first case the link between mathematics and physics was lacking and Leibniz's result was therefore unsuitable for a response to Newton. The second example, however, in which mathematics and physics are brought together, appeared in print in a prominent position.

Regarding the differential equations in *De Motu Gravium*, together with the problematic link between mathematics and physics, it is necessary to consider an obvious factor preventing publication, namely Leibniz's failure to find a link with ellipses and especially with Kepler's third law. These considerations are discussed in Section 7.3, where I examine some aspects of the theory of differential equations and the problems related to the early applications of the integral calculus. I discuss briefly also the role of elasticity and the implications of his equations for the debate on the relationships between dead and living force.

In the last three sections I outline the development of Leibniz's theory after February 1689, taking into account in particular the 'zweite Bearbeitung' of the *Tentamen* and *De Causa Gravitatis*, the *Excerpts* from the *Principia*, and the correspondence with Huygens in connection with the related manuscript *De Causis Motuum Coelestium*. These observations complement the picture of Leibniz's theory in several important respects. In the 'zweite Bearbeitung' Leibniz tried to present his ideas in a form acceptable to the Catholic Church; this revised version of the *Tentamen* shows convincingly the importance of Leibniz's contingent preoccupations in the exposition of his theory. In the *Excerpts* he expanded his exploration of the *Principia* and improved his interpretation of several passages. In the letter intended for Huygens and *De Causis* he revised his theory, showing where he thought its strengths and weaknesses lay. The last section also contains a brief reassessment of the material presented up to that point, and especially of the difficulties in interpreting Leibniz's texts.

7.2   Private research versus publication: priority and styles of writing

A comparison between the previous two chapters reveals immediately how the genesis of Leibniz's theory differed widely from its presentation. This difference is crucial in understanding the author's aims and problems. The first obvious issue is priority. The manuscripts *Inventum a me est* and *De Motu Gravium* show that the *Tentamen* is based on the transformation of projectile lines ('lineae projectitiae') into vortical lines ('lineae vorticales'); the former correspond to Newtonian central attraction, the latter to Leibnizian vortical motion. Leibniz's claim to priority, or at least to independent discovery, is based on the concealment of this fundamental proposition, which was still referred to in *Tentamen de Systemate Universi*.

Worries about priority represent only one aspect of Leibniz's response to the *Principia*. Concerning the related issue of private versus public texts, it is instructive to compare two results on orbital motion attained by Leibniz.[1] In *De Conatu* and *Inventum a me est* we find the general equation for centripetal conatus for a body moving along a curve with respect to an arbitrary centre of attraction. However, this result was exclusively a mathematical exercise with no physical interpretation. It is not surprising that Leibniz seemed to attach no particular value to his generalization in a response to the *Principia* and that his equation remained unpublished. We shall see in Chapter 9 that the mathematicians of the generation after Leibniz adopted a different attitude on this

---

[1] On the issue of public and private texts compare M. J. S. Rudwick, 'Charles Darwin in London: The Integration of Public and Private Science', *Isis*, *73*, 1982, pp. 186–206, esp. pp. 198–206.

issue. By contrast, the mathematical result involving the harmonic circulation had a direct physical interpretation and became the most representative element of Leibniz's theory. The integration of Kepler's area law, physical interpretations in terms of vortices, and mathematics, made the result interesting to Leibniz. This analysis of purely mathematical results versus physicomathematical laws constitutes a prominent feature of the *Tentamen*. Despite the hurry with which Leibniz's essay was composed, the *Tentamen* is a carefully drafted piece whose rhetorical strategy deserves to be investigated in relation to contemporary styles of writing.

The introduction to the *Tentamen* outlines the history of the vortex theory from Leucippus and Epicurus to Leibniz's own time. Similar historical introductions were relatively common in the seventeenth century, especially in the astronomical literature. Newton also provided an historical account of his theory in *De Mundi Systemate*, which was to become the third book of the *Principia*. Without knowing of each other's texts, Leibniz and Newton gave almost antithetical accounts. According to Leibniz, the ancients, especially the followers of Aristotle and Ptolemy, did not fully appreciate the splendour of nature; even the Pythagoreans had intuitions rather than rigorous demonstrations. Newton, on the other hand, presented himself as the restorer of the forgotten ancient knowledge of the Chaldeans, Anaximander, Aristarchus, and the Pythagoreans. Leibniz's list of heroes starts much later, from Copernicus, and includes Tycho and especially Kepler, as we have seen in Chapter 1. For him the merit lay in setting celestial motions in fluid orbs; for Newton in setting them through empty spaces without any resistance.[2]

Moving from the introduction to the first two paragraphs of the *Tentamen*, we find an attempt to establish the physical foundations of the essay in a rigorous fashion. Seventeenth-century attempts to render one's text uncontroversial followed different patterns. Before examining the *Tentamen*, I compare Robert Boyle's rhetorical style in his account of experimental philosophy with the strategies adopted by Newton and Leibniz. Boyle's extraordinarily detailed account of his experiments, performed in the presence of distinguished and trustworthy witnesses of high social stand, are described so as to create the impression in the reader that he is present and watching the functioning of the air-pump in all its minute operations. This technique has been aptly defined 'virtual witnessing'. It is not surprising, therefore, that for Boyle mathematics was

---

[2] I. Newton, *De Mundi Systemate* (London, 1728), was written in 1685 but remained unpublished in Newton's lifetime. J. E. McGuire and P. M. Rattansi, 'Newton and the Pipes of Pan', *NRRS, 21*, 1966, pp. 108–43. P. Casini, 'Newton: gli scolii classici', *Giornale Critico della Filosofia Italiana, 60*, 1981, pp. 7–53.

to be avoided because the presence of such a difficult discipline would restrict his gentlemanly audience. Moreover, Boyle believed that the use of mathematics was inappropriate in experimental philosophy, because the complexities of nature were likely to be hidden by mathematical idealizations. Mathematics would bring with it almost inevitably unwarranted and possibly dangerous expectations of certainty and accuracy.[3]

Newton adopted almost the opposite strategy. On the one hand he had no doubts about the perfect adequacy of mathematics for describing natural phenomena. Indeed, as we have seen in Section 3.4, the very foundations of his calculus were interwoven with kinematics and the study of motion. Further, by limiting his audience to proficient mathematicians, he seemed to be deliberately trying to avoid 'litigious' philosophy and seek a higher ground of certainty. Following the publication of a *New theory of light and colors* in 1672, Newton became involved in endless disputes concerning the execution of his experiments and their interpretations. If mathematics rather than natural philosophy was involved, however, he thought that controversies could be avoided. Although the argument of the *Philosophiae Naturalis Principia Mathematica* was bound to involve advanced mathematics to a considerable degree, Newton could certainly have presented his discourse in a less forbidding fashion, mitigating his ruthless and at times pointless mathematization.[4]

By contrast, Leibniz tried to render his ideas about planetary motion uncontroversial by having recourse to the authority of the Keplerian programme, by spelling out the premises of his reasonings, and especially by the deployment of logic. The last two aspects of his strategy, which owed much to his Aristotelian formation, were characteristic of Leibniz's philosophy. It is worth recalling here also his extensive theoretical investigations in logic. Probably the most spectacular case of his deployment of these arguments in a controversy can be found in the dispute on the conservation of living force with Denis Papin. In April 1696, 'as they could not reach an agreement about what they were discussing, Leibniz proposed reducing the arguments to a syllogistic form. But the difficulties in establishing which propositions were already demonstrated and accepted by the opponent, as well as of finding the agreed meaning of the technical words were so numerous, that they

[3] S. Shapin and S. Schaffer, *Leviathan and the Air-Pump* (Princeton, 1985), esp. pp. 25 and 60. S. Shapin, 'Robert Boyle and Mathematics: Reality, Representation, and Experimental Practice', *Science in Context*, 2, 1988, pp. 23–58.
[4] R. Iliffe, *The idols of the temple: Isaac Newton and the private life of anti-idolatory*, Ph.D. Thesis, Cambridge University, 1989. Cohen, *Newtonian Revolution*; D. T. Whiteside, 'Newton the Mathematician' in Z. Bechler (ed.), *Contemporary Newtonian Research* (Cambridge, 1982) pp. 109–27. Compare also Chapter 8 in the present work.

arrived at the sixteenth syllogism in 1697 without either of them having acknowledged defeat.' Two years later, on realizing that his own analysis of the issue was not to Papin's taste, Leibniz had sadly and scornfully to conclude—following the words of a 'certain courtier'—that 'de gustibus non est disputandibus': logic had failed miserably.[5]

In the theory of planetary motion Leibniz adopted the same strategy without greater success. We have seen that moving from the unpublished *Tentamen de Systemate Universi* to the *Tentamen de Motuum Coelestium Causis* the harmonic circulation was shifted from the opening sentence to leave space for an introduction largely based on Kepler, and to two opening paragraphs. In Chapter 1 I emphasized the significance of the choice of Kepler as an ally against Newton; it is now time to read again and analyse the two opening paragraphs. Leibniz says that on the basis of laws of nature it can be proved ('demonstrari potest') that all bodies which describe a curved line in a fluid are driven by the motion of the fluid. His reasoning is modelled on the syllogism. All bodies which describe a curve tend to escape along the tangent because of the nature of motion, therefore something constraining them is needed. Since, from the nature of bodies, no conatus is constrained except by a contiguous means in motion, and, by hypothesis, only the fluid is contiguous to the bodies, then all bodies which describe a curve are constrained by a fluid in motion. From this general statement it follows that since planets describe curved trajectories, they must be moved by the surrounding fluid. Here the term 'hypothesis' refers to the statement on which the demonstration is based, and indicates a *praecognitum*. Thus the structure of the first two paragraphs indicates that for Leibniz the opening statement is not hypothetical, but descends from laws of nature and therefore has moral certainty. His disputes with Newton and Papin bring to mind his 'calculemus!' and the utopia that one day logic will be so advanced that all disputes could be settled by an unquestionable logical calculus.[6]

The other paragraphs of the *Tentamen* can be organized in different groups. With the exception of paragraph 5, dealing with mathematical issues and especially the application of differentials, paragraphs 3–8 regard harmonic circulation, paracentric motion, and their basic properties. The mathematics employed up to this point is fairly simple; paragraphs 9–17, however, require considerable skills since Leibniz tackles the problem of paracentric motion. His preliminary analysis is

[5] A. G. Ranea, 'The *a priori* Method and the *actio* Concept Revised', *SL*, *21*, 1989, pp. 42–68, on pp. 44 and 66. This article is based on the study of the unpublished correspondence between Leibniz and Papin. A recent and broad collection of essays on Leibniz's logic is in *SL* Sonderheft *15*, 1987, *Leibniz: Questions de Logique.*

[6] G. W. Leibniz, *Dissertatio de Arte Combinatoria* (Leipzig, 1666) = *LSB*, VI, *1*, pp. 163–230, on p. 169, *LPG*, *7*, pp. 198–203, on p. 201 (without title). See also *LMG*, *6*, p. 211.

expanded in paragraphs 18 and 19 by means of his differential calculus. Indeed, the *Tentamen* is the first publication in which Leibniz applied the calculus to a physical problem. In previous essays, such as the 1682 *Unicum opticae principium* or the 1689 *Schediasma de resistentia medii,* the reader finds no differentials, despite the fact that they were used by Leibniz in his private calculations. Apparently only in the *Tentamen* did he feel sufficiently motivated to present a portion of his equations to the public. It is reasonable to see in this deployment of the calculus yet another aspect of his confrontation with Newton. The following paragraph contains the reference to Newton concerning the inverse-square law; strictly speaking Leibniz does not deny that he had seen the book, his point being merely that in the review in the *Acta* there is no demonstration of the law. Paragraphs 21–9 are worth considering in order to appreciate a further aspect of Leibniz's rhetorical strategy. Their contents consist of rather elementary statements which are nothing more than corollaries to the previous demonstrations. Yet they present to a reader with a modest competence in geometry, and especially conic sections, a series of easily understandable propositions and a convincing exposition of motion along ellipses. We can reasonably infer that a reader like Christoph Pfautz, for example, would have appreciated this clear, at times pedantic, account. Indeed, the English mathematician Charles Hayes, who followed Leibniz's mathematical theory, seemed to appreciate precisely this aspect of his exposition.[7]

The last paragraph consists of two parts. The former contains an attempt to classify the orbits, the latter an evaluation of the problems left unsolved by his theory. I outline Leibniz's classification taking into account his 1706 corrections of the factor of two for centrifugal conatus. The argument is based on the ratio between attraction and centrifugal conatus at the perihelion, where paracentric or radial velocity vanishes, for orbital velocity is equal to the velocity of circulation, as Leibniz explains in paragraph 23. Further, from paragraph 21 we know that the ratio between attraction and centrifugal conatus is as r to a:2, where a is the latus rectum of a conic section; notice that the equation of para-centric motion is valid for all conics, as we read at the end of paragraph 19. The cornerstones of this classification are the circle and parabola, from which the other cases can be easily obtained. If gravity is equal to centrifugal conatus $r = a$:2 and paracentric solicitation vanishes; since there is no cause for approach or recess, attraction and centrifugal conatus remain equal, paracentric or radial motion is nil, and the orbit is a circle. We move now to the parabola. From paragraph 18 we know that the ratio between velocity of circulation and orbital velocity is as a to

[7] Hayes, *Treatise of Fluxions,* pp. 291–305. On Hayes see Section 9.3

$2\sqrt{ra}$. Since at the perihelion the two velocities coincide, $a = 2\sqrt{ra}$ and $r = a{:}4$, therefore attraction is equal to half centrifugal conatus. If centrifugal conatus is smaller than this threshold value, but greater than attraction, the curve is an ellipse; if half the centrifugal conatus is greater than attraction, the curve is a hyperbola. If we had repulsion instead of attraction, the curve would be an opposite hyperbola whose focus lies outside its concavity, as Newton stated in *Principia*, book I, proposition 12.[8] However, it is worth recalling that Leibniz's reasoning is based on the unproven assumption that the curve is a conic section.

In the second part of paragraph 30 Leibniz mentions two unsolved problems of his theory: the mechanism which generates gravity inversely proportional to the square of the distance, and Kepler's third law. These concluding observations show that even Leibniz was aware of the problematic status of his theory; in such circumstances establishing independent discovery and priority was all the more necessary in order not to appear as an inventor who had not only come second, but was also worse than the first.

7.3 Some observations on differential equations and their interpretation

In the manuscripts described in Chapter 5 Leibniz used a variety of mathematical techniques ranging from elementary geometry to differential equations. Here I present some reflections on differentials, on the integration of differential equations, and on the interpretation of some of the equations we have seen above.

As we have seen in Chapter 3, the Leibnizian calculus is based on operations with variables including differentials of first and higher order, such as x, dx, ddx, etc. In the *Nova Methodus* Leibniz introduced dx as a straight line taken at will ('recta aliqua pro arbitrio assumta'). The *Tentamen* and its related manuscripts modify and extend this definition. In the *Lemmata Incomparabilium*, namely paragraph 5 of the *Tentamen*, he explained that differentials can be understood by considering the dimensions of the Earth with respect to the sky; in paragraph 17 the differential of time is taken as the length of a day with respect to the period of an outer planet. The preparatory manuscripts enrich this picture by showing that on several occasions Leibniz attributed very

---

[8] The equation of radial acceleration with a repulsive force, namely $d^2r{:}dt^2 = (h^2{:}r^3) + (h^2{:}r^2c)$, is indeed obtained by taking twice the derivative with respect to time of the polar equation of the hyperbolic branch whose focus is not contained within its concavity, namely $r = c{:}(\varepsilon\cos a - 1)$; here c is the parameter of the hyperbola, ε its eccentricity, r and α its polar coordinates, and $h = r^2 da{:}dt$.

small values to differentials for the sake of calculation. This way of conceiving them served several purposes. Firstly, Leibniz was able to avoid
infinitesimals as well as the related conceptual problems and philosophical objections. Secondly, he could provide a practical justification
for avoiding higher-order differentials, since they can be chosen so as to
produce errors smaller than any given quantity. Thirdly, as the physical
examples above suggest, differentials could be directly applied to the
investigation of nature. In the study of orbital motion, for example,
impetus corresponds to a displacement incomparably smaller than the
radius, conatus or solicitation to a displacement incomparably smaller
than impetus. In this way nature was investigated by means of differential
equations and represented by the curves resulting from their integration.

The theory of differential equations around 1700 is surprisingly and
regrettably little studied. The following observations concern a small
portion of Leibniz's investigations, yet at the same time they raise
broader issues and provide some general interpretive guidelines. From
the examples in Section 5.5 we have seen that in the integration of
differential equations Leibniz's technique broke down. This technique
involves procedures—such as substitution and separation of variable—
requiring specific skills. Leibniz's difficulties appear clearly: in equation
(3a) he introduced the integration constant w with the same symbol used
for the constant differential of the time variable. This problem of
notation whereby conceptually different constants are not distinguished
among themselves plagues several of his attempts. In equation (5a) he
was unable to separate the variables and this failure led him into trouble
with the integration: clearly the choice of the differential of time as a
constant blurred the distinction between parameters and variables, thus
invalidating the integration procedure. The same problem occurs in
other texts by Leibniz.[9] A very interesting case can be found in equation
(8a); since this equation is of second order, the neglect of signs and
constant factors affects the core of the solution procedure so that the
final result is completely different—compare equations (10a) and (10b).
The role of signs in geometry, and the role of signs and constant factors
in proportions, is considerably less important. Although this type of
mistake may appear to the modern reader as due to inattention, it would
be misleading to interpret it as such and to state that if Leibniz had paid
more attention, he would not have used the same symbol for two

[9] *LSB*, III, *2*, 21 June 1677, Leibniz to Oldenburg, p. 176, lines 15–17. Hess, 'Zur
Vorgeschichte', publishes several important manuscripts. It is doubtful whether Leibniz
understood the issue of separation of variables as clearly as Johann Bernoulli; see his letter
to Leibniz of 9 May 1695, *LMG*, *3*, p. 138; Johann Bernoulli, 'Modus Generalis
Construendi Omnes Aequationes Differentiales Primi Gradus', *AE* Nov. 1694, p.
435 = *JBO*, *1*, p. 123. E. Grosholz, 'Two Leibnizian Manuscripts of 1690 Concerning
Differential Equations', *Historia Mathematica*, *14*, 1987, pp. 1–37.

completely unrelated constants, or forgotten a sign or a constant. Leibniz obviously had not appreciated the importance of parameters in the new mathematics. These problems mark the transition from geometry and proportions to the new mathematics of differential equations, where a sign or a constant factor may radically change the result. The examples which we have seen illustrate that the transition from the former to the latter was by no means smooth, especially with regard to the integral calculus. Similar examples can be found in the other fields in which Leibniz employed the calculus. Commenting on a mistake concerning a constant factor in a differential equation concerning motion in a resisting medium, Eric Aiton points out that 'the difference is not material, because Leibniz is really concerned only with proportions.' Precisely this concern, however, constitutes a line of separation between two styles of practising mathematics. In this regard it is revealing that Leibniz rarely makes mistakes with variables.[10]

At the end of his calculations in Addition 2 to *De Motu Gravium*, Leibniz realized that his technique was not sufficient and tried to take into account that the radius over which he was taking the integral did not start from being infinitely small, since it ranged between the aphelion and perihelion distances. Integration techniques—which we identify as the definite and indefinite integral—were not clearly distinguished and were more complex than Leibniz had thought.

The equations above, however, testify to Leibniz's skill with the technique of substitution of variable; after equation (10a), for example, he wrote the correct substitution without hesitation. As we have seen in Section 3.4, Leibniz had developed his calculus almost from the beginning in a form allowing the selection of the most convenient progression of the variable. Moreover, the differential equations discussed in Section 5.5 are a powerful illustration of the difficulties and the importance of the transformation of language of the *Principia* into the differential calculus. The attempt to find ellipses and Kepler's third law in the second addition to *De Motu Gravium* led Leibniz to write a very general and powerful equation in which gravity depends on an arbitrary power n of the distance from the centre. In this fashion orbital motion was reduced to the problem of integrating differential equations

[10] Aiton, 'Resisting Medium', p. 264, n. 30. See also M. S. Mahoney, 'Diagrams and Dynamics: Mathematical Perspectives on Edgerton's Thesis', in J. W. Shirley and F. D. Hoeniger, eds., *Science and the Arts in the Renaissance* (Washington, 1985), pp. 198–220, on pp. 213–16; 'Beginning of Algebraic Thought', in Gaukroger, ed., *Descartes*, pp. 141–55; 'Infinitesimals and Transcendent Relations'. Cohen, *Newtonian Revolution*, sect. 1.3. E. J. Hofmann, 'Über Auftauchen und Behandlung von Differentialgleichungen im 17. Jahrhundert', *Humanismus und Technik*, 15, 1972, pp. 1–40. C. J. Scriba, 'The Inverse Method of Tangents: A Dialogue between Leibniz and Newton', *AHES*, 2, 1963, pp. 113–37.

depending on a parameter. This example is representative, I believe, of the versatility of the new methods devised and practised by Continental mathematicians. Although Leibniz's own attempts in *De Motu Gravium* were largely unsuccessful, the efforts of several generations of mathematicians in the eighteenth century followed the same lines and led to the transformation of mathematics and mechanics alike. The differential equations employed by Leonhard Euler and Alexis Clairaut around 1750 in their works on celestial mechanics and especially lunar theory bear a remarkable similarity to Leibniz's equation of paracentric motion. Thus the rediscovery of the fruitful radial equation, in a different physical and mechanical context and with more sophisticated mathematical tools, bore important fruits.[11]

The differential equations discussed in Chapter 5 have important implications for the debate on the relation between dead and living force. The left-hand sides of equations (2a) and (3a), and (7a)–(9a), clearly express the relation between dead force and living force with the factor $\frac{1}{2}$. Similar equations occur frequently in Leibniz's mathematics, but in this context they have a direct relation to mechanics; it is worth emphasizing that dead and living force are mentioned in *De Motu Gravium*. Thus for Leibniz living force could be represented as the integral of dead force times an infinitesimal distance.[12]

My interpretation of the right-hand sides of Leibniz's equations is more conjectural and is based on the role of elasticity discussed in Sections 2.4, 4.3, and 5.6. We have seen that Leibniz established an analogy between centripetal force, centrifugal force, and the force with which a compressed elastic spring begins to restore itself. Together with this association centred on the notion of dead force, Leibniz considered another analogy based on living force, the height to which a body can raise itself, and the compression of an elastic spring. Although Leibniz does not distinguish clearly between dead and living force in this context, it is worth emphasizing once again the central role played by elasticity in Leibniz's system. In the case of the differential equations above, I believe that if a physical interpretation of the terms which we identify as 'potential energy' has to be given, it would be based on the elasticity of the interplanetary aether. Although this interpretation may provide a convincing qualitative account, quantitatively it would fail to explain the dependence on distance. After 1688 Leibniz's main objective was to find explanations in terms of impacts, rather than to write equations which could not be interpreted in this way. It is in this context that he wrote the

---

[11] This issue is discussed in D. Bertoloni Meli, 'The emergence of reference frames and the transformation of mechanics in the Enlightenment', forthcoming in *Historical Studies in the Physical and Biological Sciences*. See also chapter 9 below.

[12] See the essay *De Motu Gravium*, lines 95–101.

equations above; from his point of view, however, they create new problems instead of solving old ones. With respect to the new aims dictated by the appearance of the *Principia*, Leibniz's equations were not only mathematically unsatisfactory, in that they failed to establish a link between paracentric motion and ellipses, but were also an additional puzzle in physics.[13]

This survey of Leibniz's texts sets the scene for a study of the development of his theory; as we shall see, rather than cracking new problems, Leibniz became engulfed in an ever growing number of difficulties and doubts.

7.4   The 'zweite Bearbeitung' and the cause of gravity

> Talia frustra
> Quaerite quos agitat mundi labor, at mihi semper
> Tu quaecumque paret tam crebros causa meatus
> Ut superi voluere late.
>
> Lucan, *De Bello Civili*, I. 417–9.

This quotation concludes the introduction to the 'zweite Bearbeitung' of the *Tentamen*. Leibniz claims that the inquiry into the causes of planetary motion has reached such a satisfactory stage, that a *poeta intelligens* would no longer say that the Gods wished them to remain hidden forever. In spite of this optimism, Leibniz was far from satisfied with his own theory: from the publication of the *Tentamen* in February 1689 until his letter intended for Huygens of October 1690 he changed his opinion several times on several issues.[14]

In Leibniz's mathematical papers edited by Gerhardt the *Tentamen* of the *Acta*, or published version, is printed together with a second revised version, named by Gerhardt 'zweite Bearbeitung'. The manuscripts consist of a set of notes referring to the pagination of the text as it appears in the *Acta*.[15] Its probable date of composition is between the

---

[13] See *LMG, 4*, p. 399, Leibniz to Hermann, 17 Sept. 1715.

[14] Further manuscripts on planetary motion composed in Rome include *Compositio Motus ex Circulatione Circa Aliquod Centrum, et Solicitatione Paracentrica Attractionis vel Repulsae Respectu Ejusdem Centri*, LH 35, 10, 1, f. 17; the text begins from f. 17v. and then goes on to f. 17r. It is half a folded folio. In the other half, f. 16r. contains two short texts on planetary motion, (f. 16v. is blank). The folded folio is on Roman paper. Also f. 15—a quarto sheet—is on Roman paper and deals with planetary motion; 'circulatio harmonica' and 'motus paracentricus' are explicitly mentioned. Moreover, Leibniz refers to twice centrifugal conatus in the analysis of orbital motion.

[15] *LMG, 6*, pp. 161–87. LH 35, 9, 2, f. 54–5 and 77–8 (complete set of notes); f. 62–4 and 74–6 (copies of paragraph 18 written by a secretary and revised by Leibniz); f. 60–1 (copy of paragraphs 19–30 in Leibniz's hand).

summers of 1689 and 1690; it is difficult to be more precise, because the notes forming the text were written in successive stages, beginning with a preliminary draft on the manuscript of the *Tentamen*. Additional elements in support of my dating emerge from the following discussion. The two versions are similar, the main differences being a long passage which has been added to the introduction in the 'zweite Bearbeitung' and a few other additions and recastings. It appears that Leibniz had three reasons to write a revised version: he wanted to introduce a new model for gravity; he tried to be more explicit in his definitions and calculations; he wished to present his theory in a form acceptable to the Catholic Church. The last issue was probably related to a series of attempts undertaken by Leibniz during the Italian journey to have the ban against the Copernican system lifted. Thus despite their similarity, the 'zweite Bearbeitung' shows a considerable shift of preoccupations with respect to the published version. As we are going to see presently, Leibniz's emphasis moved from the themes of the Cassirer thesis to a more 'positivist' account based on relativity of motion. These observations reinforce my interpretation of the link between the Cassirer thesis and the dispute with Newton, and at the same time prove convincingly the central role played by circumstances of composition on Leibniz.

The new passage in the introduction begins with a reference to William Gilbert's theory of magnetism, and contains reflections on the history of the Earth and its original fluid state, as well as on the cause of its sphericity.[16] The main novelty concerns the cause of gravity, which is no longer seen as generated by the same fluid which carries the planets, but by a fluid emitted from a centre and propagating according to the inverse-square law, on the example of light. This fluid penetrates the pores of bodies interposed to it; since the bodies contain less fluid receding from the centre than the unperturbed fluid, it is necessary that they are pushed back towards the centre with a force depending on the number of pores open to the fluid. On a more philosophical tone, Leibniz claims that the inverse-square law has been found first a priori, meaning that it has been assumed as a reasonable hypothesis. Subsequently, it has also been found a posteriori without hypotheses, by combining laws of nature and mathematical calculations based on astronomical data, as he shows in paragraph 19.[17]

It is useful to complement the observations on the cause of gravity discussed in the introduction to the 'zweite Bearbeitung' with another

---

[16] A commentary to the revised introduction is in Koyré *Newtonian Studies*, appendix A. Leibniz's reflections on the formation and history of the Earth were expounded in *Protogaea* (Göttingen, 1749).

[17] On Leibniz's a priori and a posteriori see Gueroult, *Dynamique*, chapters 3–5; Leibniz, 'An Introduction', in Loemker, *Papers*, pp. 280–90.

essay on the same topic. In May 1690, when Leibniz was on his way back to Hanover, a paper of his appeared in the *Acta*, *De Causa Gravitatis et Defensio Sententiae Autoris de Veris Naturae Legibus Contra Cartesianos*.[18] As the title indicates, this paper touches upon two topics, the cause of gravity and the response to a paper by Denis Papin on conservation of force which had appeared in the *Acta* the previous year. Bearing in mind Leibniz's promise at the end of the *Tentamen* of a new essay in which the cause of gravity and Kepler's third law would be better explained, we can consider *De Causa Gravitatis* as a continuation of the *Tentamen*. Leibniz provides a possible model for gravity, although the mechanism he adopts in the end is different. He considers a tube filled with mercury in which a glass sphere floats. The tube is closed at the extremities and is made to rotate in a horizontal plane around a vertical axis passing through one of the bases. Since the fluid is very dense, it pushes the lighter sphere towards the centre. This example explains why bodies are heavy as a result of their density relative to the aether. This model, however, does not account for an attraction inversely proportional to the squared distance, nor can the circulation be harmonic, since the fluid rotates like a rigid body. Further, in the case of a vortex it would be necessary to consider problems arising from its stability. Leibniz surveys alternative explanations for gravity: a vortex rotating along the parallels would be unsuitable because gravity would act towards the axis of rotation and not towards the centre; also a vortex rotating along the meridians, of the same kind as that he had envisaged in the *Hypothesis Physica Nova*, would be unsatisfactory because gravity would be much stronger in the polar regions, where all meridians meet, than at the equator, and this is contrary to experience. According to the third alternative, gravity is caused by a fluid emitted from a centre, as in the 'zweite Bearbeitung' and in a coeval letter to Antoine Arnauld.[19]

In the introduction to the 'zweite Bearbeitung' Leibniz continues with a passage often quoted by historians as an example of his 'positivism':[20]

What follows is not based on hypotheses but is deduced from phenomena by the laws of motion; whether or not there is indeed an attraction of the planets by the Sun, it is sufficient for us to be able to infer the approach and recession, that is, the increase or decrease of distance, which would occur if they were attracted by the prescribed law. And whether they do indeed circulate about the Sun, or do

---

[18] *AE* May 1690, pp. 228–39 = *LMG*, 6, pp. 193–203.

[19] At the end of his Italian journey Leibniz wrote the last letter of his correspondence with Arnauld, dated Venice, 23 March 1690. The letter and its preparatory draft contain some observations on planetary motion. Le Roy, ed., *Discours de Métaphysique*, pp. 202–3.

[20] *LMG*, 6, p. 166; Aiton, *Vortex Theory*, p. 132, from where the translation is taken; Koyré, *Newtonian Studies*, p. 136; Cohen, *Newtonian Revolution*, p. 323, n. 10.,

not circulate, it is sufficient that they change position relative to the Sun as if they were moved by a harmonic circulation.

Although at first sight this passage bears a similarity to Newton's General Scholium concluding the second edition of the *Principia* (1713), and in particular to his famous *hypotheses non fingo*, the context in which it ought to be interpreted is different. I consider two aspects. First, when Leibniz claims that his theory is not based on hypotheses, he certainly does not refer to the cause of the solicitation of gravity and to the existence of fluids; as we have seen, these are not hypothetical, but corollaries to the laws of nature, as he explains in paragraphs 1 and 2. For Leibniz the existence of attraction is hypothetical, since attraction is a purely imaginary mathematical construct to save the phenomena; reality lies in the impulsions of fluids. His interpretation is almost the opposite of Newton's *hypotheses non fingo*; for Newton the existence of attraction inversely proportional to the squared distance is a mathematically demonstrated fact, whereas the existence of subtle fluids is a hypothesis: the two players could not agree on the rules of the game. Thus it is likely that in the first part of the quotation Leibniz had Newton in mind as a target.

The second aspect concerns Leibniz's attempt to propagate his theory in Catholic countries and his defence of the freedom of philosophizing. While he was travelling through Southern Germany, Austria, and Italy, he was actively involved in a debate with several Catholics and especially Jesuits on the censorship of the Copernican system, in the hope of having the ban lifted. This is the other context in which the passage above needs to be interpreted. In the second part of the quotation Leibniz did not have Newton in mind, but Andreas Osiander's preface to *De Revolutionibus*, Cardinal Roberto Bellarmino and the anti-Copernican decree. Instead of claiming that his theory was merely a mathematical hypothesis, Leibniz defended its truth on the basis of relativity of motion and regardless of the adoption of any astronomical system. My interpretation is supported by other variant readings between the two versions. For example, in the introduction to the 'zweite Bearbeitung', references to Copernicus and Epicurus are crossed out, as well as passages such as the following: 'the entire aether with the planets is driven round by the motion of the Sun about its centre'. But the philosophical implications of the essay have been modified too. In the *Acta*, for example, Leibniz says referring to Kepler: 'For to him we owe the first indication of the true cause of gravity and of this law of nature on which gravity depends'. And in the 'zweite Bearbeitung': 'For to him we owe the first indication of the physical use of this law of nature on which either gravity depends, or at least it can be described in a wonderful way'.[21] It is likely that Leibniz revised the *Tentamen* after it had appeared in the *Acta*, having in mind a second publication in a form acceptable to the Catholic Church. In a letter from Florence of December 1689 to the Jesuit Antonio Baldigiani in Rome, Leibniz asked for his opinion on an enclosed memorandum on the Copernican system and announced his intention to publish an essay on that topic. He may have referred to the 'zweite Bearbeitung'.[22]

Further additions and recastings, apart from some minor variants, occur in paragraphs 2, 19, 27, and 30. In paragraph 2 Leibniz adds that since all planets move in the same direction and are very nearly on the same plane, and similar effects have the same cause, it follows that they are carried by fluid orbs. This seems to be a polemical usage of the second hypothesis in book III of the *Principia*, where Newton states: 'The causes of natural effects of the same kind are the same.'[23]

The additions to paragraph 19 consist in the attempt of reformulating the demonstration of the equation of paracentric motion without the differential calculus. The new proof is based on the osculating circumference, and contains the important point where Leibniz clarifies that the outward conatus ('conatus excussorius') differs from the centrifugal conatus ('conatus centrifugus'), which corresponds to the case when the curve is a circumference.[24]

Some of the most interesting variants are in paragraph 27. Leibniz stresses that the advantage of the harmonic circulation is in the agreement between physical and geometrical representations. If the circulation were not harmonic, the planet would move faster or slower than the surrounding fluid because of its *impetus conceptus*. Once again, Leibniz is concerned with the principle of inertia. Immediately below he states the equivalence between two representations of orbital motion: the trajectory can be seen as the resultant either of rectilinear motion ('ac in

[21] *LMG*, 6, pp. 147–9 and 162–3 and Bertoloni Meli, 'Censorship', sect. 4. The passage from the 'zweite Bearbeitung' reads: 'Nam ipsi primum indicium debetur usus physicis huius naturae legis, a qua vel pendet gravitas, vel saltem mirifice illustratur'.

[22] *LMG*, 6, pp. 145–7n., and Bertoloni Meli, 'Censorship', sections 3 and 4, where I date the letter and identify the addressee. Gerhardt erroneously associates this letter with the published version of the *Tentamen*, not with the 'zweite Bearbeitung'. Leibniz, *Vorausedition zur Reihe VI*, 1989, Faszikel 8, pp. 1754–62.

[23] The *Hypotheses* (*Regulae philosophandi* in the later editions), all mentioned in Pfautz's review, are discussed in Koyré, *Newtonian Studies*, chapter 6. I. B. Cohen, 'Hypotheses in Newton's Philosophy', *Physis*, 8, 1966, pp. 163–84. See also *KGW, 3, Astronomia Nova*, p. 236: 'Est siquidem axioma per universam philosophiam naturalem: eorum, quae simul et eodem modo fiunt, et easdem ubique dimensiones accipiunt, alterum alterius causam aut utrumque ejusdem causae effectum esse.'

[24] *LMG*, 6, p. 178: 'Nam conatus centrifugus, ut supra diximus [paragraph 10], est species tantum simplicissima conatus excussorii, cum scilicet motus est circularis.' It is useful to contrast this careful distinction with the less sophisticated treatment in *De Conatu* (Section 5.3 above).

vacuo') and of the attraction of gravity; or of circular harmonic motion and of paracentric motion. This equivalence accompanies Leibniz's theory from a very early stage, namely from *Inventum a me est* and especially *De Motu Gravium*. The awareness of the existence of two equivalent mathematical representations of orbital motion is one of the most characteristic aspects of Leibniz's theory. It is useful to focus on this aspect.

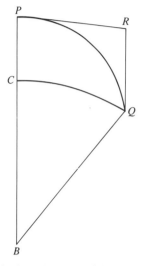

Fig. 7.1   Equivalent decompositions of orbital motion.

An infinitesimal arc *PQ* of an orbital trajectory can be decomposed in an infinite number of ways. For example one can consider the tangent *PR* to the curve, and the deviation *RQ*; or alternatively radial motion *PC* and circular motion *CQ*. In the first case the trajectory results from rectilinear inertial motion *PR* and central attraction *RQ*; in the second case the orbital path results from a circular motion *CQ* and radial motion *CP*. For Leibniz this equivalence could be resolved on a physical level by the existence of a vortex.

In the conclusion of the same revised passage, Leibniz makes an interesting remark on the relation between mathematics and natural philosophy. He states that the preceding geometrical constructions do not occur with absolute geometrical precision because of the mutual action among the planets and indeed the bodies of the universe. Nothing has ideal geometrical of dynamical properties; it is sufficient that matter does not behave too differently from our limited representations. Here the expression 'proprietates Dynamicae' contains one of the first occurrences of the new term.

Lastly, in paragraph 30 Leibniz alters the conclusion and refers to cometary paths, leaving the question whether they are closed, hyperbolic, or parabolic to further investigations. He maintained a very cautious attitude to comets—so difficult to explain in his vortex theory—for many years, until the *Illustratio Tentaminis* and *Theodicy*.[25]

7.5 The *Excerpts* from the *Principia Mathematica*

Further manuscripts dating from Leibniz's stay in Rome include the two sets of *Excerpts* from the *Principia Mathematica*. Leibniz focused on three areas: the concept of the infinitely small; the notion of force in connection with infinitesimal velocities and accelerations; the existence of the aether and of a vortex carrying the planets in connection with the cause of gravity.

With regard to the infinitely small, Leibniz reconsidered lemma 11 and added some comments which highlight the differences between his own and Newton's notions and practice. In the demonstration that the subtense of the angle of contact is ultimately as the square of the subtense of the bounding arc, Newton stated that one extreme of the arc approaches right up to the other extreme. Leibniz objected that in this case they would coincide: in his opinion it should be said that they approach until they form an arc infinitesimal of the first order which differs incomparably from a straight line. The relevant sentences in the concluding scholium were also excerpted. Newton explained that with his method of first and last ratios he did not consider velocities or quantities before they vanish, or after they vanish, but with which they vanish. Leibniz's attention was also caught by Newton's statement that whenever he mentions least, evanescent or last quantities, he does not understand them to be determinate, but always diminished without end. Whereas Newton's least quantities become vanishingly small and occur only in finite ratios, Leibniz's differentials are infinitely small but not vanishing, and behave in the realm of the infinitely small as variables do for ordinary magnitudes. His observations on lemma 11 were generally of superior quality than the corresponding ones in the *Marginalia*, where he had stated that Newton's demonstration was valid only in the special case of the circumference. The unambiguous statement in the *Excerpts* that the arc of the curve can be taken as the arc of the osculating circle,

---

[25] *LMG*, 2, p. 54, Leibniz to Huygens, autumn 1690; *Theodicy*, par. 245; in *Illustratio Tentaminis*, *LMG*, 6, p. 266, Leibniz adds the following considerations: 'Sed eo licet supposito verisimile alicui fortasse videbitur, contingere in his vorticibus, quod in aquarum circulis quod diversi lapilli faciunt simul in aquam injecti, aut diversi simul soni suis ondulationibus eundem aeris locum permeantes, ubi alter alterum non turbat.'

and other pertinent observations, show that by 1689 Leibniz had come to accept the substance of lemma 11 with its corollaries.[26]

Leibniz confronted Newton's statements on force in lemma 10 and proposition 6, which we examined briefly in Section 4.2. The former states that the spaces traversed by a body urged by a regular force at the beginning of motion are proportional to the squares of the times, and is a generalization of Galileo's law of fall. Unlike the *Notes*, where lemmas 9 and 10 were severely criticized, here they are transcribed without critical comments, as if Leibniz had come to accept them. Indeed, lemma 10 is employed in proposition 6, which was also accepted by Leibniz, though in a peculiar fashion. As we have seen in Section 4.2, this fundamental proposition proves that the centripetal force retaining a body in its orbit is directly as the vanishing deviation from the tangent to the curve, and inversely as the square of the area of the focal sector swept out by the radius; this area is proportional to the time required to traverse the corresponding arc, for Kepler's law. It is worth considering Leibniz's struggle with this passage in more detail by taking into account the evolution of his thought from the *Marginalia*. On page 45 of the *Principia*, in the demonstration of proposition 6, lemma 10 is mentioned in the first line. In the *Marginalia* Leibniz commented: 'I doubt lemma 10' ('In lemmate X adhuc haereo'). The text of Newton's demonstration reads (see Fig. 4.2): 'The nascent line-element QR is, given the time, as the centripetal *force* (by Law 2) and given the *force*, as the *square* of the time (by Lem. 10)'. Leibniz altered this passage in the following way: 'The nascent line-element QR is, given the time, as the centripetal *velocity* (by Law 2) and given the *velocity*, as the time (by Lem. 10).' (My emphases.) Subsequently he systematically changed 'quadratum temporis' into 'tempus'. This is the reason why a later reference by Newton to the corollary to proposition 6 is marked by Leibniz with the words 'error ibi'. The corollary states that, given an arbitrary curvilinear figure and a centre of force, the law of centripetal force can be found by computing the ratio of the deviation from the tangent to the square of the area swept out by the radius—or to time squared.[27] By and large, the *Marginalia* predate the *Tentamen*. We have seen that Leibniz conceived a curve as a polygon composed of infinitely many first-order infinitesimal sides, and that this representation requires instantaneous impulses at each vertex of the polygon and velocities, not accelerations. This is a first

---

[26] *Exerpts*, pp. 480–1 and 484: ' Mirum quod haec dicantur ultimo rationem habere cum coincidant, sed respondeo revera non coincidere, cum formant arcum infinite parvum primi gradus, qui a recta non differt nisi incomparabiliter. Et res redit ad meam aestimationem de circulo osculante.' See Sections 3.4 and 5.2.

[27] *Marginalia*, M 44 and M 45 A. The translation is based on *NMW*, 6, p. 133. The words 'error ibi' are in *Marginalia*, M 48, line 32.

partial explanation of Leibniz's interpretation. Secondly, although in proposition 6 Newton used a continuous curve and accelerations, he referred to the second law of motion, where force appears to be proportional to an impulse, not to acceleration. No doubt this ambiguous reference also contributed to Leibniz's reading.

The situation changed—partially, at least—after the autumn of 1688. In fact, in paragraph 19 of the *Tentamen* Leibniz proved that the second-order infinitesimal spaces traversed by a body acted upon by paracentric solicitation or gravity are proportional to the squares of the differentials of time. One may wonder at this point how Leibniz could reconcile this result, which seems to require accelerations, with his views that solicitations are as infinitesimal velocities. Although he did not discuss this matter in detail, from some brief observations he seemed to suggest that the two interpretations were compatible because the element of time dt is constant. Recalling the marginal role given to constants, we can figure out why in his mind the problem did not arise. One may add that Leibniz could write the equation of paracentric motion in a form involving time because of Kepler's area law. In that equation, however, the second-order differential of the radius ddr is equal to the difference between two terms which—taking into account the physical dimension of parameters—must also be second-order infinitesimal lengths. Thus, as in the calculation of central solicitation in Section 4.2, Leibniz's equation of paracentric motion can be reasonably understood in terms of infinitesimal velocities. In the *Excerpts* Leibniz followed an interpretation derived from the *Tentamen*. Newtonian force is taken to be proportional to infinitesimal velocity, which in its turn is proportional to time squared: 'The space QR is always as the centripetal force or velocity and (time, crossed out) the square of time.' Once again, Leibnizian mechanics is based on the refusal to take accelerations into account. Significantly, while transcribing proposition 6, he started by writing: 'Aestimat vim percussionis', swiftly substituting the last word with 'centripetam'. Thus Leibniz thought that Newtonian force could be rendered as an infinitesimal velocity which, as this slip suggests, was due to an infinitesimal impact.[28]

Leibniz hardly missed a passage on vortices and fluids. As we have seen in Section 5.2, he transcribed from the general scholium to proposition 40, book II, that the resistance to pendular motion due to the internal parts of a body is utterly negligible with respect to the resistance due to its surface; this consideration led Newton to doubt the existence of a

---

[28] *Excerpts*, pp. 481 and 485–6: 'Spatium QR semper est in ratione vis centripetae seu celeritatis et (temporis, crossed out) quadrati temporis.' In *Marginalia*, M 48, lines 33–4, Leibniz tried to reconcile his own views with Newton's by considering that the element of time is constant.

subtle fluid penetrating all bodies. In corollary 3 to proposition 6, book III, Newton claimed that a vacuum is necessarily given. Leibniz's immediate comment was that in this way gravity can only be explained in terms of incorporeal operations.

In propositions 52 and 53 and their respective scholia at the end of book II, Newton claimed that planets cannot be carried by vortices for several reasons. Newton's shaky analysis of vortical motion in proposition 52 alleged that 'If a solid sphere, in a uniform and infinite fluid, revolves about an axis given in a position with a uniform velocity . . . the periodic times of the parts of the fluid are as the squares of their distances from the centre of the sphere.' Proposition 53 states that 'Bodies carried about in a vortex, and returning in the same orbit, are of the same density with the vortex, and are moved according to the same law (as to velocity and direction of motion) with the parts of the vortex.' Therefore, since planets should follow exactly the same motion as the vortex, and the vortex ought to rotate with a period proportional to the square of the distance from the centre, planets could not obey Kepler's third law. Further, from Kepler's second law, planets move more swiftly in perihelion than in aphelion; in perihelion, however, according to the laws of mechanics, the vortex with its planets would have a slower motion, because planets move in a wider space. Newton's conclusion is that 'the hypothesis of vortices is utterly irreconcilable with astronomical phenomena, and rather serves to perplex than explain the heavenly motions.' Leibniz excerpted these passages adding an interesting remark in the margin. He claimed that at the beginning of section 11 Newton stated that he was 'considering the centripetal forces as attractions; though perhaps in a physical strictness they may more truly be called impulses.' However, since there is no impelling matter, it is not clear by what heavy bodies should be impelled.[29]

I wish to recall a further passage from book II, proposition 23, where Newton studied elastic fluids and gave a mathematical explanation of Boyles' law. The proposition states:

If a fluid be composed of particles fleeing from each other, and the density be as the compression, the centrifugal forces of the particles will be inversely proportional to the distances of their centres. And, conversely, particles fleeing from each other, with forces that are inversely proportional to the distances of their centres, compose an elastic fluid, whose density is as the compression.

The conclusion of the following scholium is also worth quoting:[30]

But whether elastic fluids do really consist of particles so repelling each other, is

[29] *Excerpts*, pp. 482 and 487; *Principia* (Motte and Cajori), pp. 164 and 387–96, esp. pp. 387, 394, 396.

[30] *Principia* (Motte and Cajori), pp. 300 an 302.

a physical question. We have here demonstrated mathematically the property of fluids consisting of particles of this kind, that hence philosophers may take occasion to discuss that question.

As I. Bernard Cohen has convincingly shown, the necessary and sufficient reasons regarding Boyle's law provided by Newton involve a hypothetical assumption: the gas must be composed of particles endowed with a repulsive—or centrifugal, as Newton called it—force. Few philosophers were better placed than Leibniz to grasp at once the difference between mathematical constructions and physical reality. The words *fingi potest* inserted in the *Excerpts* referring to Newton's repulsive force clearly highlight that Leibniz did not take the force to be real. The problems in this analysis of the elasticity of fluids are analogous to the issue concerning the cause of gravity. From Leibniz's perspective, in both cases Newton was playing the virtuoso mathematician while avoiding the real physical and metaphysical issues.[31]

Lastly, Leibniz excerpted from page 480 that the heavens must be devoid of any resistance, otherwise the motion of comets would be perturbed by planetary orbs. As we have seen in the preceding section, Leibniz referred to comets in the conclusion of the 'zweite Bearbeitung'.[32]

7.6 The final version of the theory

In a letter intended for Huygens of October 1690 Leibniz proposed further changes to his theory and announced a new essay on planetary motion. Gerhardt tentatively identified this essay as the *Tentamen de Physicis Motuum Coelestium Rationibus*, but the editors of Huygens's works have rightly disputed his identification on the grounds that letter and essay contain no common elements.[33] The similarity in contents between the letter to Huygens and the manuscript essay *De Causis Motuum Coelestium* suggests that this was the text meant by Leibniz and thus places its date of composition around autumn 1690.

In the letter Leibniz wrote that while admiring the beautiful things contained in the *Principia Mathematica*, he could not understand

[31] *Excerpts*, p. 482: 'Si fluidi elastici vis sit compressioni proportionalis *fingi potest* esse in partibus vim se fugendi distantiis reciproce proportionalem.' (My emphasis.) *Principia*, first edition, pp. 301 and 303; Motte and Cajori, pp. 300 and 302. Cohen, *Newtonian Revolution*, sect 3.3.

[32] *Excerpts*, p. 483: 'Coeli resistentia destituuntur alioqui cometae turbarentur ab orbibus planetarum.'

[33] *HOC*, 9, pp. 521–7, on p. 526, n. 16; *LBG*, pp. 611–13. Despite Huygens's insistence, Leibniz never sent this letter: *LMG*, 2, pp. 41, 46, 49, 64, 68.

Newton's ideas on the origin of gravity. According to Newton, gravity seemed to be an incorporeal virtue, and this was very different from the explanation given by Huygens in his *Discours de la Cause de la Pesanteur*, the treatise published in 1690 together with the *Traité de la Lumiere* which Leibniz had just received from its author.

After correcting a few misprints in the *Tentamen*, installing further slips in their place, Leibniz claimed that although Newton's account was satisfactory for one planet, he did not explain why all planets move in the same direction and on the same plane, exactly as the satellites of Jupiter and Saturn do. A similar argument, which is presented as one of the reasons why Leibniz wanted to maintain vortices, is in *De Causis Motuum Coelestium*.[34]

Leibniz then proposed two alternative explanations for gravity. First, the cause of gravity could be a fluid emitted from a centre that propagates according to the inverse-square law, on the analogy of light. This alternative is also mentioned in *De Causis Motuum Coelestium*, although in both texts it occupies an ancillary position.[35]

The second proposal, which Leibniz discusses in both texts at far greater length, is that each aether shell has the same quantity of power or living force, hence the square of velocity is inversely proportional to the radius.[36] In the letter to Huygens there follow two corollaries. The first states that the squares of the revolution times are proportional to the third power of the distances, in accordance with Kepler's third law. The second corollary states that since centrifugal endeavours are proportional to the square of velocity over distance, they are also inversely proportional to the square of the distance, as they should be. Both points are referred to in *De Causis Motuum Coelestium*, where Leibniz is more explicit about the second; since in the ellipse gravity is inversely proportional to the square of the distance, and gravity results from the centrifugal conatus of the rotating aether, if this were inversely proportional to the third power of the distance we would have a

---

[34] LH 35, 10, 1, f. 1 and 3; f. 3v.: 'Quoniam manifestum erat planetas in easdem partes ferri circa solem, et propemodum in eodem plano incidere ordine quodam in planorum declinationibus . . .' Similar reasonings can be found in the 'zweite Bearbeitung', *LMG* 6, pp. 166–7. Kant discussed the same problem in the *Allgemeine Naturgeschichte und Theorie des Himmels* (Königsberg and Leipzig, 1755).

[35] Ib., f. 3v., in marginal note: 'Memorabile est gravitatis vim decrescere ut lucis, perinde ac si radiis quibusdam ex centro radiante eductis corpus ad descendendum solicitaretur.'

[36] *HOC*, 9, p. 525 and LH 35, 10, 1, f. 3v.: 'Idque rationis est etiam hac lege harmonica fieri, ut gyri majores sint aequales potentia minoribus, seu ut velocitatum quadrata seu potestates sunt harmonicae hoc est disantiis a centro reciproce proportionales.' Since power is proportional to the square of velocity times quantity of matter, matter is proportional to distance in order for the powers of the aether shells to be equal.

contradiction.[37] This crucial observation is meant to amend the result stated in paragraph 13 of the *Tentamen*, where centrifugal force was set inversely proportional to the third power of distance from the centre as a consequence of the harmonic circulation. According to Leibniz, the new dependence on distance is more satisfactory because gravity and centrifugal force act together, but since gravity is inversely proportional to the squared distance, centrifugal force should also follow the same law. This erroneous correction is probably related to the *Principia Mathematica*, proposition 11, book I, where it is proved that if a body revolves along an ellipse, the centripetal force with respect to the focus is inversely proportional to the squared distance.

Leibniz was aware of the existence of two different motions for the vortex in his second explanation for gravity: one is the simple harmonic motion explaining the area law for a single planet; the other is the harmonic motion *secundum potentiam*, explaining Kepler's third law for a planetary system and the 'right' dependence on distance of centrifugal force.[38] Thus, it is necessary to consider two independent vortices: one generates gravity and is not itself heavy; the other carries the planets around the Sun. The two aethers were chosen alternatively according to the result sought.[39]

I know no other case in Leibniz's theory of planetary motion of a virtually complete identity of views as in the two texts I have compared. This suggests that *De Causis* is the text referred to in the letter to Huygens. The style of the essay resembles some of the works by Kepler, especially the *Astronomia Nova*. We find the genesis of his own ideas, the doubts and the mistakes, frustration and eventually the triumph. Leibniz's account of the evolution of his theory provides us with a retro-

---

[37] Ib., f. 3r.: 'Nam cum aetherem secundum legis secundae explicationem posuerimus homogeneum, aether autem homogeneus harmonice curculans habeat vires centrifugas cubis distantiarum a centro reciproce proportionales; sequeretur planetam detrudi gravitationibus quae sint etiam cubis distantiarum a sole reciproce proportionales. Ergo non posset fieri in ellipsi, ubi gravitationes sunt reciproce ut quadrata distantiarum.'

[38] *HOC*, 9 p. 526: 'La circulation harmonique se rencontre dans châque corps à part, comparant les distances differentes qu'il a, mais la circulation harmonique en puissance (où les quarrés des velocités sont reciproques aux distances) se rencontre en comparant des differens corps, soit qu'ils décrivent une ligne circulaire, ou qu'on prenne leur moyen mouvement ... pour l'orbe circulaire qu'ils décrivent.' LH 35, 10, 1, f. 3v.: 'Manet etiam verum absolute quidem circulationes aetheris, ipsiusque aurae planetas deferentis esse harmonicas secundum potentiam ... In orbibus tamen deferentibus seu interceptis inter aphelium et perihelium, circulationes aurae deferentis fieri harmonicas secundum celeritatem.' The same explanation is adopted in 'Illustratio Tentaminis' and 'Excerptum ex Epistola', *LMG*, 6, pp. 267–8 and 276–7 respectively.

[39] *HOC*, 9, p. 526: 'Cependant je distingue l'ether qui fait la pesanteur ... de celuy que defere les planetes, qui est bien plus grossier.' LH 35, 10, 1, f. 3v.: 'Is autem est causa gravitatis, non aura illa crassior seu spongiosior in circulis latitudinum mota quae ipsa est gravis ...'

spective description of the problems he encountered. It is unusual to find Leibniz writing in this way: 'I could assign sufficient causes to any rule [Kepler's law], but I was in trouble with their conjunction, until at last a wonderful reasoning to reconcile them appeared'; 'But the comparison between the different planets disturbed exceedingly these circles of mine when I began investigating the cause of the second rule'; 'Moreover, having worked much in vain, without being able to abandon the hope of explaining planetary motions by means of the motion of the aether, . . .'; 'But here a new difficulty appears, which might less easily drive me to despair of success whilst I am engaged in these matters'; lastly the bold opening, 'Eventually after many pains I can well exclaim the Archimedean εὕρηκα having found the *true causes* of planetary motion.' [40]

We have now charted the development of Leibniz's theory over a period of two years. The most striking feature is that although it is possible to identify a progressive line of thought and a growing awareness about the new problems involved at each stage, few solutions satisfied Leibniz in a definitive way. He seemed to be involved in a continuous process of revision in which the only constant elements were the existence of fluids generating gravity and carrying the planets, and Kepler's laws. Everything else was questioned, from the mechanism generating gravity, to the number of vortices, from the law of centrifugal force to the inverse-square law—which Leibniz abandoned in *De Motu Gravium* while trying to solve the inverse problem of central forces, and soon afterwards accepted again after a more thorough study of page 50 from the *Principia*.[41] His uncertainty about the role of centrifugal force inversely proportional to the third power of the distance was a major problem which undermined his belief in his own equation of paracentric motion.

---

[40] LH 35, 10, 1, f. 1r.: 'Cuilibet regulae separatim facile assignabam causas sufficientes, sed haesi in earum conjunctione, donec tandem conciliandi ratio pulcherrima sese aperuit'; f. 1v.: 'Sed hos circulos meos mirifice turbavit comparatio diversorum planetarum inter se quando regulae secundae causam investigare aggressum sum', f. 3r.: 'Cum autem multa frustra molitus essem, nec tamen spem deserere possem explicandi planetarios motus per motum aetheris, . . .'; f. 3r.: 'Sed hic sese aperit nova difficultas, quae minus in his versatum ad desperationem successus facile adigeret'; f. 1r.: 'Tandem post multa agitata Archimedeum εὕρηκα mihi exclamare posse videor, *veris* motuum planetariorum *causis* repertis, . . .'

[41] I. Bernard Cohen has suggested that at around 1700 Leibniz was even prepared to give up planetary ellipses and accept the revised form put forward by Giandomenico Cassini. See his 'Leibniz on Elliptical Orbits: As Seen in His Correspondence with the Académie Royale des Sciences in 1700', *Journal of the History of Medicine and Allied Sciences*, 17, 1962, pp. 72–82. However, compare *LMG*, 3, pp. 497–500:498, Leibniz to Johann Bernoulli, 7 June 1698: 'Interroga, quaeso, Dominum Varignonium de progressu Astronomiae apud ipsos, . . ., et quid sentiatur de lineis, quas Dn. Cassinus voluit substituere Ellipsibus Keplerianis, quarum tamen novarum linearum causas Physico-Mechanicas dare difficile erit, quas nobis utique facilius praebent Ellipses.'

From the texts we have considered we can identify a crucial common theme which guided Leibniz's attempts from a very early stage, namely the awareness that there exists a plurality of mathematical representations of phenomena. In *De Conatu* and *Inventum a me est* Leibniz stated that orbital motion can result from a composition either of inertial motion along the tangent and of gravity along the radius, or of rectilinear motion along a rotating ruler and circular motion of the ruler; in *Repraesentatio Aliqua* and a series of related manuscripts he adopted a different mathematical representation which I have called 'pseudo-Galilean' motion; in *De Motu Gravium* he realized that different decompositions of motion imply different paracentric endeavours; commenting on Newton's analysis of elastic fluids in proposition 23, book II, Leibniz stressed the mathematical and fictitious character of the demonstration; in the letter intended for Huygens of October 1690, *Tentamen de Systemate Universi*, and 'zweite Bearbeitung' he explicitly acknowledged the existence of a plurality of representations with regard both to mathematics and to the endeavours involved. His awareness about this aspect reinforced his belief that mathematics cannot dictate physics and metaphysics, and guided him through a successful and spectacular case of plagiarism. If the existence of several mathematical descriptions of the same phenomena was widely accepted in the seventeenth century, nobody else had considered this issue in relation to the *Principia*. The reading of Newton's masterpiece, from the *Notes*, where Leibniz immediately recorded the words *vacuum necessario datur*, to the *Excerpts*, where he missed virtually no passage on vortices, was a crucial factor in the reorganization of his priorities and shaping of his ideas and strategy. The Cassirer thesis becomes a powerful instrument in understanding Leibniz's reading of Kepler and ideas about the role of laws of nature, mathematics, and physical explanations. In the dispute about the world system between Newton and Leibniz, the issue of equivalence of their mathematical theories cannot be separated from priority claims and from Leibniz's creation of a highly ingenious new theory and adoption of the Keplerian programme.

Even from the limited selection of texts considered in this work it is clear that the Cassirer thesis occupies a central position in Leibniz's response to Newton, not necessarily, though, in other texts. Indeed, a general conclusion which can be drawn from this work is precisely the difficulty in interpreting Leibniz's texts. Of course, similar problems affect the whole spectrum of historical researches, although in this case they are magnified by Leibniz's extraordinary range of activities, by the at times unpredictable intersection between different fields of knowledge, and by his habit of adopting the terminology of his audience or correspondent. Instances of these difficulties can be found in the material presented so far, and here I wish to reconsider some of them. Take, for example, Kepler's role. The German astronomer was the unchallenged protagonist of the published version of the *Tentamen* because of his discovery of the laws of planetary motion and conflation of mathematical astronomy and physics. In the *Tentamen de Physicis Motuum Coelestium Rationibus*, however, Kepler was portrayed purely as a mathematical astronomer and his role greatly diminished. In the 'zweite Bearbeitung' of the *Tentamen* he regained a dominant position, though his three laws and explanation of gravity had to be reinterpreted following Leibniz's alternative account of the cause of gravity and attempt to avoid religious censorship. If one considers that these three essays were written in the space of a few months, and that they represent a fragment of Leibniz's production during those months, the difficulty of providing broader interpretations emerges immediately. Even the task of providing a picture going beyond some general guidelines of his views about such issues as space, time, and motion appears to be a formidable one. An adequate investigation would require a detailed study of intellectual reflections, circumstances of composition, rhetorical strategy, and audiences in very different situations. They include memoranda on relative motion aimed at having the anti-Copernican ban lifted, debates with Huygens, and attempts to win the Princess of Wales's support in the controversy with Samuel Clarke and the Newtonians. If this book were to stimulate fresh readings of those texts, one of its aims will have been achieved.[42]

---

[42] Interesting analyses of the Leibniz-Clarke correspondence can be found in S. Shapin, 'Of Gods and Kings: politics and the Leibniz-Clarke dispute', *Isis*, 72, 1981, pp. 187–215. S. Schaffer, 'Occultism and Reason', in A. J. Holland, ed., *Philosophy, Its History and Historiography* (Dordrecht, 1985), pp. 117–43, esp. sect. 6.

8

A REAPPRAISAL OF NEWTON'S ITINERARY

8.1 Introduction

The analysis of the formation of Leibniz's techniques and ideas provides us with stimulating material for re-examining Newton's itinerary towards the *Principia* and his criticism of the *Tentamen*. Their rival interpretations are mutually clarifying in many respects, such as the problem of the infinitely small in mathematics, the notion of acceleration, and the analysis of orbital motion.

The background and development of Newtonian mechanics have been investigated in great detail over the last few decades, especially after the works by A. Rupert Hall and Marie Boas Hall, John Herivel, Richard Westfall, I. Bernard Cohen, and Tom Whiteside. Although Newton retrospectively dated his discovery of universal gravity back to the 1660s, and although he did work on problems such as curvilinear motion or the motion of the Moon at that time, recent scholarship has reached a consensus in dating the crucial period during which Newton's mature ideas emerged between 1679 and 1684. In 1679 Robert Hooke, having become Secretary of the Royal Society, restarted a correspondence with Newton and stimulated him to work on curvilinear motion; in 1680 and 1681 John Flamsteed and Newton exchanged some important letters on comets; lastly, following Edmond Halley's visit to Cambridge in August 1684, Newton was stimulated to compose the fundamental tract *De motu corporum in gyrum*, the first step on the way to the *Principia*.

In the first part of this chapter I present a reappraisal of the crucial phase in the formation of Newton's theory. After reviewing in Section 8.2 some typical seventeenth-century approaches to curvilinear motion, Section 8.3 focuses on the debate between Newton and Hooke, and Newton's later views on this matter. Current historiography tends to consider Hooke's idea that a curvilinear path results from rectilinear motion combined with central attraction, without centrifugal force, as obviously correct and as the only sensible solution to the problem. My contention is that these two assumptions cannot be accepted. Hooke's interpretation had no privileged status with respect to Huygens's; Hooke himself in the 1660s and in 1680 took into account an outward tendency from the centre after he had already formulated his 'correct' theory of curvilinear motion. Indeed, after 1679 even Newton still referred to centrifugal forces, although he had allegedly already carried out the calculations following Hooke's advice and established a link between ellipses and an attractive force inversely proportional to the square of the distance. I will show that up to 1680 Newton conceived centrifugal force in orbital motion to counterbalance attraction, being at times smaller and at times greater than it. From the mid-1680s onwards centrifugal and centripetal forces in orbital motion were set equal and opposite for the third law of motion. My second objection concerns the tacit assumption that if one solution is 'right', different solutions must necessarily be 'wrong'. In particular, the idea that there is an imbalance between opposing centripetal and centrifugal forces along the radius has been criticized for no clear reason.

These observations lead us to Section 8.4, where I examine Newton's onslaught on Leibniz's theory. The *Tentamen* pointed to a different theory of curvilinear motion. As Eric Aiton has shown, Newton's attempt to demolish Leibniz's work on a mathematical ground was unsuccessful. Indeed, I hope to show that Newton's criticisms concerning the analysis of curvilinear motion involved ideas differing considerably from more modern formulations of mechanics; his investigations and ideas require a careful assessment in the context of the practices of mechanics around 1700. I recall that opening a work on lunar or perturbation theory of the late 1740s by Euler or Clairaut, it is not difficult to realize that Leibniz's equation of paracentric motion was a reasonable and fertile way of tackling the problem. A comparative investigation of the formation of Leibniz's and Newton's theories reveals more questions than one would have expected.

8.2   Continental and English approaches to curvilinear motion

In Chapters 1 and 2 we have seen that Continental mathematicians, notably Descartes, the member of the Florentine Accademia del Cimento, Borelli, and Huygens shared common ideas about curvilinear motion. Despite some differences on specific points, they believed that curvilinear motion results from the interplay of a tendency towards a centre and an outwards tendency due to the rotation of the body and to its rectilinear inertia. These explanations relied on a number of typical analogies: the bucket full of water with several bodies floating in it, used for example by Kepler, Descartes, and Huygens: the rotating sling, discussed by Descartes and several others including Newton; the rotating ruler, used by Kepler, Descartes, Borelli, and Leibniz; the pendulum, adopted by Borelli and Huygens. These explanations were not necessarily mutually exclusive, but rather complementary, and often one author discussed several of them in the same work. In his study of the Medicean planets, for example, Borelli conceived orbital motion as the resultant of a circular and a radial component. While studying the latter, he imagined radial motion to take place on a rotating lever moved by the rotation of the central body, on the example of Kepler. In the same work Borelli also tried to establish a quantitative relation between orbital velocity and distance from the centre by means of the analogy with the pendulum. Following the priority dispute about the invention of the pendulum clock between Huygens and the Accademia del Cimento in Florence, which defended Galileo's rights, pendular oscillations occupied a fairly prominent position on the Academicians' agenda. Hence it is not surprising that Huygens also employed the pendulum in his study of centrifugal force on the Earth and in orbital motion, though in a much more sophisticated way than Borelli.[1]

Descartes, Huygens, and their contemporaries did not consider centrifugal force as dependent on the choice of a rotating reference frame, as is commonly done in more modern formulations of mechanics. In order to understand the practice of mechanics in the late seventeenth century it is essential to grasp this point. Modern formulations—misleadingly named 'Newtonian'—take for granted that there exists a plurality of alternative representations of curvilinear motion: if the reference frame is inertial, and more precisely if it does not rotate, centrifugal forces do not appear; if the frame rotates, centrifugal forces arise from the art of representation, regardless of the motion which is observed. Indeed, even around 1700 these two representations—although for us they are mutually exclusive—were commonly conflated in the following way: a body moving along a curvilinear path endeavours to escape along the tangent and therefore away from the centre. Thus centrifugal force was associated with rectilinear inertia and the curvilinear motion of the body, as opposed to that of the frame of reference. As a result, inertial motion was considered together with attractive and centrifugal forces in a way which appears to be inconsistent to the modern reader. It was practice which guided mathematicians to provide case by case solutions to the problems they were investigating. Moreover, Newton and several mathematicians after him often considered centripetal and centrifugal forces to be related for the third law of

---

[1] On Kepler, Descartes, and Huygens, compare Chapter 1 and Sections 2.2 and 2.3 above. G. A. Borelli, *Theoricae mediceorum planetarum ex causis physicis deductae* (Florence, 1666), pp. 53f. and 76f. Koyré, *Revolution Astronomique*, part III. W. E. Knowles Middleton, *The experimenters: a study of the Accademia del Cimento* (Baltimore, 1971).

motion. This approach appears to us to be plagued by conceptual difficulties, such as the following two. There are several cases in which centripetal and centrifugal forces cannot be set to be equal and opposite, because the whole study is based on them being different. An example is the study of the shape of the Earth, where the ratio between gravity and centrifugal force at the equator is approximately as 300 to 1. The third law, however, is universal and cannot be switched off depending on the circumstances. The other difficulty concerns a different aspect of the application of the third law. Briefly, if body A acts on body B, also B acts on A with a reaction equal and opposite to the action. In orbital motion centrifugal force was conceived as a reaction to gravity; however, the reaction was not affecting the acting body, but was contained within the body acted upon. Considering a planet, for example, one would expect the reaction to be directed towards the body whence gravity or the action originates, such as the sun for example. If the reaction is identified with the centrifugal force, though, it remains constrained within the orbiting body: A acts on B, and B reacts on itself. Although these problems were not discussed around 1700 in this form, the present analysis has historical significance because it emphasizes the difference between Newton's and later formulations of mechanics. I believe this defamiliarization with the *Principia* to be essential for a more satisfactory interpretation of its contents.[2]

These preliminary remarks will be referred to while analysing the genesis of Newton's ideas, which were significantly greatly influenced by Descartes as well as by Borelli and Huygens. I discuss now the English approach, namely a different way of accounting for curvilinear motion which was then emerging on this side of the Channel, but was by no means dominant or even widely known. Indeed, until 1679 Newton seemed to be unaware that Christopher Wren, John Wallis, and Robert Hooke were developing an explanation according to which rectilinear inertia and a tendency towards a centre accounted for curvilinear motion, whilst the outward tendency or centrifugal force did not appear explicitly. This explanation was based on the analogy of the conical pendulum and on the study of cometary trajectories. Until 1679 Newton followed a theory, which he had already developed in the 1660s in the Waste Book, along the Continental lines. In a manuscript on circular motion dating from the late 1660s, for example, as we have seen in Section 1.2, he hit upon the inverse-square law by combining Kepler's third law with the expression for an outward or centrifugal tendency. The assumption that outward and attractive tendencies—at least for

---

[2] Bertoloni Meli, 'Relativization'.

orbits differing little from circles—were almost identical was implicit in his calculations.[3]

The years 1664–6 turned out to be crucial in the formation of the English approach in two respects. In 1664 Wallis and Wren examined some of the manuscripts left by the Liverpool astronomer Jeremiah Horrocks, who had prematurely died in 1641. As a result of their efforts, a selection of his manuscripts was published in the *Opera Posthuma* (London, 1672–8). In a letter to William Crabtree of 25 July 1638 Horrocks used the conical pendulum to explain the elliptical motion of planets. In spite of the difference between the accounts of pendular motion given by Horrocks in the 1630s and by Robert Hooke in the 1660s and 1670s, Bennett has convincingly argued for a connection between them. Moreover, Wren, who belonged to the same circle with Wallis and Hooke, had already discussed in 1661 with Christiaan Huygens the application of the conical pendulum to clocks. Hooke's ideas can be traced from the records of the debates at the Royal Society. On 9 May 1666 the Society was presented a paper by Wallis on tides. Wallis thought of modifying the Galilean hypothesis based on the combined motion of the Earth around its axis and around the Sun, by considering the motion of the common centre of gravity of the Earth and the Moon around the Sun. At the following meeting, after the paper was read, it was objected 'that it appeared not, how two bodies, that have no tie, can have one common center of gravity, upon which the whole hypothesis of Dr. Wallis is founded.' Hooke replied that celestial motions 'may be represented by pendulums', and on 23 May he submitted his paper 'concerning the inflection of a direct motion into a curve by a supervening attractive principle.' Hooke further suggested that by using two pendulums it may be possible to account for the motion of the Moon around the Earth.[4]

The other area which turned out to be of crucial importance in the formation of Hooke's ideas is cometary motion. The appearance of a comet at the end of 1664 stimulated Wren, Wallis, and Hooke to discuss the matter with respect to physics, mechanics, and mathematics. Curiously, the same comet inspired Newton to the study of astronomy in those years. Wren and Wallis discussed the problem of finding the position of a comet from a number of observations. The problem had been posed by Wren to Wallis on 1 Januray 1665; the solution by Wallis

---

[3] J. A. Bennett, 'Hooke and Wren and the System of the World: Some Points Towards an Historical Account', *BJHS*, *8*, 1975, pp. 32–61. ULC, MS add. 3958(5), fols. 87, 89, Herivel, *Background*, pp. 192–8. Westfall, *Force*, pp. 350–60.

[4] 'An essay of Dr. John Wallis, exhibiting his Hypothesis about the flux and reflux of the sea', *PT*, *1*, 1666, pp. 263–88. T. Birch, *A history of the Royal Society of London*, 4 vols. (London, 1756–7), vol. II, pp. 89–90. Bennett, 'Hooke and Wren', pp. 44–7.

is preserved at the Bodleian Library, whilst that by Wren was later
included by Hooke in *Cometa*, published in 1678 after the appearance
of the 1677 comet. Hooke's work, though, was largely composed in the
late 1660s.[5] In the seventeenth century the areas of consensus within
cometography were very limited indeed. Setting aside theories like those
defended by Galileo in *Il Saggiatore*, denying that comets were real
bodies moving through the spaces above the Moon, opinions about
cometary paths ranged from the circular to the rectilinear hypotheses.
Kepler in particular defended the idea that the path was rectilinear on
the basis of metaphysical assumptions and rather rough astronomical
observations. The rectilinear hypothesis was generally regarded as a
good approximation, though, as Hooke wrote in *Cometa*, there were two
causes of perturbations: one was the attractive power of celestial bodies
such as the Sun or the Earth, the other was the action of a fluid vortex in
which the comet was moving. Concerning the crucial problem of orbital
motion, Hooke supposed that 'the attractive power of the Sun, or other
central body may draw the body towards it, and so bend the motion of
the Comet from the straight line, in which it tends, into a kind of curve,
whose concave part is towards the Sun.' Bennett has convincingly argued
that 'it is clear that through the problem of the motion of comets, Hooke
arrived at the idea of combining a rectilinear inertial motion with a
central attractive force, and he eventually explained the motion of the
1664 comet in these terms'.[6]

Thus the years 1664–6 saw the birth of Hooke's ideas of celestial and
in general curvilinear motions. The passage from his 1674 Cutlerian
lectures with his famous three 'suppositions' appears as the result of an
English tradition which had matured over several years:[7]

First, that all Coelestial Bodies whatsoever, have an attraction or gravitating
power towards their own Centers, whereby they attract not only their own parts,
and keep them from flying from them, as we may observe the Earth to do, but
that they do also attract all the other Coelestial bodies that are within the sphere
of their activity; . . . The second supposition is this, That all bodies whatsoever
that are put into a direct and simple motion, will so continue to move forward in
a streight line, till they are by some other effectual powers deflected and bent into
a Motion, describing a Circle, Ellipsis, or some other more compounded Curve
Line. The third supposition is, That these attractive powers are so much more
powerful in operating, by how much the nearer the body wrought upon is to their
own Centers. Now what these several degrees are I have not yet experimentally

[5] Bennett, 'Hooke and Wren', p. 50, n. 103. Hooke, *Cometa*, reprinted in Gunther, *Early Science in Oxford* (Oxford, 1931), vol. 8, pp. 56–9.

[6] J. A. Ruffner, *The Background and Early Development of Newton's Theory of Comets*, Ph.D. Thesis, Indiana University, 1966, pp. 94–118 and 168–84. Bennett, 'Hooke and Wren', pp. 49–60, esp. p. 58. Hooke, *Cometa*, in Gunther, p. 229.

[7] Gunther, *Early Science*, pp. 27–8.

verified; but it is a notion, which if fully prosecuted as it ought to be, will mightily assist the Astronomer to reduce all the Coelestial Motions to a certain rule, which I doubt will never be done without it. He that understands the nature of the Circular Pendulum and Circular Motion, will easily understand the whole ground of this Principle.

Universal attraction decreasing in some ratio of the distance from the centre, the idea that unperturbed motion is rectilinear and uniform, and the circular or conical pendulum interacted in the formulation of an account of circular motion in which centrifugal force was ignored.

8.3 Newton versus Hooke

The correspondence between Hooke and Newton in 1679–80 represented the meeting of two different approaches none of which was a priori self-evident or even more convincing than the other. It is well known that on 24 November 1679 Hooke asked Newton to consider what would happen by 'compounding the celestiall motions of the planetts of a direct motion by the tangent and an attractive motion towards the centrall body'. In his letter of 28 November Newton avoided the issue and discussed instead an experiment intended to prove the diurnal motion of the Earth by dropping heavy bodies from high places and studying their trajectory. The ensuing discussion concerned the trajectory of the bodies inside the hollow sphere of the Earth as well. On 9 December Hooke claimed that he 'could adde many other conciderations which are consonant to my Theory of Circular motions compounded by a Direct motion and an attractive one to a Center.' This time, in his reply on 13 December, Newton objected that Hooke ought not to overlook that if 'gravity be supposed uniform', a body falling inside the Earth would 'circulate with an alternate ascent and descent made by its *vis centrifuga* and gravity alternately overballancing one another.' On seeing that Newton had taken the attraction to be constant, on 6 January Hooke specified that 'the Attraction always is in a duplicate proportion to the distance from the center Reciprocall.' This claim was the source of the later priority dispute with Newton. But there is another passage in the same letter which is of interest to us. While discussing Newton's experiment, Hooke stated that 'the further a body is from the centre the Lesse will be its gravitation, which I have a Long time supposed not only upon the account of the Decrease of the attractive power but upon the increase of the Indeavour of Recesse'. In a preliminary draft Hooke had begun to write 'centrifugal'; this increase was due to 'the Circular motion being swifter'. These quotations are revealing of the problem of identifying the components of curvilinear motion. Initially Hooke considered rectilinear inertia and central attraction, whilst Newton, in addition, considered a Huygensian centrifugal force opposite to gravity, at times greater and at times smaller than it. Later Hooke talked of an 'indeavour of recesse', thus following Newton's interpretation.[8] Symmetrically—as we know from his own later account—Newton seized upon Hooke's challenge, almost certainly adopting his correspondent's initial suggestion. He thus apparently established a link between Keplerian ellipses and the inverse-square law. We do not know exactly what his 1680 calculation looked like; claims that a copy of that first demonstration are extant are still controversial.[9]

Setting aside this issue and the exact details of that early proof, the question I wish to address here concerns its status for Newton at the time. My claim is not only that Hooke's original suggestion, but also Newton's 1679 demonstration of the link between ellipses on the one hand, and rectilinear inertia composed with an attraction inversely proportional to the square of the distance on the other, had an unclear status. Hooke's suggestion was problematic and at best worth investigating further with regard to physical actions and the problem of centrifugal force. I survey briefly these two aspects.

Around 1680 Newton, like Hooke and the great majority of natural philosophers, believed in the existence of an aether responsible for gravity. In *An Hypothesis explaining the properties of light*, read at the Royal Society on 9 December 1675, Newton presented his speculations about the cause of gravity. He started by claiming that 'it is to be supposed therein, that there is an aethereal medium much of the same constitution with air, but far rarer, subtler, and more strongly elastic. Of the existence of this medium the motion of a pendulum in a glass exhausted of air almost as quickly as in the open air, is no inconsiderable argument.' We have seen in Section 5.2 that in the *Principia* Newton used pendulum experiments to reach the opposite conclusion, namely that the aether penetrating the pores of solid bodies opposes virtually no resistance to motion. In *An Hypothesis*, after a discussion of the likelihood that the attraction of a rubbed piece of glass may be caused by an aethereal wind, Newton continues:[10]

So may the gravitating attraction of the Earth be caused by the continual condensation of some other such like aethereal spirit, not of the main body of

---

[8] *NC*, 2, pp. 297–313. Koyré, *Newtonian Studies*, ch. 5. J. A. Lohne, 'Hooke *versus* Newton', *Centaurus*, 7, 1960, pp. 6–52.

[9] Westfall, *Never at Rest*, pp. 387–8, n. 145. Whiteside, *Preliminary manuscripts*, pp. xiv and xx–xxi, n. 53.

[10] Birch, III, pp. 257–305, in Cohen, *Papers and Letters*, pp. 177–235; quotations from pp. 179–80 and 180–1. The pendulum experiment is also discussed in *De aere et aethere*, Hall and Hall, *Scientific Papers*, pp. 220 and 227, dated 1673 to 1675.

phlegmatic aether, but of something very thinly and subtilly diffused through it, perhaps of an unctuous or gummy, tenacious, and springy nature, and bearing much the same relation to aether, which the vital aereal spirit, requisite for the conservation of flame and vital motions, does to air.

In a letter to Robert Boyle of 28 February 1679 Newton tried to provide a causal explanation of gravity based on that aether. His mechanism involved the pressure of finer and grosser particles of the aether moving through the pores of solid bodies.[11] Lastly, in a series of propositions relating mainly to comets and dating from approximately 1681, Newton stated that 'the matter of the heavens is fluid' and that it 'revolves around the centre of the cosmic system in the direction of the courses of the planets.' Whether and how this matter affected the motion of celestial bodies was not clear.[12]

These reflections followed shortly after the important correspondence with the Astronomer Royal John Flamsteed on the great comet of 1680–1. This exchange contains an interesting observation on centrifugal force. It is well known that initially Newton, like most of his contemporaries, believed in the existence of two different comets. Their correspondence ranged from observational astronomy to the analysis of the effects of the solar vortex, from orbital dynamics to the study of the magnetic properties of the Sun in relation to its extraordinarily high temperature. In a draft of a letter probably intended for James Crompton, a friend of Flamsteed and Fellow of Jesus College, Cambridge, Newton criticized Flamsteed's theory that the comet was attracted while approaching the Sun and repelled while receding:[13]

But all these difficulties may be avoyded by supposing ye comet to be directed by ye Sun's magnetism as well as attracted, and consequently to have been attracted all ye time of its motion, as well in its recess from ye Sun as in its access towards him, and thereby to have been as much retarded in its recess as accelerated in its access, and by this continuall attraction to have been made to fetch a compass about the Sun ... the *vis centrifuga* [in Perihelion] overpow'ring the attraction and forcing the Comet notwithstanding the attraction, to begin to recede from ye Sun.

[11] Cohen, *Papers and Letters*, p. 253.

[12] ULC, Add 3965(14), f. 613r.: '2. Materiam caelorum fluidam esse ... 3. Materiam caelorum circa centrum systematis cosmici secundum cursum Planetarum gyrare.' These propositions were first made known by J. Ruffner in his doctoral dissertation, *Newton's theory of comets*, pp. 310–13. The statement in proposition [X] that 'the curve is an oval if the comet returns in an orbit, if not [the curve] is nearly a hyperbola' (p. 312), indicates that Newton's manuscript follows his correspondence with Flamsteed. Moreover, prop. 16, stating that 'The comet descended below the sphere of Mercury', clearly refers to the 1680–1 comet.

[13] *NC*, 2, pp. 358–62, probably dating from April 1681, on p. 361. Cohen, *Newtonian Revolution*, sect. 5.4.

The analogy with the ideas expressed in the letter to Hooke strongly suggests that around 1680 Newton was convinced that a body moving along a curvilinear path endeavoured to escape from the centre. This centrifugal force had been given a quantitative expression by Huygens in the unpublished *De vi centrifuga* of 1659 and later in the *Horologium Oscillatorium* of 1673, and by Newton himself in some reflections dating from the mid-1660s in the Waste Book and in other manuscripts. In the Waste Book Newton calculated the endeavour from the centre of a body moving inside a hollow sphere by considering the pressure of the body against the inner surface of the sphere. He first considered a trajectory along a regular inscribed polygon with a given number of sides. In this way the pressure could be calculated at the vertices of the polygon, where the impacts of the ball with the sphere occur; then the number of sides was increased to infinity, and the result calculated in this limiting case. In later manuscripts Newton adopted different methods and found, as Huygens, that for a body moving along a circumference the tendency to escape from the centre was as the square of the velocity over the radius. Its measure for an arbitrary curve and especially for the ellipse, however, had not been calculated, nor was it immediately clear how this should be done. In the continuation of his discussion in the Waste Book Newton referred to this problem: 'If the body *b* moved in an Ellipsis then its force in each point (if its motion in that point bee given) will bee found by a tangent circle of Equall crookednesse with that point of the Ellipsis.'[14] Whether the force was directed along the radius of the 'tangent circle', or the radius to the centre of the ellipse, or that to one of its foci, was not specified. Before asking this question, one may even query whether centrifugal force can be measured in this way. Alternatively, one may fix a centre and a radius, and then take the component of motion along the ellipse—or any other curve—perpendicular to that radius; in this way the measure of centrifugal force would be equal to the square of that component of velocity over the radius. As it happens, this is the standard procedure in modern mechanics, since centrifugal force is usually measured with respect to a fixed centre. In addition, as Newton's objection to Hooke reveals, it was by no means clear how the outward tendency should effect curvilinear motion, or, more precisely, how the components of curvilinear motion should be represented.

Thus around 1680 Newton's views were broadly similar to Leibniz's at the end of the decade. Following Hooke's suggestion, Newton had

[14] Herivel, *Background*, p. 130; I have taken into account Whiteside's improved transcription in *NMW*, *1*, p. 456. Aiton, *Vortex Theory*, pp. 115–18. Bertoloni Meli, 'Relativization', sect. 2. See also the recent reappraisal of Newton's mechanical investigations by D. T. Whiteside, 'The prehistory of the *Principia* from 1664 to 1686', *NRRS*, *45*, 1991, pp. 11–61.

probably provided a 'successful' purely mathematical explanation which seemed to bear no relation either to shared ideas about mechanics, such as the existence of centrifugal force, or to physical explanations concerning the aether. In such circumstances it is not surprising that in 1680 Newton's demonstration 'failed to seize his imagination':[15] indeed, one may well ask how it ever did.

Newton's purely mathematical demonstration and, even more problematically, Hooke's famous observation in the Cutlerian Lectures that curvilinear motion results from rectilinear inertial motion combined with a central attraction, are all too easily retrospectively judged by the historian to be the 'correct' proof and suggestion respectively. At the time, however, they were both surrounded by uncertainty as to their status and significance.[16] Several historians, and Newton himself in his 1686 letter to Halley, emphasized the distance between Hooke's suggestion and Newton's own mathematical proof.[17] In addition, here I wish to stress also the distance between Newton's 1680 alleged demonstration and the tracts *De motu* of 1684–5. Besides the emergence of the notion of universal gravity by the end of 1684, there were other issues Newton had to address. In order to render his mathematical proof acceptable first to himself, and then to his readers, Newton had to tackle the problems related to mechanics and physics as well. Take, for example, the problem of curvilinear motion.

In the augmented tract *De motu corporum* of winter 1684–5 he tried to account for centrifugal force in terms of the *vis insita* and of the reaction to the force deflecting a body from its curvilinear path. A preliminary rejected definition of *vis exercita* reads:[18]

The exercised force of a body is that by which it attempts to preserve that part of its state of rest or motion which it gives up instantaneously and it is proportional to the change of its state or to that portion of its state given up instantaneously,

---

[15] Westfall, *Never at Rest*, p. 390.

[16] Westfall, *Force*, p. 426: 'In the mechanics of circular motion, Hooke's suggestion was of capital importance, slicing away as it did the confusion inherent in the idea of centrifugal force and exposing the basic dynamic factors with striking clarity.' Ib. p. 430: 'Under the tutelage of Hooke, he bid centrifugal force adieu and turned down the path that would lead him eventually to the *Principia*.' A more satisfactory account is in Cohen, *Newtonian Revolution*, sect. 5.4. It will become clear in the following that Newton never bid centrifugal force adieu in the study of curvilinear motion.

[17] *NC*, 2, p. 438, 20 June 1686. Cohen, *Newtonian Revolution*, sect. 5.4.

[18] ULC, Ms Add. 3965 (5), f. 25v. and 26r. (for the two following quotations, respectively). Herivel, *Background*, pp. 317 and 320, definition 14; I have improved Herivel's translation. The following quotation is from *Background*, pp. 306 and 311, definition 12; I have taken into account Whiteside's improved reading in 'Newtonian Dynamics', p. 115, n. 1, and *NMW*, 6, p. 191. Compare also Whiteside, *Preliminary Manuscripts*, pp. 30–1.

and not improperly is said to be the reluctance or resistance of the body, of which one species is the centrifugal force of rotating bodies.

The revised version, which is not crossed out by Newton, established an even clearer link between *vis insita*, or 'internal', and centrifugal force:

The internal and innate force of a body is the power by which it preserves in its state of rest or of moving uniformly in a straight line. It is proportional to the quantity of the body, and is actually exercised proportionally to the change of state, and in so far as it is exercised it can be said to be the exercised force of the body, conatus or reluctance, of which one species is the centrifugal force of rotating bodies.

Here centrifugal force appears to be related to rectilinear inertia and is seen as a reaction proportional to the force which bends the body's orbit. No hint can be found of a distinction between inertial and rotating frames. We notice also the difficulty about the link between centrifugal force and third law discussed in the previous section: centrifugal force does not act on the cause bending the body's orbit, but appears as a reaction contained within the orbiting body.

In his maturity Newton interpreted centrifugal force in orbital motion as a reaction to centripetal force: as such, they were considered to be equal and opposite—whilst in the past they appeared to be only opposite but not necessarily equal—and then centrifugal force was ignored in the calculations. The clearest statement in this regard can be found in two memoranda against Leibniz which we are going to examine in the following section. Here I survey other sources dating from 1687 onwards. In the manuscript 'The Elements of Mechanicks', which according to its editors 'is certainly later in date than the *Principia*', Newton stated:[19]

Bodies circulating in concentric circles have centrifuge forces proportional to ye radii of ye circles directly and the squares of the times of revolution reciprocally, and are kept in their Orbs by contrary forces of the same quantity.

The *Principia* contains several passages consistent with this interpretation. Although from my reading Newton's views on this matter did not change from 1687 to 1726, I prefer to quote from the second and third editions, where he became more explicit. In book I, scholium to proposition 4, Newton referred to the *Horologium Oscillatorium*, where Huygens had compared the force of gravity with the centrifugal force of revolving bodies, and to the motion of a body inside a hollow sphere, which he had studied in the Waste Book. The scholium ends with the

---

[19] ULC, Ms Add 4005, f. 23–5, published by Hall and Hall, *Scientific Papers*, pp. 165–9, on pp. 165–6. Further references to a *conatus recedendi a centro* and *vis centrifuga* after 1684 can be found in the letters to Halley of 20 June and 27 July 1686, *NC, 2*, pp. 436 and 446, respectively.

words: 'This is the centrifugal force, with which the body presses upon the circle; and to which the contrary force, wherewith the circle continually repels the body towards the centre, is equal.'[20] This quotation gives the impression that Newton is using the third law: note in particular the words *huic aequalis est vis contraria*. In a passage related to definition 5, and omitted in the first edition, Newton follows Descartes in explaining curvilinear motion with the example of the sling. The representations involving rectilinear inertia and an outward tendency are conflated:[21]

A stone, whirled about in a sling, endeavors to recede from the hand that turns it; and by that endeavor, distends the sling, and that with so much the greater force, as it is revolved with the greater velocity, and as soon as it is let go, flies away. That force contrary to this endeavor, and by which the sling continually draws back the stone towards the hand and retains it in its orbit, because it is directed to the hand as the centre of the orbit, I call the centripetal force. And the same thing is to be understood of all bodies, revolved in any orbits.

Also in this passage Newton seems to have the third law in mind. Lastly, in the third edition, scholium to proposition 4, book III, while discussing the motion of a hypothetical little moon rotating very close to the surface of the Earth, Newton states:[22]

Therefore if the same little moon should be deserted by all the motion which carries it through its orb, because of the lack of centrifugal force with which it had endured in the orb, it would descend to the Earth,

In cases different from orbital motion, such as the study of the shape of the Earth, centrifugal force was neither set equal and opposite to gravity, nor was it neglected in the calculations; in book III, propositions 18, 19 and in the corollary to proposition 36, the effects of centrifugal force were carefully evaluated. This difference of treatment emphasizes the difficulty in establishing the components of curvilinear motion. Newton's approach was shaped by the particular problem under investigation; this emphasis on practice gives the impression that his theory was based on a case by case analysis and that the solution in one specific area could not easily be generalized to other fields. Interestingly, even Robert Hooke in a manuscript of 1667, after having already found the 'correct' approach to curvilinear motion, stated that 'In all circular motion the indeavour of *receding* from the center is in a duplicate proportion to the velocity.' This passage and his 6 January 1680 letter to Newton show

---

[20] The word 'centrifugal' is used only in the last two editions. I have slightly improved the translation of the *Principia* by Motte and Cajori, p. 47. See also *NMW, 6*, pp. 200–1.

[21] *Principia*, definition 5. I have slightly improved the translation by Motte and Cajori, pp. 2–3. The crucial passage reads: 'Vim conatui illi contrariam . . . centripetam appello.'

[22] *Principia* (third edition), p. 398, lines 22–4 (my translation).

that the alleged self-evidence of Hooke's 'correct' approach is not easy to defend.[23]

The problems related to the role of the celestial aether in Newton's world system have been thoroughly investigated by historians and require no extensive discussion here. From the mid-1680s onwards Newton denied that the heavens oppose any resistance to the motion of celestial bodies, planets, and comets alike. In this context the motion of comets in all directions was crucial in the establishment of his views. Concerning the issue of a mechanical or physical cause for gravity, he produced no quantitative account and left the issue to further investigations. Unlike Roger Cotes in the preface to the second edition of the *Principia*, Newton repeatedly denied that gravity was a primary or essential property of matter. As he wrote at the end of the third rule in the second and third editions of the *Principia*, 'I do not affirm at all that gravity is essential to bodies.'[24]

I believe that these observations show not only that there was no a priori way to establish the 'correct' approach to the study of curvilinear motion, but even a posteriori the transition from the 1680 demonstration to the tracts *De motu* involved important decisions on Newton's part and did not follow automatically. Without underestimating the intrinsic—though not necessarily conclusive—value of Newton's mathematical demonstration, it is also worth recalling the public support he thought he would command at the Royal Society from Wren, for example, and especially Halley, who had been 'struck with joy and amazement' on hearing of Newton's result involving inverse-square law and elliptical orbits.[25] In the case of celestial matter Newton may have been led to strip the heavens of any resistance by the consideration that the motions of celestial bodies are remarkably regular. Thus the surprising agreement between mathematics and astronomical data turned out to be crucial.

---

[23] P. Pugliese, 'Robert Hooke and the dynamics of motion in a curved path', M. Hunter and S. Schaffer, eds., *Robert Hooke. New Studies* (Cambridge, 1989), pp. 181–205, on p. 204, my emphasis.

[24] Among the many sources on the role of the aether in Newton's thought see: H. Guerlac, 'Newton's Optical Aether', *NRRS*, 22, 1967, pp. 45–57. J. L. Hawes, 'Newton's Revival of the Aether Hypothesis and the Explanation of Gravitational Attraction', *NRRS*, 23, 1968, pp. 200–12. J. E. McGuire, 'Active Principles and Newton's Invisible Realm', *Ambix*, 15, 1968, pp. 154–208; 'Body and Void and Newton's *De Mundi Systemate*: Some New Sources', *AHES*, 3, 1966, pp. 206–48. A. Thackray, *Atoms and Powers* (Cambridge, Mass., 1970), chapters 2 and 3. R. W. Home, 'Force, Electricity and the Powers of Living Matter in Newton's Mature Philosophy of Nature', in M. J. Osler and P. L. Farber, eds., *Religion, Science and Worldview* (Cambridge, 1985), pp. 95–117. I. B. Cohen, *Newtonian Revolution*, sect. 3.8. See also the correspondence with Richard Bentley in Cohen, ed., *Papers and Letters*, pp. 271–312, esp. pp. 298 and 302–3. Koyré and Cohen, eds., *Third edition with variant readings*, p. 555.

[25] Westfall, *Never at Rest*, pp. 402–6.

With regard to mechanics and the role of centrifugal force he had a similar convincing tool: success. The extraordinary fertility of his approach concerning all possible features of planetary and cometary paths, as well as of the shape of celestial bodies, convinced Newton, more strongly at each step, that he was 'right'. It was not the individuation of the 'correct' analysis of curvilinear motion that led Newton to succeed. Rather, success in accounting mathematically for celestial motions led him to redefine, against the views he had previously held, how curvilinear motion should be analysed. Had Newton explained only circular motion, the impact of his approach would have been incomparably less significant. Widening the range of his theory was a way of strengthening its uncertain premises. As Kepler had stated in the *Mysterium Cosmographicum*, the 'outcome of false premises is fortuitous, and that which is false by nature betrays itself as soon as it is applied to another related matter.' In an analogous fashion, the extraordinary success of the method Newton was developing supported its credibility.

His approach hinged on the application of the three laws of motion: an undisturbed body moves in a straight line with a uniform velocity; a deviation from this state of inertial motion is due to a force, either impulsive as in an impact, or continuously acting as in the case of centripetal force; centrifugal force is the reaction to centripetal force and is equal and opposite to it from the third law. In the *Principia* the third law is used to prove that attraction must be mutual: if body *A* attracts body *B*, the reaction is the equal and opposite force with which *B* attracts *A*. This is made clear in proposition 69, book I. Therefore, in my interpretation the third law has a double role for Newton, as an explanation of centrifugal force and of the reciprocity of attraction.[26]

Newton's interpretation differs widely from more modern versions of 'Newtonian' mechanics to the extent that his views appear to the modern reader as inconsistent. Centrifugal force is set equal and opposite to gravity for the third law. Despite the universality of the laws of motion, however, on the rotating Earth centrifugal force and gravity were different. Moreover, in the mathematical analysis of orbital motion centrifugal force was switched off for no apparent reason. As we are going to see in the following chapter, Newton's contemporaries found his explanation of centrifugal force in terms of the third law highly convincing. Newton himself used it in his attack on the *Tentamen*.

[26] B. Latour, *Science in Action* (Milton Keynes, 1987), esp. p. 12. Aiton, 'Mathematical Basis', p. 222. I. B. Cohen, 'Newton's third law and universal gravity', *Journal for the History of Ideas*, 48, 1987, pp. 571–93.

8.4   Newton's onslaught on the *Tentamen*

In the 1710s, at the climax of the priority dispute over the invention of the calculus, Newton carefully engineered an onslaught on the *Tentamen*. If the final broadside came from John Keill, there is no doubt that ammunition had been provided by Newton himself. We possess two memoranda in his hand spelling out the attacking lines; several further unpublished manuscripts with criticisms and calculations concerning the *Tentamen* are preserved at the University Library, Cambridge.[27] A fresh analysis of this controversy not only provides us with the most detailed contemporary evaluation of the *Tentamen*, but also illuminates Newton's views. As Eric Aiton has shown, although Leibniz had considerable difficulties in accounting for Kepler's third law, Newton's attacks were not as devastating and self-evident as previous commentators had supposed.

The issue of proficiency in second-order infinitesimals was an important theatre of the war between Newtonians and Leibnizians. Although this controversy affected Newton's reading, his criticisms were not restricted to mathematics, but involved other aspects as well. In particular they concerned the cause of motion with regard to a corporeal agent versus the human mind or God, the incongruences following from the idea that gravity is due to a vortex rotating harmonically, the mathematical treatment of the theory, and finally the analysis of curvilinear motion in connection with the role of centrifugal force.

The first and more philosophical point concerned the opening two paragraphs of the *Tentamen*. By stating that it was absurd to believe that 'a body is only moved by a corporeal agent, not by the human mind (unless it be corporeal) nor by God (unless he be corporeal)', Newton was deploying a typical argument of the time. Similar observations can be found in *De gravitatione et aequipondio fluidorum* and in the correspondence between Leibniz and Samuel Clarke.[28]

The problems related to Cartesian vortices were that gravity would

---

[27] The memoranda are published in Edleston, *Correspondence*, pp. 307–14, and *NC*, 6, pp. 116–22. Other related manuscripts are ULC, Add 3968.4, f. 15r., 16v., 17r., 19v. (remarks on the *Tentamen*); Add 3968.12, f. 176r.; Add 3968.41, f. 124v. (where Newton checked Leibniz's calculation of the equation of paracentric motion).

[28] *De Gravitatione et Aequipondio Fluidorum* (written around 1670), is published by Hall and Hall, *Scientific Papers*, pp. 90–121, on pp. 138–9 (English translation, pp. 121–56, pp. 141–2). Leibniz claimed several times that everything in the world happens mechanically, but the laws of mechanics depend on God's choice for the best. Compare the letter to Herman Conring of 1678, *LSB*, II, *1*, p. 400; *Discours de Métaphysique*, paragraph 18; the letter to Nicholas Malebranche of 1687, known as 'Principium Quoddam Generale', *LMG*, 6, pp. 129–35:135; the correspondence with Samuel Clarke, Leibniz's fifth paper, paragraphs 92 and 124. These themes were standard arguments against Descartes and Hobbes. See also Section 1.1.

tend towards the axis of rotation rather than towards the centre, either of the Sun or of other celestial bodies. More crucially, as Leibniz knew only too well, and as David Gregory and George Cheyne had already pointed out, the harmonic vortex could account for the area law or, with some modifications, the third law, but he had great difficulties in explaining the two together.[29] Newton also claimed that according to the *Principia*, corollary 1 to proposition 4, book I, centripetal forces in a circumference are proportional to the square of velocity over the radius; if the circulation is harmonic, centripetal forces would be inversely proportional to the third power of the distance. This criticism is not correct because harmonic circulation and inverse-square law are perfectly compatible. The harmonic circulation imposes no constraints on the dependence of force on distance.[30] A further criticism implied that Leibniz's reasoning was absurd because, following his vortex theory, the velocity of planets and comets at equal distances from the Sun would have to be equal. Indeed, already in the 1685 *De mundi systemate* Newton had employed the rule whereby, at the same distance from the Sun, the speed of a comet is to the speed of a planet as the square root of 2 is to unity. This approximate rule was based on the assumptions that the orbit of the comet is nearly parabolic, and the orbit of the planet is circular. Although Newton's criticism would be even more cogent if he referred to the component of orbital velocity perpendicular to the radius, his objection was still very powerful, provided one accepted his theory of comets. As we are going to see in the next chapter, however, at the beginning of the eighteenth century the status of cometography on the Continent differed greatly from that in England, and none of the main mathematicians working on celestial mechanics had any specific interest in astronomy and especially in cometography.[31]

We move to the third point. Some of Newton's criticisms about Leibnizian mathematics have been analysed in masterly fashion by Eric Aiton and need therefore not detain us too long. They regard the notion

---

[29] D. Gregory, *Astronomiae Physicae et Geometricae Elementa* (Oxford, 1702), pp. 99–104. A copy with marginal annotations by Leibniz is at the NLB (Marg. 124). In a letter to Johann Bernoulli Leibniz said that this book contained nothing new. He replied to Gregory's criticisms, however, in the 'Illustratio Tentaminis', *LMG*, 6, p. 254. G. Cheyne, *Philosophical Principles of Natural Religion* (London, 1705), pp. 37–42. I owe this reference to Eric Aiton.

[30] *NC*, 6, p. 116 and 118; Edleston, *Correspondence*, pp. 310–11. See also *Principia*, corollary 6 and scholium to proposition 4, book I.

[31] On cometography around 1700 see J. A. Ruffner, *Background*; chapters 7–9 are on Newton. S. Schaffer, 'Newton's Comets and the Transformation of Astrology', in P. Curry, ed., *Astrology, Science and Society. Historical Essays* (Woodbridge, 1987), pp. 219–43. *NMW*, 5, pp. 210–13, 298–303, 524–31, and 6, pp. 57–61, 81–5, 481–507. See also *Principia Mathematica*, first edition, pp. 474–510.

and measure of centrifugal force, and two equalities in paragraph 15 of the *Tentamen*. The two equalities are $N_2M = G_2D$ and $NP = {}_2D_2T$.[32]

$\odot$ is the centre of the circulation; ${}_1M$, ${}_2M$, ${}_3M$ are the positions occupied by the planet at three successive equal instants of time $\theta = dt$; ${}_1MP$, ${}_2M_1T$, and ${}_3M_2T$ are three circular arcs with radii $\odot{}_1M = r + dr_1$, $\odot{}_2M = r$, and $\odot{}_3M = r - dr_2$, respectively, where $dr_1 = {}_1T_1M$ and $dr_2 = {}_2T_2M$. Further, ${}_2D_3M$ and N_1M are perpendicular to $\odot{}_2M$; ${}_3MG$ is parallel to ${}_1M_2M$, ${}_1M_2M$ is equal to ${}_2ML$, and L_3M is parallel to $\odot{}_2M$. Despite Newton's criticism, Leibniz's first equality is correct if the tangent ${}_1ML$ is the prolongation of the chord ${}_1M_2M$, as Leibniz specified in 1706 and John Keill critically observed in his 1714 attack on the *Tentamen*. Newton's criticism of the second equality, that second-order differences are neglected, is not correct; if θa is equal to twice the area of triangles ${}_1M\odot{}_2M$ and ${}_2M\odot{}_3M$, which is constant because ${}_1M_2M$ and ${}_2M_3M$ are traversed in equal times, the segment $PN = a^2\theta^2:2(r + dr_1)^3$; the segment ${}_2D_2T = a^2\theta^2:2(r - dr_2)^3$; their difference is proportional to $\theta^2(dr_1 + dr_2)$, which is a third-order infinitesimal. In conclusion, by 1706

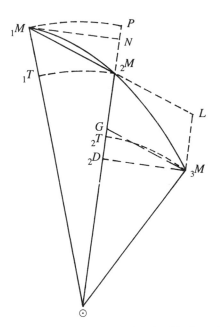

Fig. 8.1   Leibniz's analysis of orbital motion.

[32] Aiton, *Vortex Theory*, pp. 144–5. I have introduced indices to the differential *dr* of the radius, although they were not employed by Leibniz.

Leibniz's emended version was not easy to challenge from a purely mathematical standpoint: the main problem involved matters of definition—such as the notion of tangent—and these were linked to deep philosophical commitments. We move now to the last issue on the agenda, namely mechanics and the analysis of curvilinear motion.

With regard to the measure of centrifugal force, Newton objected to the mistake concerning the factor of 2, which was corrected by Leibniz in 1706 thanks to the correspondence with Varignon, and to the very definition of centrifugal force or conatus as well. For Leibniz centrifugal conatus pertained only to circular motion; when motion was not circular, he considered that component of the orbital trajectory pependicular to the radius, or the circular component of the curvilinear trajectory: centrifugal force was as the square of that component over the radius. By contrast, Newton used the expression 'centrifugal force' for any curve, and set it equal and opposite to gravity. Newton's objections are particularly interesting because Newton himself until 1680 held views almost identical with Leibniz's. In the controversy Newton was induced to spell out his theory more explicitly than anywhere else in his work. It is worth noticing Newton's reluctance to put forward his ideas on centrifugal force in print: his criticisms were only the basis for Keill's public attack. The strategy of pushing his champion Keill forward may suggest a degree of doubt on Newton's part on this issue. In the following discussion I consider the 1706 version of Leibniz's theory, ignoring for simplicity the mistake by a factor of 2. Newton, to be sure, was much less charitable on this issue.

In both memoranda against the *Tentamen* referred to at the beginning of this section, Newton stated that paracentric solicitation, or centripetal force, is equal and opposite to centrifugal conatus, or force. In the memorandum named *Ex epistola* he explained further that this equality derived from the third law of motion, namely action equals reaction. His objection to the *Tentamen* was that since for Leibniz paracentric solicitation and centrifugal conatus were in general different, the third law of motion was violated. Far from being merely a polemical device employed against Leibniz, the explanation of centrifugal force in terms of the third law had been implicitly endorsed by Newton already in the first edition of the *Principia*. I have pointed out above that the variant readings among the three editions show that he became progressively more explicit on this issue. It is tempting to see this process as revealing of Newton's initial uncertainty on this matter, and as part of the war with Leibniz involving the attack on the *Tentamen* in particular. Moreover, Newton was able to show a further consequence of Leibniz's reasoning: 'If the curvature of the curve be diminished until the curve coincides with its tangent, paracentric solicitation and centrifugal conatus will cease, but

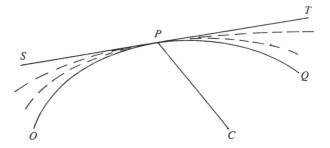

Fig. 8.2   Newton's criticisms of the *Tentamen.*

the paracentric velocity and its element will not cease.' Newton's reasoning can be illustrated with the following diagram.[33]

Let us imagine a curve *OPQ* and its tangent *SPT* in *P*. If a body moves along the curve, for Leibniz the body will have both a centripetal and centrifugal conatus. If we now imagine the curve to be progressively straightened, as indicated by the dotted lines, until it coincides with the tangent *SPT*, motion becomes rectilinearly uniform, therefore centripetal and centrifugal force ought to vanish from the first two laws of motion. Indeed, this is what happens following Newton's approach. Following Leibniz, however, paracentric or radial motion along *CP* and its element *ddr* do not vanish, and this appeared to be a crucial fault to Newton, because the first two laws of motion were violated too. Carrying out the calculations, one would find $ddr = a^2\theta^2/r^3$. In conclusion, a direct consequence of Leibniz's mathematical representation of curvilinear motion violates all three laws of motion. For Newton there was only one reasonable representation of curvilinear motion: different representations were inevitably leading to inconsistencies. From this analysis of Newton's reasoning, it seems difficult to defend the view that for Newton centrifugal force was 'fictitious' or that it was related to the rotation of the frame of reference as opposed to the rotation of a body. On the contrary, for Newton centrifugal force in orbital motion appears to be a 'real' force, as real as a reaction to a centripetal force. These observations emphasize the difference between Newton's own formulation of mechanics and later formulations of 'Newtonian' mechanics.

This evaluation of contrasting views bears little historical value unless one considers the reception of Newtonian and Leibnizian theories of curvilinear and especially planetary motion; this is our following task.

---

[33] The diagram is mine: see *NC*, 6, pp. 117–18.

9

THE RECEPTION OF NEWTONIAN AND LEIBNIZIAN THEORIES

9.1 Introduction

Although in the seventeenth century several philosphers including René Descartes, Giovanni Alfonso Borelli, Robert Hooke, and Christiaan Huygens studied celestial phenomena using the laws of mechanics, with the appearance of Newton's *Principia* the mathematical study of the system of the world reached a new level of sophistication and a broader horizon. If physical causes had been left at the margin of Newton's account, mechanics and astronomy were brought together to an unprecedented degree of cohesion. Not only planetary motion and Kepler's three laws, but also the motion of comets, tides, the shape and mutual perturbations of celestial bodies were studied on the basis of the same principles and laws of motion. This unified treatment of celestial and terrestrial phenomena set a number of problems in mathematics, mechanics, and astronomy and established a new field within the mathematical disciplines, namely celestial mechanics. It is essential at this point to permit some remarks about disciplinary boundaries. Newtonian and Leibnizian celestial mechanics had a location on the map of knowledge different from that which became standard several decades later. With regard to thcology, for example, the order and constitution of the cosmos and the status of its laws with respect to the Creator were a major concern for the two contenders. From the references to God in the first edition of the *Principia* to the general scholium concluding the later editions, theology was interwoven with Newton's ideas about the system of the world. As the series of annual Boyle lectures begun in 1692 shows, theology and politics played a central role in the reception of Newtonianism. Likewise Leibniz, in many letters and pamphlets, in his main published work significantly named *Théodicée*, up to his correspondence with Samuel Clarke, stressed that laws of nature were an integral part of his philosophical and theological system. Although theology probably marks the most important difference with respect to modern perceptions, other disciplines too figure in the arena of celestial mechanics. Cometography, for example, was an area where Earth history, biblical hermeneutics, and alchemy interacted in Newton's theory and influenced his belief in the periodicity of comets. Their tails in particular were endowed with remarkable properties, as the following quotation from the *Principia* shows:[1]

So for the conservation of the seas, and fluids of the planets, comets seem to be required, that, from their exhalations and vapors condensed, the wastes of the planetary fluids spent upon vegetation and putrefaction, and converted into dry earth, may be continually supplied and made up; for all vegetables entirely derive their growths from fluids, and afterwards, in great measure, are turned into dry earth by putrefaction; and a sort of slime is always found to settle at the bottom of putrefied fluids; and hence it is that the bulk of the solid earth is continually increased; and the fluids, if they are not supplied from without, must be in a continual decrease, and quite fail at last. I suspect moreover, that it is chiefly from comets that spirits come, which is indeed the smallest but the most subtle and useful part of our air, and so much required to sustain the life of all things with us.

Within the several contexts of the reception of the *Principia*, I focus here on the mathematical study of celestial motions—especially on the Continent—and on the competitive interaction with the *Tentamen* from the late 1680s to approximately Leibniz's death in 1716. In those years not only Newton's masterpiece, but also Leibniz's response were known to the main mathematicians.[2] I restrict my field of inquiry to advanced research, setting aside the problem of how celestial mechanics entered university teaching.[3]

The awareness of the wealth of staggering results contained in the *Principia* grew slowly and induced a response which can be analysed with the help of a periodization linked to the attitudes of different

---

[1] Newton, *Principia*, first edition, p. 506; Motte and Cajori, pp. 529–30. Notice the references to 'vegetation and putrefaction'. On alchemy and cometography in Newton see S. Schechner Genuth, 'Comets, teleology and the relationship of chemistry to cosmology in Newton's thought', *Annali dell'Istituto e Museo di Storia della Scienza di Firenze*, 10, 1985, pp. 31–65. Cohen, *Introduction*, pp. 155–6 and 240–5. M. C. Jacob, *The Newtonians and the English Revolution* (Ithaca, 1976).

[2] Johann Bernoulli referred to the *Tentamen* in a letter to Leibniz in *LMG*, 3, p. 250, 22 Febr. 1696. P. Varignon, 'Des Forces Centrales', *MASP*, 1700, pp. 218–37, on p. 224. On David Gregory and Cheyne see Section 8.2 above. J. Hermann, 'Metodo d'Investigare l'Orbite de' Pianeti', GLI, 2, 1710, pp. 447–67, on p. 450. C. Wolff, *NC*, 6, pp. 216–18, 24 April 1715. G. Poleni, *De vorticibus coelestibus dialogus* (Padua, 1712), pp. 34–5. *Celestino Galiani—Guido Grandi. Carteggio (1714–1729)*, eds. F. Palladino and L. Simonutti (Firenze, 1989), p. 34, Grandi to Galiani, 19 May 1714. For a broader picture of vortex theories including those not directly related to the *Principia* and *Tentamen* see Aiton, *Vortex Theory*, chs. 7–8.

[3] On this topic see L. W. B. Brockliss, *French higher education in the seventeenth and eighteenth centuries* (Oxford, 1987), ch. 7; N. Guicciardini, *The development of the Newtonian calculus in Britain, 1700–1800* (Cambridge, 1989) provides a useful survey. H. J. Waschkies, *Physik und Physikotheologie des jungen Kant* (Amsterdam, 1987), chs. 7 and 18.

communities in Britain and on the Continent. The locations and approximate dates of activity of the scholars engaged in the practice of celestial mechanics can be seen in the following diagram.

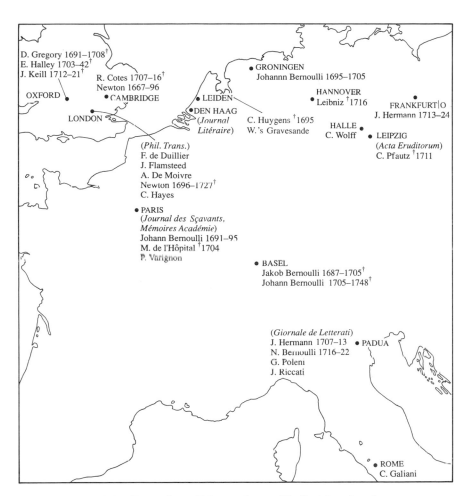

D. Gregory 1691–1708[†]
E. Halley 1703–42[†]
J. Keill 1712–21[†]   R. Cotes 1707–16[†]
Newton 1667–96

OXFORD   ● CAMBRIDGE

LONDON

● GRONINGEN
Johann Bernoulli 1695–1705

● LEIDEN
● DEN HAAG
(*Journal
Litéraire*)

HANNOVER
● Leibniz [†]1716

C. Huygens [†]1695
W.'s Gravesande

HALLE ●
C. Wolff

FRANKFURT|O
J. Hermann 1713–24

● LEIPZIG
(*Acta Eruditorum*)
C. Pfautz [†]1711

(*Phil. Trans.*)
F. de Duillier
J. Flamsteed
A. De Moivre
Newton 1696–1727[†]
C. Hayes

● PARIS
(*Journal des Sçavants,
Mémoires Académie*)
Johann Bernoulli 1691–95
M. de l'Hôpital [†]1704
P. Varignon

● BASEL
Jakob Bernoulli 1687–1705[†]
Johann Bernoulli 1705–1748[†]

(*Giornale de Letterati*)
J. Hermann 1707–13   ● PADUA
N. Bernoulli 1716–22
G. Poleni
J. Riccati

● ROME
C. Galiani

Fig. 9.1   Reception of Newtonian and Leibnizian theories.

On the Continent the emergence of a community of practitioners of the Leibnizian calculus created at the same time a competent audience for the *Principia*. This paradox, whereby Newton was read and admired thanks to his rival, can be documented at several centres. The presence of Johann Bernoulli in Paris in the 1690s led to the formation of a circle of mathematicians whose most distinguished members were the Marquis de l'Hôpital and Pierre Varignon.[4] Johann's career continued later at Basel, where he counted Leonhard Euler among his students. Still at Basel Johann's brother Jakob taught Jakob Hermann, who later became professor of mathematics at Padua. At Padua we find a distinguished circle of mathematicians practising the Leibnizian calculus and studying the *Principia*: Jacopo Riccati and Giovanni Poleni were among them. In Germany the Leibnizian Christian Wolff, professor of mathematics at Halle, contributed several important reviews to the Leipzig *Acta* and wrote didactic treatises which formed the backbone of German higher education for several years. Once again, in his reading of the *Principia* he had recourse to the differential calculus. On the whole, knowledge of higher mathematics on the Continent was based on a common set of texts, mainly articles from the *Acta Eruditorum*, Paris *Mémoires*, Venice *Giornale*, and textbooks such as l'Hôpital's *Analyse des infiniment petits*. Continental mathematicians shared a conceptual framework, language, and set of problems: a small but significant portion of them concerned the motions and shape of heavenly bodies.[5]

The immediate reception of the *Principia*, which is the subject of Section 9.2, centred on the cause of gravity. This aspect is singled out immediately in the titles of the most important works of the early period, Leibniz's *Tentamen de motuum coelestium causis*, Huygens's *Discours de la cause de la pesanteur*, and Fatio de Duillier's unpublished essay *De la cause de la pesanteur*. Another important reader of the *Principia* was David Gregory, who carried out an extensive and accurate study of the text. His annotations were to become an important resource for his later book on the elements of astronomy. After the three works mentioned above, we have to wait until the new century to find other contributions to the mathematical study of celestial motions.

In the new century the *Principia* inspired several investigations, notably by Pierre Varignon and Johann Bernoulli. The adoption of some guiding lines from Newton's mathematical treatment of celestial motions, however, did not imply the acceptance of his ideas about vortices and the cause of gravity. Their work poses again the problem of equivalence, though in a less dramatic way than with Leibniz. Despite the obvious

[4] On the Parisian circles see A. Robinet, 'Le groupe malebranchiste introducteur du calcul infinitésimal en France', *RHS*, *13*, 1960, pp. 287–308; 'La philosophie male-branchiste des mathématiques', *RHS*, *14*, 1961, pp. 205–54. P. Costabel, ed., *Oeuvres de Malebranche*, vol. 17.2, *Mathematica* (Paris, 1968). See also A. R. Hall, 'Newton in France: a new view', *History of Science*, *13*, 1975, pp. 233–50. H. Guerlac, 'Some areas for further Newtonian studies', ib., *17*, 1979, pp. 75–101. Much useful material is in *JBB*, *1* and *2*.

[5] A. Robinet, 'Lan conquête de la chaise de mathématiques de Padove par les leibniziens', *RHS*, *44*, 1991, pp. 181–201. I have been unable to see A. Robinet, *L'Empire leibnizien* (Trieste: Lint, 1991). Further relevant literature can be found below. For some interesting guidelines on the period see M. S. Mahoney, 'On Differential Calculuses', *Isis*, *75*, 1984, pp. 366–72.

relations to the *Principia*, their essays contained several innovative features and led to a decisive algebraization of the science of motion. Although geometrical diagrams were still widely used, they no longer occupied such a prominent role as in the *Principia*. On the Continent it is possible to trace a dual response regarding mathematics and physics. Several mathematicians such as Johann Bernoulli or Leonhard Euler later in the century, adopted the mutual attraction of celestial bodies while remaining highly sceptical if not altogether hostile towards Newton's ideas about the heavenly matter. This situation resembles in some respects the state of pre-Keplerian astronomy, when Aristotelian cosmology could not produce a predictively adequate system able to compete with those based on fictitious mathematical constructions. The natural philosophy of the heavens set out in *De Caelo* and *Metaphysica, XII*, allowed only concentric orbs, while mathematical astronomy was based on epicycles, eccentrics, and equants. Besides this analogy involving the dichotomy of mathematical versus physical explanations between the sixteenth and the eighteenth centuries, however, an important difference needs to be singled out. In the sixteenth century mathematical devices designed to save the phenomena were accompanied by serious doubts about the possibility of attaining knowledge of the heavens. By contrast, at the beginning of the eighteenth century the inverse-square law was accepted as a genuine discovery about the motions of planets and satellites. The propositions of the *Principia* had to be explained, complemented and extended, rather than subverted.[6]

By and large, Continental and British readers showed different interests in their reading of Newton's masterpiece. Although the influence of theology on Continental mathematicians has not been satisfactorily investigated, at present is appears that authors like Varignon, Hermann, Johann and Niklaus Bernoulli showed no great theological preoccupations.[7] The main developments in the new century are surveyed in Section 9.3. Whilst on the Continent we find important mathematical contributions, in Britain one witnesses a stronger interaction between mechanics and astronomy. John Flamsteed, Edmond Halley, and Roger Cotes were concerned with lunar motions, the theory of comets, and tides. By contrast Continental astronomers, notably Giandomenico Cassini, largely ignored the *Principia*.

[6] R. S. Westman, 'The astronomer's role in the sixteenth century: a preliminary study', *History of Science*, *18*, 1980, pp. 105–47. Jardine, *Birth*, chs. 6–9; 'Scepticism in Renaissance astronomy'.

[7] Regarding l'Hôpital see *JBB*, *1*, pp. 305–7, 22 Aug. and 3 Sept. 1695. The situation in Italy has been investigated in V. Ferrone, *Scienza, Natura, Religione. Mondo newtoniano e cultura italiana nel primo settecento* (Napoli, 1982). H. Guerlac, *Newton on the Continent* (Ithaca, 1981). E. A. Fellmann, 'The *Principia* and Continental Mathematicians', *NRRS*, *42*, 1988, pp. 13–34.

At the beginning of the eighteenth century an important context of the reception of the *Principia* and *Tentamen* was the priority dispute over the invention of the calculus. The inverse problem of central forces, which emerged around 1710, was interwoven with the war between Newton and Leibniz. Section 9.4 focuses on this issue and especially on the works by Newton, Johann Bernoulli, Varignon, and Hermann. At the end of the section I survey the priority dispute between Johann Bernoulli and John Keill. The very notions of 'demonstration' and 'solution' became controversial; this dispute emphasizes the differences between British and Continental mathematics, especially about geometrical and analytical representations of curves.

The last section examines Keill's attack on the *Tentamen* and the reasons for the final defeat of Leibniz's theory. Despite the polarization of the mathematical community into rival groups, the defeat of the *Tentamen* depended overwhelmingly on intellectual factors and on the fertility of the *Principia* as a source for further researches. The main difficulty for Leibniz, namely the combination of Kepler's third law and the harmonic vortex, involved arguments that we would still perceive as valid. Other factors, however, such as the link between centrifugal force and third law in orbital motion, or the belief that centrifugal force could not decrease inversely as the third power of the distance, are alien to our interpretation. Thus the reasons I invoke have to be located in the historical practice and intellectual horizon of baroque mechanics, mathematics, and natural philosophy. While early eighteenth-century mathematicians investigated celestial motions in a purely mathematical fashion, Leibniz's strategy shifted from the attempt to produce a theory able to compete with Newton's, to the formulation of theological and metaphysical objections typical of the correspondence with Samuel Clarke in 1715 and 1716. Thus Leibniz abandoned the Keplerian programme.

9.2   The early response to the *Principia* and *Tentamen*

As I. Bernhard Cohen has shown, the *Principia* was reviewed in the main journals of the late seventeenth century, namely the *Philosophical Transactions*, *Acta Eruditorum*, *Bibliothèque Universelle* and *Journal des Sçavants*. In addition, we can rely on a number of private documents to trace the first responses to Newton's masterpiece both in Britain and on the Continent. Together with Edmond Halley, who edited the book and wrote the review for the *Philosophical Transactions*, and Robert Hooke, who made an accusation of plagiarism with respect to the inverse-square law, we know of the reactions of readers such as Fatio de Duillier, David Gregory, John Locke, and Gilbert Clerke.[8]

In the late 1680s the two outstanding figures on the European philosophical scene were Huygens and Leibniz. Although they shared many ideas on physics and mathematics, their reaction to the *Principia* also presented different features. Their correspondence, some annotations by Huygens, and his *Addition* to the *Discours de la cause de la pesanteur*, allow us to reconstruct Huygens's complex reaction to the *Principia* and *Tentamen*. Although Huygens wondered why Newton would have gone through all the trouble of so many difficult mathematical demonstrations when the whole system was based on the 'absurd' principle of attraction, it appears that those demonstrations left a deep mark on him. In particular Huygens accepted the inverse-square law as a genuine discovery, although he believed that its cause remained to be investigated. With regard to physical explanations he thought that Newton had demolished Cartesian vortices, in which Huygens himself had believed until the late 1680s. Despite the rejection of deferent vortices, Huygens still looked for a mechanical cause for gravity; he justified the existence of an aether with the propagation of light through celestial spaces. Notoriously, he compared this propagation to that of waves in a medium. The rejection of Cartesian vortices was the main point of departure from Leibniz: this is precisely the issue on which Huygens pressed his German friend in their correspondence, containing the only early opinion on the *Tentamen* I am aware of.[9]

With regard to the analysis of curvilinear motion and centrifugal force, current historiography tends to see a major break between Huygens and Newton. Briefly, Huygens is presented as the last outstanding representative of the old school which focused on outwards tendencies and stopped on the threshold of the correct formulation. Newton, on the other hand, is portrayed as a modern who resolutely passed that threshold and provided the definitive solution to the problem, at least within classical mechanics. Thus his move from centrifugal to centripetal forces would represent the decisive step from error to truth. Contrary to this interpretation, and despite the differences between their approaches and the new results attained by Newton, I see a continuity between

---

[8] The main source for the reception of the *Principia* is Cohen, *Introduction*, pp. 145–61. *NC*, *2*, p. 431 (Halley to Newton, 22 May 1686), pp. 435–41 (Newton to Halley, 20 June 1686). Lohne, 'Hooke *versus* Newton', R. S. Westfall, 'Hooke and the Law of Universal Gravitation', *BJHS*, *3*, 1967, pp. 245–61, and *Never at Rest*, pp. 382–3, 446–53, 471–2, with additional bibliography.

[9] *LMG*, *2*, pp. 41 and 57, Huygens to Leibniz, 8 Febr. and 18 Nov. 1690. *HOC*, *10*, pp. 147–55, with Cohen's commentary in *Introduction*, pp. 186–7. *HOC*, *9*, pp. 167–71; 190–1; 357–60; 381–9; 391–3; 407–12 (correspondence with Fatio De Duillier). *Addition* to the *Discours*, *HOC*, *21*, pp. 466–88. Cohen, *Newtonian Revolution*, section 3.4.

Huygens's and Newton's analyses of curvilinear motion. My interpretation is supported by the material presented in the previous chapter and by Huygens's own statements. In the 1690 *Addition* he wrote:[10]

But seeing now by the demonstrations of M. Newton that, supposing such a gravity toward the sun, and that it diminishes according to the said proportion, it counterbalances so well the centrifugal forces of the planets and produces precisely the effect of the elliptical motion that Kepler had guessed and proved by observation, I cannot doubt the truth either of these hypotheses concerning gravity or of the System of M. Newton.

The same opinion about centrifugal force and gravity counterbalancing each other was expressed in the letter to Leibniz of 8 February 1690 referred to above. Huygens probably had in mind corollaries 1, 6, 7, and the scholium to proposition 4. Corollary 1 states that the centripetal forces of bodies moving uniformly along circles tend to the centres of the circles and are as the squares of the velocities over the radii. Corollary 6 states that if the squares of the periods are as the third powers of the radii, centripetal forces are inversely as the squares of the distances. Corollary 7 generalizes these results to other curves. Lastly, the scholium refers to the propositions at the end of the *Horologium Oscillatorium* and establishes the relation between centripetal and centrifugal forces that we have seen in Section 8.3. Thus the quotation above and the 8 February letter to Leibniz were not the product of an old mind unable to follow the latest developments in mechanics, or the result of misunderstanding. Quite on the contrary, Huygens provided a reasonable interpretation which was shared by many of his contemporaries and which coincided to a large extent with Newton's own. We shall see in the following section that one of the most influential readers of the *Principia* on the Continent and even Newton's mouthpiece—Pierre Varignon and John Keill respectively—expressed similar views on this matter.

From 1690 onwards Leibniz referred to the *Principia* on several occasions. In the *Specimen Dynamicum,* just to mention one example, he criticized Newton's theory of attraction and ideas about absolute motion. The preparatory manuscripts, whose critical edition unfortunately is not always reliable, reveal the existence of further references to the *Principia.*[11] Another line of action can be traced through the contacts sought by Leibniz with Newton and the Royal Society. In October 1690 Leibniz wrote to his old Paris friend Henri Justel, by then a royal librarian in London, praising the *Traité de la Lumiere* and *Principia*

---

[10] *HOC, 21*, p. 472; transl. in Cohen, *Newtonian Revolution*, p. 80.

[11] Leibniz, *Specimen Dynamicum* (Hamburg, 1982), pp. 22–4 and 58, and the essay-review by E. Knobloch, *AS, 40,* 1983, pp. 501–4. Translations in Ariew and Garber, *Philosophical Essays*, pp. 125 and 136.

Mathematica as the most important works of their kind since Descartes. However, he criticized Newton's attraction and his rejection of vortices. In the same letter Leibniz also advertised the main features of the *Tentamen* and of his recent findings about planetary motion, probably in the hope that his ideas and essay would be discussed in England. Justel's reply, in which he defended Newton with the claim that in physics all attempts are justified because *tentare non nocet*, probably induced Leibniz to pursue other routes.[12] In 1693 he wrote directly to Newton, urging him to continue 'to handle nature in mathematical terms', while suggesting a new interpretation of his results:

You have made the astonishing discovery that Kepler's ellipses result simply from the conception of attraction or gravitation and trajection in a planet. And yet I would incline to believe that all these are caused or regulated by the motion of a fluid medium, on the analogy of gravity and magnetism as we know it here. Yet this solution would not detract from the value and truth of *your* discovery. (My emphases.)

In his reply Newton defended the interpretation of the *Principia*:

For since celestial motions are more regular than if they arose from vortices and observe other laws, so much so that vortices contribute not to the regulation but to the disturbance of the motions of planets and comets; and since all phenomena of the heavens and of the sea follow precisely, so far as I am aware, from nothing but gravity acting in accordance with the laws described by me; and since nature is very simple, I have myself concluded that all other causes have to be rejected and that the heavens are to be stripped as far as may be of all matter, lest the motions of planets and comets be hindered or rendered irregular. But if, meanwhile, someone explains gravity along with all its laws by the action of some subtle matter, and shows that the motion of planets and comets will not be disturbed by this matter, I shall be far from objecting.

By throwing the ball into Leibniz's court, Newton challenged him to provide not a partial explanation of this or that phenomenon, but to account for 'all phenomena of the heavens and of the sea'.[13]

In the following year Leibniz wrote again to London. On 31 October 1694 a letter of his was read at the Royal Society, 'wherein he recommends to the Society to use their endeavours to induce Mr Newton to publish his further thoughts and emprovements on the subject of his late book Principia Philosophiae Mathematica, and his other Physicall and Mathematicall discoverys, least by his death they should happen to be lost.' Possibly Leibniz was hoping that new results by

[12] *LSB*, I, 6, pp. 263–7 and 300–2; Aiton, *Leibniz. A Biography*, (Bristol, 1985), pp. 171–2.

[13] *NC*, 3, pp. 257–60, on p. 258, 17 March 1693, and Guerlac, *Continent*, p. 53, n. 41.; *NC*, 3, pp. 285–9, on p. 287, 15 Oct. 1693.

Newton might have led to a reconciliation with physical interpretations.[14]

In England, soon after publication of the *Principia*, rumours and plans about a revised edition were under way. Within weeks of the appearance of the book Halley urged Newton to improve lunar theory. This attitude was typical of the reception of the *Principia* in Britain, where emphasis was laid on the links between theory and astronomical observations.[15]

Plans for a new edition were associated with Fatio de Duillier and David Gregory. The Genevan mathematician, who had come to England in 1687 and had become a very close friend of Newton, felt confident enough to undertake the task and started drafting a list of errata and improvements. More interesting from our point of view is Fatio's attempt to provide a physical and mechanical explanation for gravity. His theory was based on a quantitative attempt to calculate the effect of the pressure of a very rare aether on the bodies floating in it. Although an account of Fatio's theory was presented at the Royal Society on 26 February 1690 and was known outside Britain to influential mathematicians such as Huygens and Jakob Bernoulli, his ideas remained unpublished and exerted only a limited influence. Initially Newton himself believed that Fatio's hypothesis was the only possible physical explanation for gravity, but he later retracted his support; as reported by Gregory, 'Mr Newton and Mr Hally laugh at Mr Fatios manner of explaining gravity'.[16]

The Scottish mathematician David Gregory started his study of the *Principia* in Edinburgh in September 1687 and completed it in Oxford in 1694, where he had become Savilian Professor of Astronomy in 1691. His *Notae* reveal him as one of the most careful readers of Newton's masterpiece. Gregory, however, did not develop any particular branch of Newton's work. Together with his *Memoranda*, recording his conversations with Newton, the *Notae* will become an important resource for the *Astronomiae physicae et geometricae elementa* (Oxford, 1702), the first textbook of astronomy based on Newton's *Principia*.[17]

---

[14] *NC, 4*, p. 24, n. 5.

[15] *NC, 2*, p. 482, Halley to Newton, 5 July 1687. Cohen, *Introduction*, p. 173.

[16] *NC, 3*, p. 191. Cohen, *Introduction*, pp. 177–87. B. Gagnebin, 'Memoire de Nicolas Fatio De Duillier. De la Cause de la Pesanteur. Presente à la Royal Society le 26 Fevrier 1690', *NRRS, 6*, 1949, pp. 105–60. Hall and Hall, *Scientific Papers of Isaac Newton*, pp. 312–17. K. Bopp, 'Die wiederaufgefundene Abhandlung von Fatio de Duillier: De la Cause de la Pesanteur', *Schriften der Strassburger Wissenschaftlichen Gesellschaft in Heidelberg*, 1929, Neue Folge, *10.* Heft, pp. 19–66. H. Zehe, 'Die Gravitationstheorie des Nicolas Fatio de Duillier', *AHES, 28*, 1983, pp. 1–23.

[17] W. P. D. Wightman, 'Gregory's *Notae in Isaaci Newtoni Principia Philosophiae*', *Nature, 172*, 1953, p. 690 and 'David Gregory's Commentary on Newton's *Principia*', *Nature, 179*, 1959, pp. 393–4. C. M. Eagles, *The Mathematical Work of David Gregory, 1659–1708*, Ph.D. Thesis, University of Edinburgh 1978, pp. 26–34.

9.3   Developments in the new century

The appearance of Pierre Varignon's work on central forces in 1700 marks a crucial date in the reception of Newton's and Leibniz's theories. Unlike Fatio, Huygens, and Leibniz, Varignon focused on mathematics and mechanics, ignoring the problem of physical explanations. Varignon was the first of a series of Continental mathematicians whose main concern was not the reconciliation of gravity with vortices, but the public demonstration of the power and versatility of the differential calculus and the deployment of their mathematical skills. The *Principia Mathematica* was used as a battleground for the new mathematics. In the works of Continental mathematicians after 1700 astronomy was largely neglected; none of the leading figures of the new generation, notably Varignon, Johann Bernoulli, and Hermann, was very proficient or interested in it.

Varignon was a member of the Paris Academy. He was indirectly instructed in the calculus by Johann Bernoulli and, following Fontenelle's *Éuloge*, studied the *Principia* in the late 1690s. The recently established *Mémoires* of the Academy provided an ideal arena for his publications. His series of essays have been judged differently by historians: some have treated him as a mere translator of Newtonian mechanics into the language of the calculus, while others have presented him as a more original thinker. A puzzling historical insight can be gained by considering a debate between the Marquis de l'Hôpital—one of the leading mathematicians at the end of the century—and his mentor Johann Bernoulli. When the latter posed a problem involving centrifugal forces in 1695, the Marquis had to confess his lack of a clear understanding of what was meant with 'force centrifuge'. Johann in his reply referred to the propositions contained in the appendix of the *Horologium Oscillatorium*.[18] This exchange, dating only one year before the appearance of l'Hôpital's *Analyse des infinement petits* in 1696, emphasises the gulf between pure and applied mathematics. This seems to suggest that common mechanical notions were not necessarily known even to the best mathematicians of the time. Although Johann Bernoulli was a more talented mathematician than Varignon, the French had developed a particular skill in mechanics.

Before attempting an evaluation of Varignon's contributions, at this point I wish to introduce some general reflections. The process of bring-

---

[18] *JBB, 1*, p. 236, l'Hôpital to Johann, 19 Febr. 1695; pp. 270–1, Johann to l'Hôpital, 5 Mar. 1695. The Marquis later published a 'Solution d'un problème physico-mathématique', *MASP*, 1700, pp. 9–21, on the original problem posed by Johann Bernoulli.

ing together the most advanced areas of mathematics and mechanics of the time bore fundamental results which are not captured by the term 'translation', at least in its most immediate meaning of rendering a passage from one language into another, such as Greek into Latin. Other examples may turn out to be more appropriate. An extreme case emphasizing the importance of notations and style is the calculation of the square of a number, say 86, written in Roman numerals as 'LXXXVI'. Even simple multiplication becomes highly problematic, since our positional notation allows the execution of several operations by means of simple techniques which cannot be 'translated': notation and manipulation of equations cannot be separated. In the general introduction I argued that translating a set of operations between two computer languages may require deeper changes than translating between two Indo-European languages. Even if the two programmes are designed to perform the same operations, the skills required to manipulate them may differ considerably. Thus subsequent modifications and developments may follow different routes, and this is precisely what happened in Britain and on the Continent in the eighteenth century: despite the initial 'equivalence' of fluxions and differentials, British mathematicians remained closer to geometry and tended to conceive the increments of a variable with respect to time. By contrast, Continental mathematicians developed an algebraic approach in which variables and parameters were clearly distinguished by means of a standard notation: x, y, etc. for variables, a, b, etc. for parameters. Differential equations became the focus of attention, and a wealth of integration techniques was developed. By the middle of the century, whilst British mathematicians were still struggling with diagrams and a cumbersome geometric notation, Continental mathematicians were beginning to investigate equations in several independent variables, partial differentiation, and even simple but effective new representations involving rotating axes. As a result, entire fields such as the study of elasticity, celestial mechanics, and the theory of rigid bodies were transformed by Daniel Bernoulli, Alexis Clairaut, Jean d'Alembert, and especially Leonhard Euler. In 1757 the distinguished British mathematician Thomas Simpson could state: [19]

And it appears clear to me, that, it is by a diligent cultivation of the *Modern*

[19] Bertoloni Meli, 'Emergence'. See the collection of essays by Truesdell, *Essays in the History of Mechanics* (New York, 1968), and his important contributions in Euler, *Opera Omnia*, II, vols. *11*, (second part) to *13*. D. T. Whiteside, 'Newton's lunar theory: from high hope to disenchantment', *Vistas in Astronomy*, *19*, 1976, pp. 317–28. A different interpretation of British mathematics in the eighteenth century is in Guicciardini, *Development*. Further observations on this topic can be found in the following section. T. Simpson, *Miscellaneous Tracts* (London, 1757), preface, quoted in Guicciardini, *Development*, p. 84.

Analysis, that *Foreign Mathematicians* have, of late, been able to push their *Researches* farther, in many particulars, than Sir Isaac Newton and his *Followers* here, have done.

In Varignon's case the problem of equivalence appears again in a form different from that we have seen for the *Tentamen*. Leibniz's transformation of some results of the *Principia* involved mathematics and physics at the same time. Varignon's work, by contrast, is purely mathematical. By wanting to prove the versatility of the differential calculus, he was setting the scene for the algebraization of Newtonian mechanics, deepening the understanding of the link between mathematical representations and motion, and beginning to develop new practices and results. Varignon did not prove new fundamental theorems in mechanics; for example, nothing in his work compares in originality to any one of Huygens's major achievements a few decades earlier, such as the impact laws, centrifugal force, or the isochronism of the cycloid. Varignon's importance must be evaluated in different terms.

His first essay on mechanics in the Paris *Mémoires*, read in January 1700, sets out a general correlation between mathematics and mechanics involving spaces, times, speeds, and forces. At the end of his essay Varignon was able to express in algebraic form proposition 39 of the *Principia*, which states: 'Supposing a centripetal force of any kind, and granting the quadratures of curvilinear figures; it is required to find the velocity of a body, ascending or descending in a right line, in the several places through which it passes . . .' Whereas Newton's demonstration was entirely geometrical, Varignon was able to write an elementary differential equation: $\int y dx = 1/2 \ vv$, where y is the acceleration, v velocity, and x the space traversed. Propositions 39–41, as we shall see in the following section, became a focal point of interest for the mathematicians involved in the formulation of an algebraic mechanics. Varignon's second memoir deals with central forces developing mathematical examples, such as motion along ellipses and the circle in particular, logarithmic and hyperbolic spirals, and the case of forces acting along parallel lines. The third essay is particularly interesting since it follows the outbreak of the controversy with Rolle about the calculus. In this memoir Varignon focused on astronomical examples, starting again from the ellipse, and then moving to the Cassinoid, and to the eccentric. Each example was studied with the correct form of the area law, and with the empty focus rule which had been commonly adopted in the seventeenth century as a computational device. With these examples Varignon was more interested in emphasizing the universality of his method and possibly gaining Cassini's support against Rolle, than in seeking a collaboration with astronomers, as Newton was with Flamsteed in the

1680s and 90s. The fourth essay introduces the osculating circumference and generalizes some of the preceding results. It is interesting to notice that analogous results with the osculating circumference had been obtained, though not published, by Leibniz in *De Conatu* of late 1688. In a later essay Varignon was stimulated by Leibniz to tackle the three-body problem, an area in which he attained no significant result.[20]

In Varignon's memoirs there are two points in particular which deserve attention. In the second essay he referred to central forces—both centrifugal and centripetal—as being the foundation of the *Principia*. The suspicion that he interpreted Newton's masterpiece as concerning centripetal as well as centrifugal forces is confirmed by a further memoir in the Paris Academy for 1706, in which Varignon explicitly stated that centrifugal forces are equal and opposite to centripetal forces. This interpretation largely corresponds to Huygens's and shows once again the gulf between interpretations around 1700 and more recent ones. Even Johann Bernoulli followed partly this approach. Unlike Varignon, however, Johann applied the third law only to the component of gravity which is perpendicular to the curve, attaining the expression v^2/p, where v is orbital velocity and p the osculating radius.[21]

The other issue concerns the link between mathematical representations of curves and dynamics. As we have seen in Section 4.2, conceiving a curve as an infinitangular polygon affected the dynamics of curvilinear motion because accelerations do not arise in the elements. By and large, Leibniz preferred to dispense with accelerations and adopted the polygonal representation of curves. As Eric Aiton has shown, at the beginning Varignon represented curves as polygons, but at the same time he employed accelerated motion. Awareness of this mistake emerged in his debates with Leibniz and Johann Bernoulli, and is due to Varignon. He explained the matter to Leibniz, who in his turn passed it on to Johann. Later the Swiss mathematician communicated the result to the

---

[20] P. Varignon, 'Maniere generale de déterminer les forces, les vitesses, les espaces, et les temps', *MASP*, 1700, pp. 22–7, esp. p. 27. Newton, *Principia* (Motte and Cajori), p. 125. Varignon's memoir was preceded by several essays on the mathematical analysis of motion; compared M. Blay, 'Quatre memoires inédits de Pierre Varignon consacres a la science du mouvement', *AIHS*, *39*, 1989, pp. 218–48. On Varignon and the *Principia* see M. Blay, "Varignon ou la theorie du mouvement des projectiles 'comprise en une Proposition generale'", *AS*, *45*, 1988, pp. 591–618. Varignon's second memoir is 'Du mouvement en general', *MASP*, 1700, pp. 83–101; the third is 'Des Forces centrales', *MASP*, 1700, pp. 218–37; the fourth 'Autre regle generale des forces centrales', *MASP*, 1701, pp. 20–38. P. Varignon, 'Des courbes décrites par le concours de tant de forces centrales qu'on voudra', *MASP*, 1703, pp. 212–29.

[21] P. Varignon, 'Comparaison des forces centrales avec les pesanteurs absolues des corps', *MASP*, 1706, pp. 178–235, on pp. 178–9. Johann Bernoulli's relevant passage is quoted in a letter by Varignon to Leibniz, *LMG*, 4, pp. 136–8, 9 Oct. 1705. Bertoloni Meli, 'Relativization', sect. 3.

Huguenot refugee in England Abraham De Moivre. Although Varignon acknowledged that the polygonal model led to the correct result, he believed that centripetal force acts continuously and that motion under a centrifugal force is accelerated; hence he preferred the continuous approach. Johann Bernoulli and Hermann also preferred to use accelerations despite their allegiance with Leibniz. The correspondence between Hermann and Leibniz contains several debates on this issue. At one point, for example, Hermann claimed that the 'causa agens' inducing in the body m the infinitesimal velocity dc in the element of time dt, is equal to $mdc:dt$, a statement promptly denied by Leibniz, who saw in it a contradiction with his own metaphysics as well.[22]

In a *Remarque* in his third memoir, Pierre Varignon introduced as a purely mathematical exercise an expression resembling Leibniz's equation of paracentric motion, namely:

$$y = \frac{dz^2 - rddr}{rdt^2},$$

which can be written as $ddr = (dz^2/r) - (ydt^2)$, where y is the central force, r the radius, dz the component of the orbital trajectory perpendicular to the radius, and dt a constant element of time. Thus for a particular choice of the progression of the variables Varignon wrote the central force as the difference between two terms. However, he did not interpret his mathematical formulation as relating either to centrifugal force, or to physical actions.[23] This detachment of the mathematical formulation from the interpretation in physics and mechanics can be witnessed in another interesting case in England. In 1704 the Deputy Governor of the Royal African Company Charles Hayes published *A Treatise of Fluxions*, one of the first textbooks on the new calculus. In the preface he clearly admitted that he had borrowed freely from virtually all the main mathematicians of the time, and especially from Newton, Leibniz, the Bernoulli brothers, l'Hôpital, and John Craig. The section on astronomy in his book is particularly curious because it juxtaposes propositions from the *Principia* and from the *Tentamen*. Although he does not refer to vortices, the Leibnizian mathematical treatment of planetary motion is clearly exposed. Hayes started from a version of

[22] T. L. Hankins, *Jean d'Alembert* (Oxford, 1970), pp. 225–57 and *Science and the Enlightenment* (Cambridge, 1985), pp. 22–5; Aiton, 'Celestial Mechanics', pp. 75–82; *NMW*, 6, pp. 540–1, n. 8. K. Wollenschläger, ed., 'Der mathematische Briefwechsel zwischen Johann I Bernoulli und Abraham de Moivre', *Verhandlungen der Naturforschenden Gesellschaft in Basel*, 43, 1933, pp. 151–317, on pp. 281–2, 290, 296–7. *LMG*, 4, p. 384, Hermann to Leibniz, 22 Dec. 1712; p. 388, Leibniz to Hermann, 1 Febr. 1713. See also Chapter 4 above.

[23] Varignon, 'Des forces centrales', *MASP*, 1701, p. 234. In the equation in the fourth line of the *Remarque* the numerator should read $dz^2 - rddr$.

proposition 1 of the *Principia*, and then, following Leibniz's private itinerary, he attained the law of the harmonic circulation expressed in a purely mathematical form. The reviewer in the *Acta* did not fail to notice that his source was the *Tentamen*.[24] Moreover, although in his figure Hayes called the latus rectum of the planetary ellipse *XN*, in the text he wrote *XW*, which corresponds to Leibniz's figure! Once again, a mathematical formulation was detached from its interpretations.

Hayes was not very representative of the work carried out in Britain, which hinged on a closer integration of celestial mechanics with astronomy. The Astronomer Royal John Flamsteed, despite the heated controversy with Newton over the publication of his own observations in *Historia Coelestis* (London, 1712) without his own consent, provided crucial data for lunar theory.[25]

If Flamsteed was eminently an observational astronomer, the Savilian Professor David Gregory was almost exclusively a theoretical or 'closet astronomer', as Flamsteed called him. His *Astronomiae Physicae et Geometricae Elementa*, probably used for teaching to final year students, was largely based on the *Principia* and contained the preface and an essay on lunar theory by Newton himself. Gregory's work was structured in six books on the Newtonian system, spherical astronomy and refraction, planetary theories and the shape of rotating bodies, the astronomy of satellites including lunar theory, cometography, and comparative astronomy, namely the study of the universe viewed from different celestial bodies. In book I Gregory examined the physical theories proposed by Kepler, Descartes, and Leibniz. Interestingly, originally Gregory had been very appreciative of Kepler, but Keplerian physics was eventually dismissed, it appears, after a meeting with Newton in 1698. Gregory criticized the *Tentamen* for the lack of an explanation of the inverse-square law, the incompatibility between the harmonic vortex and Kepler's third law, and the theory of comets. In a comment made known by Christina Eagles in her doctoral dissertation, Gregory dismissed Leibniz's 1706 additions and emendations to the *Tentamen* with the words: 'It is a very poor paper, and does not so much as touch the main difficultys, but acknowledges all it touches'.[26]

[24] C. Hayes, *A treatise of fluxions or an introduction to mathematical philosophy* (London, 1704), pp. 291–305; *Acta Eruditorum* (1705), pp. 474–6. Guicciardini, *Development*, pp. 15–17.

[25] A. Chapman, *The Preface to John Flamsteed's 'Historia Coelestis Britannica'*, Maritime Monographs and Reports, 52, 1982. Cohen, *Introduction*, pp. 172–7.

[26] D. Gregory, *Astronomiae Physicae et Geometricae Elementa*, pp. 99–104. A copy with marginal annotations by Leibniz is at the NLB (see p. 187 n. 29). Leibniz replied to Gregory's criticisms in the 'Illustratio Tentaminis', *LMG*, 6, p. 254. Eagles, *David Gregory*, pp. 477–577 (on *Astronomia*); pp. 499–504 (on Kepler); pp. 505–510, esp. p. 509 (on Leibniz); pp. 574 and 597. Cohen, *Introduction*, p. 184. *Isaac Newton's 'Theory of the Moons Motion' (1702); with a bibliographical and historical introduction by I. Bernard Cohen* (Folkestone, 1975).

Gregory's colleague at Oxford Edmond Halley, since 1704 Savilian Professor of Geometry, worked extensively on comets. His seminal *Synopsis* is based on the study of 24 comets dating from 1337 to 1698.[27]

Probably the most talented mathematician in Britain after Newton was Roger Cotes, since 1707 Plumian Professor of Astronomy, and editor of the second edition of the *Principia* (Cambridge, 1713). In his celebrated preface he ascribed gravity among the primary qualities of bodies, repeated all the standard arguments against vortices, and attacked the *Tentamen* in particular without mentioning Leibniz. Indeed in a letter to his patron, the Master of Trinity College Richard Bentley, Cotes singled out the *Tentamen* as a particularly good example of a work 'which deserves a censure' because of Leibniz's 'want of candour'. His outstanding editorial work involved virtually all aspects of the *Principia* and, with regard to celestial mechanics, especially lunar theory.[28]

The different roles played by astronomy in celestial mechanics in Britain and on the Continent are exemplified by the story of the data of the early sighting of the great comet of 1680–1. In the first edition of the *Principia* the first none too accurate observation of the great comet was that of 17 November in Rome. Gottfried Kirch, however, since 1700 astronomer at the Berlin Akademie der Wissenschaften, had made reliable observations in Coburg on 4, 6, and 11 November. The existence of these observations was known to Leibniz around 1700, though he was unable to make any use of them. In the early summer of 1703 Edmond Halley stopped in Hanover on the way South to the Adriatic, and the matter of Kirch's observations arose in a conversation with Leibniz. In a letter of 14 July 1703, Leibniz promised Halley to ask Kirch about his observations. As Leibniz wrote to Halley on 8 December 1705, he hoped that Kirch's observations might bring some light to cometography. Although those data were received in England, Newton omitted them from the second edition of the *Principia*, probably because he did not want to thank Leibniz. In the third edition, however, the observations by 'Mr Kirk' were mentioned not only in book III, but also in the few lines of the preface as one of the notable additions.[29] Few examples could

---

[27] E. Halley, 'Astronomiae Cometicae Synopsis', *PT*, **24**, 1705, pp. 1882–99. A. Armitage, *Edmond Halley* (London, 1966). *Standing on the shoulders of giants*, ed. N. J. W. Thrower (Berkeley, 1990). S. Schaffer, 'Newton's comets and the transformation of astrology'.

[28] R. Gowing, *Roger Cotes. Natural Philosopher*, (Cambridge, 1983). Cohen, *Introduction*, pp. 239–40. *NC*, 5, p. 389, Cotes to Bentley, 10 Mar. 1713. *NC*, 5, Intr., pp. xxxi–xxxiv.

[29] *Correspondence and Papers of Edmond Halley*, ed. E. F. MacPike (Oxford, 1932), pp. 200–1, and p. 219. On Halley's journey see Armitage, *Halley*, p. 155, and esp. A. H. Cook, 'Halley, Surveyor and Military Engineer: Istria, 1703', in Thrower, ed., *Giants*, pp. 157–70. The whole episode is narrated in Waschkies, *Physik* pp. 277–90; the dates of cometary sightings follow the old style, as in the *Principia*. 'Observationes quaedam accuratae

emphasize more spectacularly the different roles of astronomical data in Britain and on the Continent. The close links established by Newton with the community of astronomers, mirroring the close links between his own celestial mechanics and astronomy, remained without parallel on the Continent until the time of Jean d'Alembert and Pierre Charles Le Monnier, Alexis Clairaut and Joseph Delisle, Leonhard Euler and Tobias Mayer.

At least two influential Italian mathematicians produced an early response to the *Principia* and *Tentamen*. The Marquis Giovanni Poleni was from 1709 Professor of Astronomy and Meteorology at the University of Padua, where he later occupied the chair of mathematics. In 1712 he published *De vorticibus coelestibus dialogus*, an extremely well documented survey on the current debates on vortex theories in celestial mechanics. In his characteristic style, Poleni presented the arguments for and against the various versions of the vortex theories then available, including Leibniz's harmonic vortex, without taking side with any of them or with the Newtonians. In those years Padua was the main mathematical centre in Italy, following the conquest of the chair of mathematics in 1709 by the Swiss Jakob Hermann, a pupil of Jakob Bernoulli at Basel.[30] A very interesting document about the reception of the *Principia* has been recently published by Vincenzo Ferrone. In a letter to Gregorio Caloprese of March 1714, the Celestine father Celestino Galiani attacked Cartesian vortices and implicitly Leibniz's *Tentamen*, which he had certainly read. His criticisms were based on the inconsistency between Kepler's laws and vortical motion and contained several mathematical arguments. The letter, originally intended for the Venetian *Giornale de' Letterati*, remained unpublished but seems to have been widely known in Italy.[31]

9.4   The inverse problem of central forces

The whole dispute will concern not the solution to the problem, which is not controversial, but the manner of solving it; following the spirit of this very delicate

insignis Cometae sub finem anni 1680 visi, Coburgi Saxoniae a Domino Gottfried Kirch habitae', *PT*, *29*, 1715, pp. 169–72.

[30] Poleni, *De vorticibus*, pp. 15, 34–5, 126–8, 138–42. On Poleni compare B. Dooley, 'Science teaching as a career at Padua in the early eighteenth century. The case of Giovanni Poleni', *History of Universities*, *4*, 1984, pp. 115–51, esp. p. 125f. P. Casini, *Newton e la conscienza europea*, (Bologna, 1983), pp. 179–81. Ferrone, *Scienza*, passim.

[31] V. Ferrone, 'Celestino Galiani e la diffusione del newtonianesimo. Appunti e documenti per una storia della cultura scientifica italiana del primo settecento', *Giornale Critico della Filosofia Italiana*, *61*, 1982, pp. 1–33; Ferrone, *Scienza*, passim. P. Casini, *Newton*, pp. 192–4.

century, which having set together Analysis and Metaphysics, does not value as much the truth which has been discovered, as the method with which this truth is found and made overt.[32]

A major mathematical issue in celestial mechanics tackled successfully before the generation of Leonhard Euler and Alexis Clairaut was the inverse problem of central forces. The direct problem consists in finding the force given the trajectory and is easier than the inverse problem, which requires the trajectory given the force. Before the issue became the focus of attention around 1710, the two problems were not sharply distinguished. As the opening quotation shows, what was at stake was not so much the result, as the method of demonstration.

The leading Continental mathematician at that time was Johann Bernoulli. Together with his brother Jakob, they were among the first to study Leibniz's publications on the calculus and to develop new results. Despite his prominent role in pure mathematics, however, Johann entered the arena of mechanics quite late. In the context of the priority dispute over the invention of the calculus, he seized upon the inverse problem of central forces as a further element in his attack on the English. In corollary 1 to proposition 10 of the *Principia*, book I, Newton stated that if a body moves along an ellipse, the force towards the centre is proportional to the distance, and conversely. In propositions 11, 12, and 13 he proved that if a body moves along an ellipse, a hyperbola, or a parabola, respectively, the force towards the focus is inversely proportional to the square of the distance. In corollary 1 to proposition 13 he stated that also the inverse of the previous three theorems was true, though without proving it. At the very beginning of the work for a second edition of the *Principia*, in 1709, Newton wrote to Rogers Cotes about the need to add a few lines at the end of corollary 1 to proposition 13, in order to prove the inverse problem. The additional outline of the proof, which appeared in the 1713 edition, reads:[33]

---

[32] J. Riccati, 'Risposta ad alcune opposizioni fatte del Sig. Giovanni Bernulli', *GLI, 19*, 1714, pp. 185–210, on p. 187 (the rather free translation is mine): 'Tutta dunque la disputa verserà, non sopra il problema sciolto, che non é soggetto ad opposizione; ma sopra la maniera di scioglierlo: conforme al genio di questo delicatissimo secolo, che avendo insieme congiunte l'Analisi, e la Metafisica, non fa tanto conto della verità ritrovata, quanto del metodo, con cui si scopre, e si manifesta.'

[33] *NC, 5*, p. 5, 11 Oct. 1709, Newton to Cotes. *Principia*, transl. by Motte and Cajori, p. 61. The reviewer of the second edition in the *Acta Eruditorum* missed this point. The editors of the *Acta* wrote the names of the authors into the volumes in their possession. The year 1714 (March, pp. 131–42) of the set at the University Library in Leipzig indicates Christian Wolff as the author. E. Ravier in *Bibliographie des Oeuvres de Leibniz* (Paris, 1937, reprinted in 1966), pp. 90–1, states that after 1705 Leibniz and Wolff worked together on several reviews. The antinewtonian context of Johann's work emerges clearly from a letter to Leibniz of 12 Aug. 1710, *LMG, 3*, pp. 853–4.

For the focus, the point of contact, and the position of the tangent, being given, a conic section may be described, which at that point shall have a given curvature. But the curvature is given from the centripetal force and velocity of the body being given; and two orbits, touching one the other, cannot be described by the same centripetal force and the same velocity.

This proof relies on the direct problem: Newton starts by trying to establish a possible solution with a conic section, and later provides a unicity requirement implying that the curve thus selected is the only possible one.

In 1710, thus before the appearance of the second edition, this was an ideal topic for Johann Bernoulli: unlike cometography or lunar theory, which required specific skills in astronomy, the inverse problem of central forces involved analytic skills with quadratures, one of the fields where he was strongest. Johann convincingly stated the need to prove the inverse theorem with a counter-example: if a body moves along a logarithmic spiral, central attraction is inversely proportional to the third power of the distance from the centre. However, starting from a force decreasing in this fashion, it does not follow that the trajectory will be a logarithmic spiral, because other curves, such as the hyperbolic spiral, can be described too. Hence Johann's statement that although Newton's claims were correct, they had not been demonstrated in the first edition of the *Principia*. Johann's solution was published in the Paris Mémoires for 1710, together with those by Hermann and Varignon. The problem was conceived in purely mathematical terms, with no implications on physics or mechanics. Their solutions and Johann Bernoulli's subsequent polemics with Newton's champion John Keill and with Hermann highlight a number of problems in the practice of early eighteenth-century mechanics. These problems hinged on the process of algebraization of mechanics. Newton's cumbersome geometical treatment of central forces in propositions 39–41 of book I was rendered into analytic form and developed. This process is related to several issues, such as the relationships between mathematics and its physical interpretations, the problem of what constitutes a satisfactory solution in relation to geometric versus algebraic techniques, and the notion of adequate proof in relation to auxiliary lemmas and constant factors. I discuss these problems in turn.

Regardless of their beliefs concerning the cause of gravity and vortices, mathematicians employed accelerations and continuous curves rather than uniform motions and infinitangular polygons. Thus the correlation between mathematical representations and physical interpretations established by Leibniz collapsed.[34] A particularly interesting case

[34] Hermann, however, mixed the polygonal representation with accelerated motion, as Eric Aiton had pointed out in 'The Inverse Problem of Central Forces', *AS*, *20*, 1964, pp. 81–99, on p. 94, n. 62.

can be found in a memoir by Pierre Varignon, who provided several solutions to the inverse problem involving different progressions of the variables. On the hypothesis that ydx is constant, where y is the radius and dx the height of the infinitesimal triangle swept out by the radius in the time dt, he started from an expression similar to Leibniz's equation of paracentric motion, perfectly analogous to the equation we have seen in the previous section.[35] Leibniz provided his mathematical formulation together with an intepretation in mechanics involving centrifugal conatus and the solicitation of gravity, and a physical account in terms of vortices. Paradoxically, Varignon used Leibniz's formulation, originally intended against Newton, in a purely mathematical fashion: neither Leibniz's theory of centrifugal conatus, nor his physical interpretation were given any consideration. This specific case seems to point to a separation of mathematics from its immediate physical interpretation and thus to a collapse of the Keplerian programme. However, the issue is considerably more complex and requires a broader investigation covering several fields over a longer period. Elasticity, for example, is a promising area for investigating the relationships between mathematical modelling and physical interpretations. Johann Bernoulli is a particularly interesting case: he followed and developed the Leibnizian calculus; moreover, philosophically he was in many respects close to Leibniz and believed in the conservation of living force and vortices of subtle matter. In the 1730s he tried to develop a vortex theory of gravity in several essays presented to the Paris Academy, which awarded him two prizes.[36] Although the specific form of the Keplerian programme defended by Leibniz did collapse, surely the same cannot be said for the more general spirit of his attempts. These preliminary observations set the agenda for a broader study of the status of mathematical representations in the Enlightenment.

The second issue concerns the notion of 'solution' of a mathematical question. The controversy between Johann Bernoulli and Keill highlights the disagreement about this point. As we have seen in Section 3.3, representation of curves in the late seventeenth and early eighteenth centuries was based on a plurality of methods. A curve was known not only or necessarily through its equation, but also through its properties or method of construction. With the introduction of the infinitesimal calculus the problem was further complicated by the presence of

[35] P. Varignon, 'Des forces centrales inverses', *MASP,* 1710, pp. 533–44; *LMG, 4,* pp. 170–4, Varignon to Leibniz, 4 Dec. 1710. In his reply on 17 Febr. 1711, pp. 174–6, Leibniz refers to the utility of the calculation in the study of lunar motion.

[36] Aiton, *Vortex Theory,* pp. 214–19 and 228–35. On elasticity compare Truesdell, *Euler, Opera Omnia,* II, *11* (second part) and *12.* J. Bernoulli, *Nouvelles pensées sur le système de M. Descartes* (Paris, 1730) = *JBO, 3,* pp. 131–73; *Essai d'une nouvelle physique céleste* (Paris, 1735) = *JBO, 3,* pp. 261–364.

differential equations: a solution could be such an equation, 'granted the quadratures', or its integral. If the integral was transcendent, the problem could be said to have no 'solution'. Since the inverse problem of central forces consists in finding the curve traversed by a body in special conditions, it is not surprising that the issue of representation of curves affected the notion of solution. Curious equivocations emerged from a passage by Jakob Hermann. On the same page of an article in the Venetian *Giornale de' Letterati*, he stated that probably the general solution to the problem of central forces will never be found, and that Newton had given an 'erudita soluzione' to it. Giuseppe Verzaglia, a patrician from Cesena who had studied mathematics in Basel with Johann Bernoulli, started a controversy with Hermann involving, among other things, the notion of solution and the acceptability of transcendent or mechanical curves.[37]

In the *Principia*, proposition 41 of book I, Newton provided a general geometrical method for finding the curve given an arbitrary central force. The proposition states: 'Supposing a centripetal force of any kind, and *granting the quadratures of curvilinear figures*; it is required to find as well the curve in which bodies will move, as the times of their motions in the curves found' (my emphasis). At first sight Newton's solution appears to be very general. On second reflections, however, one may wonder whether it is a satisfactory solution at all, since in practice the identification of the trajectory of the orbiting body was not provided.[38] From proposition 41 it was by no means clear that if the force is inversely proportional to the square of the distance, the only possible trajectories are conic sections. Indeed, Newton's own attempt at the inverse problem was not based on proposition 41. This issue attracted the attention of Continental mathematicians: Johann Bernoulli in particular attained the general equation of a conic from the inverse-square law by integrating a first-order differential equation. His proof begins with a preliminary lemma and then moves on to the differential equation of the orbit. The method he employed was an algebraic version of Newton's geometrical treatment in propositions 40 and 41, book I. Their equations can be shown to be 'the same' if one considers that Newton's A corresponds to

[37] J. Hermann, 'Metodo d'investigare l'orbite dei pianeti', *GLI*, 2, 1710,p. 460; G. Verzaglia 'Modo di trovare l'orbita, che descrivono i pianeti, qualunque siasi la loro forza chiamata centrale', *GLI*, 3, 1710, pp. 495–510. J. Hermann, 'Soluzione generale del problema inverso delle forze centrali', *GLI*, 5, 1711, pp. 312–35. G. Verzaglia, 'Considerazioni sopra l'articolo nel quale si tratta del problema inverso delle forze centrali nel voto', *GLI*, 6, 1711, pp. 411–40. J. Hermann, 'Riflessioni geometriche', *GLI*, 7, pp. 173–229. Verzaglia, who had a paper rejected by the Venetian *Giornale*, published a book on the issue, *Esame delle riflessioni geometriche pubblicate da un ultramontano professore in Italia* (Bologna, 1714), esp. p. 22.

[38] Newton, *Principia*, transl. Motte and Cajori, p. 130; *NMW*, 6, pp. 336–51.

Bernoulli's x, IN to dx, CX to a, YX to dz, etc. Trivial as this translation may appear, its advantage can already be grasped: Newton's notation relies entirely on the geometrical figures which one must look at at each stage, and hides the difference between variables and constants. By contrast, this difference is immediately clear in Bernoulli's work: by looking at his equation one can assess the difficulties of the integration, devise a suitable strategy, think of appropriate substitutions of variables. Hence Johann's claim that Newton's proof was 'trop embarassée' may have been ungenerous, but was not completely unjustified. The two 'equivalent' equations differed in another fundamental respect: Newton's represented the completion of his proof, Bernoulli's was the starting point of an original and skilful development from the *Principia*. By means of a series of substitutions the Swiss mathematician could attain the equation of a conic section depending on a parameter whose values identified ellipses, parabolas, or hyperbolas. Johann went even further and provided an alternative solution based on a second-order differential equation. Also in this case he was able to find the integral, this time for different progressions of the variables.[39]

The ensuing controversy with Keill highlights the conflict between two mathematical styles. Keill claimed that Bernoulli had merely translated Newton's theorem into the language of the calculus, and that the two texts differed as the Latin and Greek versions of the same passage. When Bernoulli replied that he had provided an analytical solution showing that the only possible curves in the hypothesis of central forces inversely proportional to the square of the distance were conic sections, Keill answered that Newton had provided a general solution, which his rival had applied only to a special case. Further, he claimed that Newton's solution was simpler, being three lines long, while Bernoulli had employed seven pages. However, as mathematicians knew well, the special case was not a trivial corollary of the general one: Bernoulli's integration was indeed magisterial.[40]

We move now to the third issue, namely the notion of adequate proof. Newton's own proof, which I have quoted above, relies on proposition 17, book 1, which states: 'Supposing the centripetal force to be inversely

---

[39] J. Bernoulli, 'Solution du problème inverse des forces centrales', *MASP*, 1710, pp. 521–33 = *JBO*, *1*, pp. 470–80; the equation corresponding to Newton's is on p. 475, lines 8–9. It is worth recalling that unlike first-order differential equations, higher-order equations depend on the progression of the variable: see Bos, 'Differentials', pp. 26, 29–30, and 35.

[40] J. Keill, 'De inverso problemate virium centripetarum', *PT*, *29*, 1714, pp. 91–111; 'Défense du Chevalier Newton', *Journal Litéraire*, *8*, 1716, pp. 418–33, esp. p. 420; 'Lettre de M. J. Keill', *Journal Litéraire*, *10*, 1719, pp. 261–87. J. Bernoulli was the author of the anonymous 'Epistola pro eminente mathematico ... contra quendam ex Anglia antagonistam scripta', *AE*, July 1716, pp. 296–315.

proportional to the squares of the distances of places from the centre, and that the absolute value of the force is known; it is required to determine the line which a body will describe that is let go from a given place with a given velocity in the direction of a given right line'. Newton's intention in that proposition was not to prove the inverse problem, since in the construction he assumed that the curve is a conic section, but only to provide the scale or size of the conic. However, his proof could be modified in such a way that the inverse problem results as a corollary. In 1716 Keill claimed that proposition 17 contained a demonstration of the inverse theorem. In his reply through his student Johann Kruse, however, Johann Bernoulli objected that proposition 17 assumes the result rather than proving it. Surprisingly Johann Bernoulli's most talented pupil and possibly the most gifted mathematician of the Enlightenment sided with Keill: Leonhard Euler, in his 1736 *Mechanica*, claimed that the inverse problem of central forces could be solved on the basis of proposition 17, book I.[41]

The change in perception of mathematical notions can be grasped by comparing proposition 41 in the *Principia* with a modern solution within 'Newtonian' mechanics. While Newton's geometric treatment was based on the ratio between variable areas, modern formulations are based on constant parameters such as energy and angular momentum: the importance of constant factors has grown considerably in the way a problem is conceived, tackled, and solved. The role of constants emerged in the controversy between Johann Bernoulli and Jakob Hermann. The proof provided by the latter involved two integrations: the first, however, was carried out without adding the arbitrary constant factor. Johann promptly pointed out that this omission affected the rigour of the demonstration, since it was conceivable to suspect that curves which are not conic sections might have been excluded. His article in the same 1710 issue of the Paris *Mémoires* starts from an improved version of Hermann's proof, where Johann takes into account the integration

---

[41] J. Keill, 'Défence du Chevalier Newton', p. 427. J. H. Kruse, 'Responsio ad Cl. Viri Johannis Keil', AE, Oct. 1718, pp. 454–66, on p. 461. My understanding of these events differs somewhat from the account in *NMW*, 6, pp. 146–9, n. 124 and pp. 349–50, n. 174. L. Euler, *Mechanica* (St. Petersburg, 1736) = *Opera Omnia*, II, *1*, pp. 221–2; see also J. B. Biot's review of Edleston, *Correspondence*, *Journal des Savants*, 1852, p. 532. E. J. Aiton, 'The contributions by Isaac Newton, Johann Bernoulli and Jakob Hermann to the inverse problem of central forces', *SL*, Sonderheft *17*, 1987, pp. 47–58. In recent years there have been lively debates as to the adequacy of Newton's proof of the inverse problem. In this rapidly growing literature see: R. Weinstock, 'Dismantling a Centuries-old Myth: Newton's *Principia* and Inverse-square Orbits', *American Journal of Physics*, *50*, 1982, pp. 610–17; 'Long-buried Dismantling of a Centuries-old Myth: Newton's *Principia* and Inverse-square Orbits', ib., *57*, 1989, pp. 846–9. E. J. Aiton, 'The solution to the inverse problem of central forces in Newton's *Principia*', *AIHS*, *38*, 1988, pp. 271–6. B. H. Pourciau, 'On Newton's proof that Inverse-Square Orbits must be Conics', *AS*, *48*, 1991, pp. 159–72.

constant neglected by his colleague. Hermann had to publish an appendix in which he considered constant factors in the first integration. A few years later Jacopo Riccati, countering Johann Bernoulli's further accusation that Hermann's solution was artificial and *ad hoc*, attained the same integral by means of a general elegant technique involving the substitution of variable. This brief example testifies to the specific skills involved in the application of the differential calculus and in the transition from geometry to algebra. The inverse problem marks a significant stage in the process of algebraization of mechanics.[42]

9.5 Epilogue

As we have seen in the previous chapter, in 1714 Keill published a ruthless attack on the *Tentamen* in Willem 'sGravesande's *Journal Littéraire*, claiming that Leibniz's essay was the most incomprehensible piece of philosophy which had ever appeared. Selecting from the ample material provided by Newton, the Savilian Professor of Astronomy focused on the issues of second-order infinitesimals and centrifugal force. Keill accused Leibniz of incompetence concerning the first issue because of the mistake by a factor of two in the calculation of centrifugal conatus. Turning to Leibniz's 1706 corrections, Keill objected to the polygonal method of calculation employed by Leibniz. Unlike Varignon, however, who accepted that the polygonal method leads to the correct result, Keill argued that the force produces a continuous deflection which cannot be represented with straight lines.[43] Concerning mechanics, Keill claimed that the outward tendency or centrifugal force is clearly related to the inertia of a body; since centrifugal force is nothing but the body's reaction to the force which makes it deviate from the rectilinear path, Keill concluded that centripetal and centrifugal forces are always equal and opposite. As we have seen, these views had been provided by Newton, and were in agreement with the ideas expressed by Huygens, Varignon, and in slightly different forms by the mathematical community at that time.

---

[42] J. Hermann, 'Metodo d'investigare l'orbite dei pianeti', *GLI*, 2, 1710, pp. 447–67, on p. 459f.; the problem had been posed by Johann Bernoulli in a letter in which he had not given the solution; J. Hermann, 'Breve aggiunta', *GLI*, 6, 1711, pp. 441–9, and 'Extrait d'une lettre', *MASP*, 1710, pp. 519–21. J. Bernoulli, 'Solution du problème', *MASP*, 1710, pp. 521–33. J. Riccati, 'Risposta ad alcune opposizioni'; N. Bernoulli, 'Annotazioni sopra lo Schediasma del Sig. Conte Jacopo Riccati', *GLI*, 20, 1715, pp. 316–51.

[43] Keill's mathematical criticisms are discussed in Aiton, 'Mathematical basis', pp. 223–4.

Leibniz's champion Christian Wolff sent him Keill's article and urged his master to reply, but Leibniz was adamant in his refusal:[44]

I wish to have no dealings with a man of that sort ... Those who look into the paper on celestial motions will see that there is no error there, only a phrase concerning centrifugal force which I later rendered more suitable. For what I have demonstrated, that the Keplerian ellipses result from a combination of the harmonic circulation with gravity, certainly is true and will remain so.

Significantly, Leibniz omitted any reference to Kepler's third law, and had nothing to say about the relations between centripetal and centrifugal forces. As we have seen in Section 7.6, Leibniz himself had come to doubt his own theory of centrifugal force already in 1690. In the correspondence with Samuel Clarke in 1715–16, and indeed already in the *Specimen Dynamicum* of 1695, Leibniz had abandoned the project of presenting a theory capable of competing with Newton's. Despite his subtle philosophical and theological objections, in the eighteenth century Leibniz had left Newton master of celestial mechanics.

This chapter presents a picture of Newtonian and Leibnizian celestial mechanics as they were conceived and practised around 1700. The rival intellectual constructions in the *Tentamen* and in the relevant portions of the *Principia* were perceived differently from what a modern reader would a priori expect. Leibniz's difficulties in reconciling Kepler's laws with the harmonic vortex, for example, are fully consonant with our judgement. The interpretation of centrifugal force given by Newton, Huygens, Varignon, and several other mathematicians at the beginning of the eighteenth century, however, is not; likewise, the belief that centrifugal force could not decrease following the inverse-cube law, is alien to our perception. And yet this interpretation undermined the consistency of the *Tentamen*. In this regard mechanics and the three laws of motion played an important role, though again not in a way a modern reader would a priori expect. Newton's analysis of orbital motion depended on the first law for the body's tendency to fly away along the tangent, on the second for the deviation from the tangent, and on the third for the explanation of centrifugal force and the reciprocity of attraction. Thus if Newton's account did not provide physical or causal explanations for gravity, it incorporated the laws of mechanics more satisfactorily than Leibniz's.

Addressing the question of the reception of the *Tentamen* with respect to Newtonian celestial mechanics it is useful to establish a comparison with a major debate of the same period, namely the *vis viva* controversy.

[44] *NC*, 6, pp. 179–80, Wolff to Leibniz, 22 Sept. 1714; pp. 216–18, Wolff to Leibniz, 24 April 1715; pp. 222–3, Leibniz to Wolff, 7 May 1715; Aiton, 'Mathematical basis', p. 225.

Thanks to the works by Carolyn Iltis, we have come to appreciate the interplay of rational and social explanations alike. Iltis has stressed the lack of mutual understanding of rival communities entrenched in the defence of their respective positions.[45] As a consequence of the priority dispute over the invention of the calculus, the mathematical community was radically polarized, and yet, despite differences in mathematical notions and physical theories, we find considerable common ground between Newtonians and Leibnizians. Keill understood perfectly well that the *Tentamen* was based on the polygonal representation of curves; Johann Bernoulli worked in the framework of the *Principia* rather than of the *Tentamen*; Varignon, occupying a crucial middle position between the opposing camps, grasped the subtle differences between the Newtonian and Leibnizian approaches, though he followed more decidedly the *Principia* and relegated the *Tentamen* to a mathematical exercise.

These observations on the reception of Newtonian and Leibnizian theories lead to some final considerations. At the beginning of the eighteenth century the interplay of mathematics, mechanics, and physics, acted not simply statically in the purely intellectual evaluation of the pros and cons of each theory, but also dynamically in the practice of the new problems and solutions which were emerging. The *Principia* constituted an extraordinarily fertile field for further researches on the inverse problem of central forces, cometography, the shape of rotating bodies, perturbation theory, lunar theory, and tides. Although the *Tentamen* was known to the most prominent mathematicians at the beginning of the eighteenth century, it failed to elicit a response and to serve as a basis for further investigations, at least with regard to the reconciliation of mathematics and physics. Johann Bernoulli's decision not to develop the *Tentamen* despite his general agreement with Leibniz in mathematics and physics is revealing of the difficulties he perceived in his ally's essay. Interest in it vanished with the generation following Newton and Leibniz. However, many of the researches stimulated by the *Principia* were carried out on the Continent using more and more sophisticated versions of the differential calculus.

By the beginning of the eighteenth century Leibniz's worries over priority in celestial mechanics had become less and less important. If the reasons for his defeat were mainly intellectual, the particular form of the reception and success of Newton's theory was largely shaped by the skills and interests of the scholars who had worked on the *Principia*. In the years covered by my research I have identified an astronomical

[45] C. Iltis, 'The Leibnizian-Newtonian debates: natural philosophy and social psychology', *BJHS*, *6*, 1973, pp. 343–77; 'The decline of Cartesianism in mechanics: the Leibnizian-Cartesian debates', *Isis*, *64*, 1973, pp. 356–73.

tradition in Britain and a mathematical one on the Continent. Hermann divided his *Phoronomia* into two books on the motion of solid bodies, and on the motion of fluids respectively; a third book on the system of the world, as one finds in the *Principia*, was missing. With respect to the problem of physical causes, his works, together with those of Varignon and Johann Bernoulli, mark the collapse of the specific form of the Keplerian programme defended by Leibniz and the emergence of a more pragmatic approach in which the relationships between mathematics and physical interpretations had to be redefined. At the beginning of the new century celestial mechanics was already showing autonomous features from the works by Newton and Leibniz, despite the central role played by the two contenders.

Appendix 1
TEXTS AND COMMENTARIES

Before presenting the most important and representative texts, I wish to specify my general editorial conventions. Each manuscript or group of manuscripts is introduced by a brief characterization including the analysis of papers and watermarks, a description of the documents, my dating criteria and other features relevant to the intelligibility of the material. All manuscripts in this selection were private drafts and were not written with the intention of publication. Each text is reproduced in its entirety. My aims differ from those of the *Akademie* edition, hence my decision to adopt different editorial conventions. These conventions are specifically devised for a limited selection of essays rather than for a complete edition of approximately one hundred thousand manuscripts.

Initially my transcriptions included all passages crossed out by Leibniz, but later I decided to omit slips or stylistic changes, whereas portions showing a significant change of thought or which may include interesting information are retained. If these are short, they are enclosed in angled brackets and marked with a horizontal bar; longer passages are indicated with two vertical bars in the margin. This system does not reproduce all stages of composition of a text; however, it has two desirable features. First, full justice is done to important cancellations which appear in the text rather than in the apparatus; secondly, the elaborate calculations in many of those cancellations are not printed in small character at the bottom of the page. Interpolations are indicated between corners, thus ⌊. . .⌋. As with cancellations, I have tried to select those variants which in my opinion contain useful information, rather than aiming at completeness. For reasons of clarity I occasionally insert in angled brackets punctuation marks and words omitted by Leibniz; square brackets and dots are used by him and are not a part of the editorial apparatus unless otherwise specified. The symbol ///// denotes a damaged or missing portion. Capitalization is set according to modern criteria; accents are neglected. Contractions, like a horizontal bar over a letter such as '$\bar{u}$' (um), or '&', '$\mathcal{J}$' (con), '$\mathcal{L}$' (pro) and $\overline{pt}$ (potest), are expanded; abbreviations, like 'triang.', are kept as in the original. Concerning mathematical notation, indices are lowered and fractions have been written on one line when convenient; $a/b \cdot c$ means $a/(bc)$, whereas $a/b\ c$ means $(a/b)c$. I recall that a horizontal bar over an expression is equivalent to its enclosure in brackets; commas are

used as separation symbols, both at the beginning (reversed) and at the
end of an expression.

1   Leibniz's *Notes* to the *Principia Mathematica*

LH 35, 10, 7, f. 32–5 (compare Figs. (a) and (b))

The manuscripts reproduced here consist of two sheets folded in quarto
which appear to have been written in continuous succession. The first
sheet bears no watermark, the second has one identical with that
numbered 695 in the catalogue at the NLB, letters 'MR'. Leibniz used
the same kind of paper in Vienna in 1688. The analysis of orbital motion
is considerably less sophisticated than in the *Tentamen*, and the *circulatio
harmonica* or *motus paracentricus* is not mentioned. This suggests that
the present manuscripts were written in Vienna during the autumn of
1688.

Leibniz's observations are often enclosed within brackets and
asterisks; sometimes he provided the reference to the portion of the
Principia to which his notes refer. When this is not the case, I indicate in
the margin in angled brackets the portion of the text concerned.
Whenever this is possible, these references are used in the commentary
as a substitution for line-numbers. Not all passages require explanation,
therefore indications in the margin do not necessarily imply that there is
a corresponding section in my commentary.

My aim in the commentary is to explain Leibniz's fragmentary and
extemporaneous observations considering them together with *Marginalia*, *Excerpts*, and with relevant texts such as *Dynamica* and *Specimen
Dynamicum*. I also cite some of the relevant secondary material on
Newton.

Neutoni philosophiae naturalis principia mathematica

⟨Def. 1⟩   Densitatem materiae ponderi proportionalem esse reperi per
experimenta pendulorum accuratissime instituta uti posthac
docebitur. Def. 1 (pag. 1). Scilicet se non habere rationem
fluidi interlabentis notat.   5

⟨Def. 3⟩   Materiae vis insita, seu vis inertiae, est potentia resistendi
seu perseverandi in suo statu (p. 2) eaque semper proportionalis
est suo corpori, licet quiescat, tamen impetum habet, quatenus

⟨Def. 4⟩   corporis in ipsum agentis statum mutare conatur. Corpus quod
semel novam vim impressam accepit, postea illam retinet per   10
solam vim inertiae.

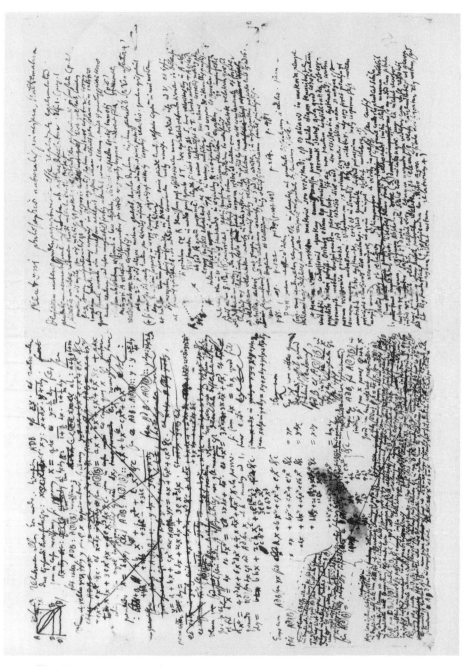

Fig. (a)   LH 35, 10, 7, f. 32r.–33v., Leibniz's *Notes* to Newton's *Principia*.

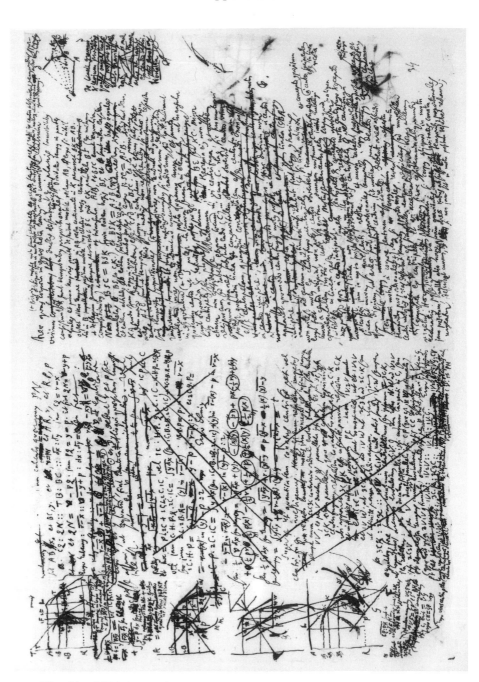

Fig. (b) LH 35, 10, 7, f. 34r.–35v., Leibniz's *Notes* to Newton's *Principia*.

⟨Def. 6⟩   Unum centrum (ad instar magnetis) fortius attrahit quam
⟨Def. 7⟩   alterum. Major est gravitas propiorum terrae.
                  Motum absolutum aestimat a loco absoluto, scilicet respectu
                  spatii immoti (p. 6).   15
⟨Pag. 9   Motus veri et relativi distinguuntur per vires, v.g. quibus
     11⟩   corpora tenderent fila, seu conatus recedendi ab axe motus
                  circularis.
                  Lex 2. Si vis aliqua motum generet, dupla duplum, ⟨tripla⟩
                  triplum generabit. (+Non assentior+) (+Si impressio sit   20   O2
                  conatus tantum seu celeritas infinite parva, admitti,
                  opinor poterit. +)
                  Lex 3. Actioni corporis in corpus semper aequalis est
                  alterius corporis reactio; quantum quis premit et trahit,
                  tantum premitur et trahitur.   25
⟨Cor. 5⟩   Si moveatur navis uniformiter in directum, omnia perinde
                  erunt, respectu corporum in navi motorum ac si navis
                  quiesceret, ob additionem communem motus navis.
⟨Pag. 21⟩   Si pendulum ex R demissum post oscillationem unam
⟨fig. 1⟩   redeat usque ad V, et ipsius RV sumatur in medio   30
                                 pars quarta ST, haec exhibebit retardationem in

                                 C descensu ab S ad A quum proxime, itaque si cadat
                                 corpus ab S, tunc velocitas eius in A absque
                                 errore sensibili tanta erit, ac si in vacuo
                  descendisset a T. Chorda autem TA celeritatem descensus in   35
⟨Pag. 22⟩   arcu TA in vacuo, metitur. Hac ratione correctis motibus
                  pendulorum, rem in ⟨pedum⟩ 10 pendulis tentando, in
                  aequalibus et inaequalibus, sic ut corpora de intervallis
                  amplissimis pedum 8, 12, 16 concurrerent, reperi semper
                  actionem reactioni aequalem, et servatam summam vel   40
                  differentiam motus, ita ut nunquam fuerit error trium
⟨Pag. 23⟩   digitorum in mensuris. In imperfecte elasticis vis
                  elastica imperfecta, facit corpora redire ab invicem cum
                  velocitate relativa, quae sit ad relativam velocitatem
                  concursus in data ratione. Idque in pilis ex lana arte   45
                  conglomerata et fortiter constricta, tentavi, erat semper
                  in variis casibus velocitas relativa separationis, ad vim
                  relativam concursus, ut 5 ad 9 circiter; in vitreis autem
                  proportio erat 15 ad 16.
⟨Pag. 20⟩   Vocat Wrennum, Hugenium et Wallisium huius aetatis   50
                  geometrarum facile principes.

                  p.85 p.105 p.122 p.139 (p.141 .142) p.264 p.481
                  p.411 vacu⟨u⟩m necessario datur

                  (+NB vim centripetam reciproce esse in planetis, ut quadrata
                  distantiarum a centro. Jam radiorum lucis densitates sunt   55
                  autem reciproce ut quadrata distantiarum a sole. +)

(+ praeter ⟨vim⟩ inertiam materiae seu resistendi, est et
aliä vis in materia, nempe unitatis seu cohaesionis, quatenus
omnis massa pro uno fluido aliqua tenacitate praedito haberi
potest. Oriturque etiam ex principio servandi status, quia   60
quod cohaesionem abrumpit, motum conspirantem perturbat. Itaque
non tantum consideranda est corporum resistentia privata, sed
et cohaesio sive unio, seu resistentia systematis; contra
omnem magnam mutationem. Videndum an ex privata sequatur
publica, nam si magna est mutatio in toto systemate, erit et   65
in singulis. Atque ita videtur, nec opus foret peculiari
principio cohaesionis sive unitatis. Illud generale sufficit
principium motus corporum sese invicem magis magisque
accomodare, ut quam minime obstent, mutenturque. +)

⟨Pag. 9⟩   Motus verus et relativus distinguuntur a viribus impressis   70
seu causis. Ex fune contorto pendeat situla, impleatur aqua,
libertas detur funi se relaxandi₍,₎ initio situla magnum habebit
motum relativum, exiguum absolutum, (𝒞₁)⟨donec vi impressa⟩(+ an quod
cum situla initio non agitatur satis +) donec motu situlae communicato
aquae, cum ea incipit revolvi, tunc decrescit relativus, crescit   75
verus, et aqua quando non videtur cum vase revolvi sed eundem cum eo
locum servat, movetur realiter, et elevat se versus extremitates
ut concava sit eius superficies (+ Experiendum an initio aqua in
situla non sequatur eius motum, sed magnum habeat motum
relativum +).   80

⟨v.⟩   Actio semper contraria est reactioni, et corporum duorum
⟨Lex 3⟩   actiones inter se semper sunt aequales et in partes contrarias
diriguntur. Siquis digito lapidem premit, premitur digitus a
lapide, si equus lapidem trahit lapis equum retrahit, nam funis
utrinque tensus suo relaxandi conatu aequaliter equum agit versus   85
lapidem, et lapidem versus equum tantumque impedit progressum
unius, quantum promovet progressum alterius. Hinc mutationes
velocitatum in contrarias partes factae, sunt corporibus
reciproce proportionales.

⟨Cor. 1⟩   Corpus si agatur vi AB, et AC ibit vi AD (malo dicere   90
conatu). (+ Sed non explicat quomodo moveatur corpus, si
debeat ire vi AB, AE, AC. Nempe ibit per centrum
gravitatis. Operae pretium erit, investigare qua linea
moveatur corpus, si simul conetur in quolibet radio
ex A ad curvam BEC ducto. ⟨Patet ex principiis meis,   95

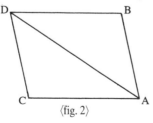

⟨fig. 2⟩   ⟨fig. 3⟩

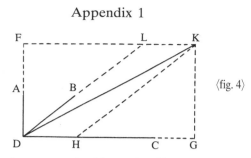

⟨fig. 4⟩

~~conatum esse in recta angulos bisecantium bisecante.~~
~~Nam quae recta basin trianguli per verticem bisecat~~
~~etiam angulum verticis bisecat.⟩~~ Ostendi scilicet
alia scheda separata directionem ex omnibus compositam
tendere ad centrum gravitatis curvae, aliaque pulchra   100
ea occasione inveni.

Dubito autem an nostris demonstrationibus intellectis
defendi possit ratiocinatio Neutoni quod mobile ex A, tendens
ad D, aeque ⟨ve⟩lociter ad DB perveniat ut ante, quia vis in AC
et parallelis nihil immutet. Nescio enim an hoc sit necesse,   105
neque apparet quomodo hoc applicabile ad compositionem
ex tribus conatibus. Illud considerandum, si tres
conatus DA, DB, DC compositi dent lineam DK, posse duci
per K parallelam ipsi DC occurrentem ipsi DA in F, et
aliam per K parallelam ipsi DA, occurrentem ipsi DC in G.   110
Idemque ergo est ac si duo essent conatus DF et DG.
Similiter si KH parallela DB, et KL parall. DH, et H sit
in DC, et L in DB si opus productis, perinde erit ac
si conatus esset compositus in DH, et DL. Sed ex conatibus
in DA, et DB sumtis non potest componi conatus in DK quia   115
non cadit DK inter DA et DB. At in extremis res semper
succedit. +)

⟨Cor. 3⟩ Quantitas motus quae colligitur capiendo summam motuum, in
easdem partes et differentiam ad contrarias, manet eadem
(rectius dicerem: summam progressus esse eandem, non motus.)   120
Nam cum actioni aequalis sit reactio, aequales in motibus
efficiunt mutationes versus contrarias partes. Ergo summa ad
easdem partes eadem manet quae prius (+ ostendendum est
prius ⟨aestimandam progressus vim a quantitate⟩ vim
actionis et reactionis se per quantitates progressuum   125
vel regressuum exerere, revera se exerit per quantitatem
recessus a se invicem. +)

⟨r.⟩   Neutoni lib. 1. De motu corporum sect. 1 De methodo rationum
primarum et ultimarum, cuius ope sequentia demonstrantur.
Lemma 1. Quantitates ut et quantitatum rationes (+ hae etiam   130
sunt quantitates +) quae ad aequalitatem dato tempore
constanter tendunt, (+ an aequabiliter? +) et eo pacto
propius ad se invicem accedere possunt, quam pro data quavis

differentia, fiunt ultimo aequales. Si negas sit earum
ultima differentia D, ergo non possunt propius ad aequalitatem   135
O 2   accedere quam pro data differentia D contra hypothesin.
(+ Dubitari potest an sit aliqua ultima differentia +).

Lemma 2 (+ hoc ita meo more effero +): parallelogramma curvae
particulis, si latitudine (lateribus) minuantur, numero
augeantur in infinitum₍,₎ figura inscripta, circumscripta et   140
curvilinea sunt aequales, nam differentia inscriptae et
circumscriptae figurae est summa parallelogrammorum, quae
minor quavis data fit. Ergo adhuc multo minor differentia
inter inscriptam vel circumscriptam et curvilineam.
(+ idem applicari potest etiam ad polygona   145

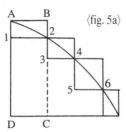

A   B   ⟨fig. 5a⟩

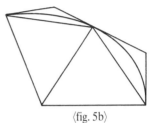

⟨fig. 5b⟩

ubi differentiae inscriptae et circumscriptae
figurae est summa triangulorum. Supponit
Neutonus in demonstratione bases
parallelogrammorum 12, 34, 56, esse
aequales et ita summam omnium esse aequalem parallelogrammo   150
ABCD sub eadem basi et summa omnium altitudinum; verum
⟨Lem. 3⟩   et si sint inaequalia manet tamen eadem demonstratio,
sumendo maximam omnium basium, tanquam ea semper maneret,
quae cum adhuc sit inassignabilis, etiam eo casu summa
omnium erit infinite parva, multo magis ergo si quaedam   155
⟨Cor. 2⟩   sint minora. Idem est de summa triangulorum polygoni,
quae ⟨aequatur⟩ summae omnium chordularum seu basium
(non majori quam est curva) in maximam altitudinem,
minor est, vel aequalis huic summae in eandem altitudinem,
si ubique aequalis demonstrari etiam poterit differentiam   160
inter ambitum polygoni interni et externi minorem fieri
quavis data. ⌊(+ Video post et Neutonum hoc usum. +)⌋
Nam si angulus sit infinite obtusus, differentia inter
basin ad summam laterum infinitesimam rationem habet ad

⟨fig. 6⟩

basin, nam sit triangulum ABC, basis ⟨AC⟩   165
centris A, C radiis AB, CB describantur

arcus BD, BE, erit ED differentia baseos et summae laterum,
demittatur perpendicularis BF secans ED in F, et sit BG
normalis ad AB occurrens ipsi AC in G. Patet BG majorem
esse quam FG, ergo et quam FD. Jam BG est infinities minor   170
quam AB (alioqui infinities majus foret triangulum ABG,
quam ABC pars toto,) ergo et FD est infinities minor, quam
AB, ergo et duplum FD, seu ED. +)

    Lemma 6. Angulus comprehensus tangente unius
puncti curvae, et chorda ex illo puncto ad   175
aliud punctum continue minuitur in infinitum

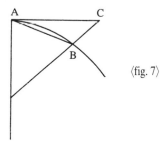

⟨fig. 7⟩

et evanescit si alterum punctum puncto
tangentis satis accedat. Hinc differentia inter tangentem
⟨et chordam evanescit⟩. Hinc ob angulum BAC infinite parvum,
⟨Lem. 7⟩   differentia inter BAC evanescet. Ergo et inter chordam   180
et arcum medium (+ multa adhuc demonstranda differentia
manet inter latera trianguli cuius angulus infinite parvus
lateribus comprehensus, sed ea est infinite parva respectu
laterum. +)

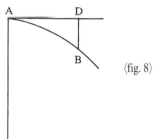

⟨fig. 8⟩

⟨Schol.⟩   Si DB sit ut AD^2, AD^3, AD^4 etc tunc primo casu ubi est ut   185
AD^2 angulus contactus est ⟨infinities⟩ ⟨aequalis circulari⟩
ejusdem generis cum circulari, sequenti ubi ut AD^3, est infinities
minor circulari, et sequenti AD^4, adhuc infinities minor ipso
AD^3 et ita porro in infinitum ⟨(+ si ut AD erit infinities
minor circulari. Quid si medium inter AD et AD^2 ut $AD^{3:2}$⟩.   190
Sed si ut AD, $AD^{3:2}$, $AD^{4:3}$, etc erit infinities major
circulari, et sequens semper infinities major praecedente
(+ non probat, pendet ex meis principiis de aestimando angulo

contactus. Videndum an ang. contactus infinite parvus conferendus
cum eo qui restat si ⟨ordinarium⟩ circularem subducas ab hoc   195
ipso, qui ei censetur homogeneus, seu circuli v.g. ab angulo
contactus parabolae in vertice. +)(+ Hinc videtur rationi
consentaneum metiri genera angulorum contactus loco circuli, et
ovalium per parabolas et paraboloeides₍.₎ videndum quod de
hyperboloeidibus. Nimirum videndum an inter curvam et rectam   200
tangentem possit duci parabola, et si non videndum an
paraboloeides aliqua, et quae parabolae proxima. Forte et
inveniri poterunt ovales, seu quasi circuli his
paraboloeidibus respondentes, ut ⟨x³ + y³ = a³⟩
$ax^2 - x^3 = y^3$ ubi posito $x = 0$ coincidit cum $ax^2 = y^3$. +)   205

     Lemma 9 Si AB, A(B) inassignabiles, erunt triangula ADB in
ratione laterum duplicata (+ nempe intelligendum est

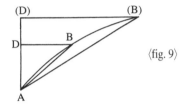

⟨fig. 9⟩

triangula esse similia, adeoque angulum BA(B) ad angulum
DAB habere rationem infinite parvam, sed non videtur
verum, quia in triangulo AB(B) latus B(B) assignabilem habet   210
rationem ad AB vel A(B) adeoque ad AD, et DB. Licebit investigare
⟨v.⟩   relationem illam hoc modo. Triangulum ADB est
$xy/2$, et ratio inter duo ejusmodi triangula est:

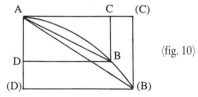

⟨fig. 10⟩

$xy : \overline{x + dx} \cdot \overline{y + dy}$. Jam in casu novissimo, si
ponamus $x = a/b\,dx$ et $y = 1/m\,dy$ fiet: 215
$al/bm\,dx\,dy : \overline{1 + a/b} \cdot dx \cdot \overline{1 + 1/m} \cdot dy$. Seu fiet ratio ADB : A(D)(B)
novissima; $al : \overline{a + b} \cdot \overline{1 + m}$. Verum, ut relatio quoque ipsius y
per x exprimatur, scribamus ⟨y = ax + bx² + cx³ + ex⁴ etc et fit ADB = ax⟩
$y = a + bx + cx^2 + ex^3$ etc. fietque xy seu
ADB = $ax + bx^2 + cx^3 + ex^4$ etc. et $\overline{x + dx}$ in $\overline{y + dy}$ seu   220
A(D)(B) = $ax + bx^2 + cx^3 + ex^4$ etc $+ adx + 2bxdx + 3cx^2dx + 4ex^3dx$ etc.
jam in eo casu quo x est prima aliqua seu aeque inassignabilis
quam dx, eique homogenea, tunc ponendo $x = h\,dx$, fiet
ADB : A(D)(B) :: $a + bx + cx^2 + ex^3$ etc: $+ a \quad + bx \quad + cx^2 \quad + ex^3$ etc,
                               $+ 1ha + 2hb .. + 3hc .. + 3he$ 225

~~(nam generaliter)~~ ~~(seu ADB:A(D)(B)::y:y + x/dx dy)~~ seu
$ADB:A(D)(B)::ydx:y\ dx + x\ dy$ quod generale est etiam
eo casu quo x et dy sunt heterogenea, nam
$y = a + bx + cx^2$ etc et ADB seu $yx = ax + bx^2 + cx^3$ etc.
pono communiter $dy = b\ dx + 2cx\ dx + 3ex^2dx$.   230

Et $\overline{x + dx}$ in $\overline{y + dy}$ erit: $ax + bx^2 + cx^3$ etc $+ a\ \overline{dx} + bx\ dx + cx^2dx$ etc

$\phantom{Et \overline{x + dx} in \overline{y + dy} erit: ax + bx^2 + cx^3 etc + a\ \overline{dx} +}$ $bx\ dx + cx^2dx$ etc

$+ b\ \overline{dx}^2 + 2cx\ \overline{dx}^2 + 3ex^2\overline{dx}^2$ et rejiciendo $\overline{dx}^2$ fiet:
$A(D)(B)$ seu $\overline{x + dx}$ in $\overline{y + dy} = ax + bx^2 + cx^3$ etc $+ adx + 2bx\ dx + 3cx^2dx$
etc. Sed haec differentialibus procedit tantum
Verum hoc loco, quia ad casum respicimus quo dx et x possunt   235
fieri inter se assignabiles, non contemnenda est decurtata
differentialibus, et fit $\overline{dx} = dx$ et $\overline{dx}^2 = 2x\ dx + \overline{dx}^2$ et
$\overline{dx}^3 = 3x^2dx + 3x\ \overline{dx}^2 + \overline{dx}^3$, et
$\overline{dx}^4 = 4x^3dx + 6x^2\overline{dx}^2 + 4x\ \overline{dx}^3 + \overline{dx}^4$, et ita porro.   240
Si jam $dx = hx$, quod fit quando $D(D)$ seu dx est ad
AD seu x, ut h numerus ad 1, tunc erit
$dy = bhx + 2ch\ x^2 + 3eh\ x^3$ etc.

$ch^2x^2\quad 3eh^2x^3\qquad$ jam $x + dx$ in $y + dy = xy + xdy + ydx + \overline{dxdy}$

$1eh^3x^3\qquad\qquad$ Ergo cum ADB   245

seu yx sit $ax + bx^2 + c\ x^3 + cx^4$ etc

fiet A(D)(B) | $ax + bx^2 + c\ x^3 + ex^4$ etc $= yx\qquad$ Ex his non video

$ahx + hbx^2 + ch\ x^3 + ehx^4$ etc $= ydx\qquad$ quomodo defendi

$+ bhx^2 + 2ch\ x^3 + 3ehx^4$ etc $= xdy\qquad$ possit quod

$1ch^2\qquad 3eh^2\qquad\qquad$ statuit Neutonus,   250

$1eh^3\qquad\qquad$ ADB et A(D)(B)

$+ bh^2x^2 + 2ch^2x^3 + 3eh^2x^4\quad = dxdy\qquad$ in casu initii,

$1ch^3\qquad 3eh^3\qquad\qquad$ seu cum x

$1eh^4\qquad\qquad$ ipsis dx homogenea

 est$_{(,)}$ esse ipsis   255

seu $A(D)(B) = $ | $ax +\quad bx^2 + 1\ cx^3 + 1\ ex^4$ | etc. Ergo ex ADB, fit A(D)(B)

$h +\qquad\quad 3h\qquad 4h\qquad$ in initiis, si pro x

$2h\qquad 3h^2\qquad 6h^2\qquad$ ponas $\overline{1 + hx}$ quod jam tum

$h^2\qquad 1h^3\qquad 4h^3\qquad$ poterat praevideri.

$1h^4\quad$ | Posito scilicet $h = \overline{dx}:x$,   260

 seu $D(D):AD$

Hoc loco igitur subtilitas ingeniosissimi Neutoni deliquium
aliquod passa est. Dum putat semper ultimo areas ADB in
duplicata ratione laterum AD ut asseruerat Lemmate 9 cuius
⟨Lem. 10⟩ demonstrationem perplexiorem examinare non vacat. Unde nec   265
admitti potest Lemma 10 quod inde infert tale: spatia quae
corpus urgente quacunque vi regulari describit, sunt ipso
motus initio in duplicata ratione temporum, exponendo enim
tempora per AD, velocitates per DB, erunt spatia ut ADB
(ipse pro DB, (D)(B) scribit DB, EC). Sed vel hoc ipsum   270
corollarium admonere erroris potuerat, cum aliqua initia
alterius naturae esse possint, et omnia initia certa aliqua

regula possint continuari. Adde quod interdum x et $\overline{dx}$ in
ipsis initiis manent inassignabilis rationis. ⌊Imo

⟨Lem. 11⟩ forte et hoc nihil est.⌋ At solidius est Lemma XI quod 275
primae abscissae AD sunt in duplicata ratione chordarum
AB, scilicet quando, ut ego loqui soleo circulus describi
potest, qui curvam in A osculatur₍,₎ nempe circulus per puncta
A, B, (B) transiens est ipse osculans, cuius centrum est in axe
AD(D) producto, in circulo autem AD sunt ut $\overline{AB}^2$. Haec breviter 280
meo more demonstrandi. Convincitur autem hinc ingeniosissimum
Neutonum, falsum eius esse Lemma 9, et quomodo se habeant ADB;
nam prima y sunt ut $\sqrt{2x - xx}$. Ergo prima xy sunt ut $\sqrt{2x^3 - x^4}$,
et quadrata primorum ABD sunt in composita ratione x^3 et
$\overline{2 - x}$. Videamus quid fiat parabolam substituendo pro circulo. 285
Vult etiam CB esse ultimo in duplicata ratione ipsarum AC,
sed hoc non consentit ei quod AD sunt in duplicata ipsarum
AIG, nec licet fingere parabolam quae circulum tangat in
vertice simul habere puncta B(B) duo cum eo communia adeoque
in summa sex. Unde mihi videtur nimia subtilitate sibi 290
nonnihil excidisse. Postremo omnibus consideratis suspicor
totam hanc subtilitatem inanem esse, et in ipsis initiis
progressus infinitis modis posse assignari. Videndum ubi
reali problemati haec applicabuntur.

⟨f. 34r.⟩ ⌊(+ Brevius: si mobile motu simul centripeto et concepto feratur, 295
areas radii ex centro abscindent temporibus proportionales,
vel si corpori jam in motu posito superveniat vis centripeta
perpetua, nullaque alia accedat, actio vim habet quod diximus. +)⌋

⟨Prop. 1⟩ Areae quas corpora in gyros acta radiis ad immobile centrum
virium ductis describunt et in planis immobilibus consistunt, 300
et sunt temporibus proportionales ⌊et vicissim⌋.
Dividatur tempus in partes aequales et prima

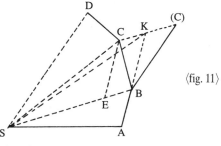

⟨fig. 11⟩

temporis parte describat mobile rectam AB,
ergo si nihil obstet altero tempore in AB
recta continuata in K percurret BK = AB. Sin 305
in B positum mobile impellatur versus
centrum, ita ut deflexum secundo tempore
percurrat BC erit CK parallela BS (ex

De lineis gravium et
levium projectitiis,
quales omnes lineae

motuum compositionibus) et triangula ASB,   <u>esse possent respectu</u>
et BSC erunt aequalia. Nam BSC = BSK, (quia   <u>centri gravium</u>   310
eadem basis BS, et sunt inter easdem   <u>dati, si modo</u>
parallelas BS et CK.) et BSK = ASB (quia   <u>sint tempora</u>
bases aequales et eadem altitudo ab eodem   <u>areis per radios</u>
scilicet vertice, S, in eandem rectam KA   <u>ex centro descriptis</u>
productam si opus in qua ambae bases)   <u>proportionalia</u>   315
ergo BSC = ASB etc. Quod erat dem.
Eodem modo demonstrabitur esse CSD = BSC, et ita porro.
Itaque areae ASD sunt ut tempora. Porro in eodem plano

⟨Cor. 1⟩ esse etiam satis patet. Hinc in mediis non resistentibus,
si areae non sint temporibus proportionales, ⟨(+ hoc est   320
~~si nullum reperiri potest punctum respectu cuius areae~~
~~sint temporibus proportionales +)⟩~~ vires non tedunt ad

⟨Cor. 2⟩ concursum radiorum. Et rursus in mediis omnibus si
arearum descriptio acceleratur vires non tendunt ad
concursum radiorum, sed inde declinant in consequentia.   325
(+ Posita quacunque curva et sumto quocunque centro ut S,
potest ita curva secari in elementa, ut triangula sint
aequalia, et definiri potest ~~⟨vis centripeta⟩~~ nisus
centripetus, seu progressus ipsurum DE seu KC. Miror
etiam non considerasse Neutonum theorema eius non esse   330
reciprocum, sed habere etiam locum in vi centrifuga,
quod apparet substituendo (C) in locum C. Itaque sic
pronuntiari potest, quamdiu linea concavitatem obvertit
centro vim esse centripetam, quamdiu convexitatem esse,
centrifugam. ⟨~~Hinc patet etiam mirabile, vim centrifugam~~   335
~~huiusmodi~~ non ~~posse a centro incipere, itemque vim centrifugam~~
~~non posse in infinitum durare imo, cavendus error.⟩ ⟨Hinc~~
~~patet centrifugum talem motum non posse a centro incipere~~
~~quod non capio.⟩~~ Hinc dubium non apparet quomodo centrifugus
motus possit a centro incipere, ~~⟨neque enim recta AB(C) ad~~   340
~~centr⟩ ⟨et semper centrifugus manere nisi alicubi habeat~~
~~flexum contrarium.⟩~~ Nam linea (C)BA non potest ad centrum
usque continuari nisi alicubi fiat flexus contrarius, ergo
fiet motus ex centrifugo centripetus. Et si linea aliqua
curva a centro incipiat, non potest centro obvertere   345
~~⟨concavitatem⟩~~ convexitatem initio. Unde sequitur omnem
motum ex centrifuga vi ⌊et concepto⌋ compositum a centro
incipientem esse in linea recta⌋. Motus autem centripetus
intelligi non potest qui a centro incipiat. ⟨~~Cum BC sit aequalis~~
~~B(C)⟩~~ Corpus quod ~~⟨motu centripeto move⟩~~ impetu centripeto   350
et concepto movetur in aliqua linea, non potest in illa
linea regredi motu centrifugo et concepto. Si corpori
jam in motu posito ⌊extra centrum constituto⌋ superveniat vis
centripeta vel centrifuga linea sit curva, si quiescenti,
recta. Et curva habebit proprietatem quam diximus, ut areae   355

sint proportionales temporibus. ⌊Corpus quod impetu composito
ex concepto et centripeto vel centrifugo movetur dicatur
<u>centrum respicere.</u>⌋ Si centrum sit mobile, tunc demto eius
motu a corpore centrum respiciente, debent areae abscindi

⟨Prop. 3⟩　　temporibus proportionales +). Itaque si corpus ⟨~~centrum~~　　　　　360
~~respiciens~~⟩ respectu centri mobilis areas describat
temporibus proportionales, revera ⟨~~non movetur motu illo~~
~~centripeto~~⟩ non est simpliciter centrum respiciens sed
movebitur vi composita ex centripeta, et vi acceleratrice
qua corpus alterum urgetur. Haec ita Neutonus (+ haec　　　　　365
consideranda aliquando accuratius, posset clarius primarium
theorema ita pronuntiari: si corpus aliquod in motu jam positum,
nullo alio novo quam gravitatis vel levitatis conatu urgeatur,
describet curvam, cuius areae radiis abscissae erunt temporibus
proportionales, si non jam sit in motu positum describet rectam;　　　　　370

⟨v.⟩　　et linea curva non alio differt a parabola Galilei quam quod
apud Galilaeum centrum intelligitur infinite distans, et
gravitas vel levitas ubique agit uniformiter. Neutonus autem
rem generalissime considerans, nam communem omnibus
proprietatem invenit, quam observavit Keplerus. ⟨~~Videndum an~~　　　　　375
~~datis quotcunque~~⟩ ⟨~~Quocunque motu assignato non semper poterit~~⟩
⟨~~Omnis motus non potest reduci ad simplicem gravis~~⟩ Itaque
generaliter <u>lineae a gravi aut levi projecto descriptae,</u>
<u>si omnis alia actio vel resistentia sequestretur, sunt lineae</u>
<u>planae, unde radiis ex centro gravium ductis abscinduntur</u>　　　　　380
<u>areae temporibus proportionales</u> (nec refert ⌊levitas vel⌋
<u>gravitas sit ubique uniformis an immutetur.) Et haec</u>
<u>propositio etiam reciproca est.</u> Quod si durante motu novae
superveniant projectiones, vel resistentiae, jam alia linea
orietur. Examinandum quomodo proposita aliqua linea et data　　　　　385
relatione inter tempus et areas ab aliquo dato puncto
abscissas₍,₎ sciri possit an detur aliud punctum cuius
respectu areae sint temporibus proportionales. Hoc datum
punctum ponere possumus infinite distans, ⟨~~si igitur sint~~
~~tempora ut abscissae, constat incrementa arearum esse ut~~　　　　　390
~~ordinatas~~⟩ Sit curva C.C. cuius axis B.B. ordinatae BC, quasi
radii respicientes centrum infinite distans, et rectangula
$_1$⟨M⟩$_1$B$_2$B sunt arearum incrementa, tempora sint
ut BE (ordinatae curvae EE) utcunque, quibus
scilicet mobile percurrit $_1$C$_2$C, $_2$C$_3$C, etc.　　　　　395

⟨fig. 12⟩

Quaeritur punctum P, tale, ut triangula P_1C_2C (P_2C_3C,) sint
datis BE proportionalia ⟨Sit ~~AB, x, et BC, y, et BE~~⟩ ⟨sit
~~AR = r, et RP = p et $_2$CP ipsi axi occurrat in 2q~~⟩ Ducatur CT
curvam tangens in C, cui occurrat ex P ipsa PN parallela
ipsis BC, cui in BC sumatur aequalis $_1B_1M$, utique spatia   400
M_1B_2B sunt ipsis triangulis proportionalia, itaque ipsae
BM demta vel addita recta constante debent esse ipsis BE
proportionales. Itaque tempora insumta in curvis projectariis
a gravi vel levi descriptis, sunt ut PN resectae e centro
gravium P, (auctae forsitan certa quantitate vel diminutae.)   405
Sed cum augmenta nihil immutet, cum a detracta respondenti
parte computari tempus possit, patet ⟨~~tempora procedere ut~~
~~resectas. Porro patet etiam nihil referre utrum corpus~~
~~positum in P~~⟩ ⟨~~verum si praecise velimus aequalibus temp~~⟩
sufficere ut dicamus aequalibus temporibus aequales esse   410
differentias resectarum, ut linea projectitia esse intelligatur.
Ubicunque autem P. ponatur in recta RP nihil refert, eaedem enim
manent differentiae, demta semper communi RP. Sed si P ponatur
altius quam R vel inferius, videndum an hinc varietas

⟨r.⟩   vel res quae uno casu non succedit alio possit. Et videtur   415
quod non, quia axem AB pro arbitrio assumsimus, unde cum
in uno axe assumto res nihil referat in qua ipsi AB parallela
locatum sit puctum B, etiam in alio nihil referet⟨,⟩ sed
hinc videtur absurdum nasci quod sequeretur quodvis punctum
assumi posse, si unum succedit.   420
Corpus incremento temporis BE percurrit elementum curvae $_1C_2C$
et ita porro; et proinde, quia $_1B_2B$, et $_2B_3B$ aequales poni
possunt, patet spatium $_1C_3C$ percurri tempore $_1E_1B_3B_3E_1E$.
   Jam quaeritur ⌊an detur⌋ punctum P tale, ut si spatium interceptum
inter $_1B_1E$, et $_2B_2E$, sit aequale spatio intercepto inter $_2B_2E$ et   425
$_3B_3E$; sint etiam aequalia trilinea P_1C_2C, et P_2C_3C. Ponamus
curvas esse infinite parvas, adeoque $_1E_2E$, $_2E_3E$ item $_1C_2C$, et
$_2C_3C$ esse rectas. Ergo si trapezia $_1E_1B_2B_2E$ et $_2E_2B_3B_3E$ sunt
aequalia, etiam triangula P_1C_2C, et P_2C_3C erunt aequalia.
Patet autem crescentibus BE, necessario decrescere intervalla   430
$_1B_2B$, $_2B_3B$, etc. Ducatur $_1T_1C$ curvam tangens in $_1C$ quam per P
ducta ordinatis parallela secet in $_1N$, et resecta P_1N transferatur
in $_1B_1M$. Patet (ex alibi a me demonstratis) rectangulum $_1M_2B$
aequari duplo triangulo P_1C_2C. Et proinde existentibus aequalibus
trapeziis $_1E_1B_2B_2E$ (adeoque triangulis P_1C_2C) erunt etiam   435
aequalia trapezia $_1M_1B_2B_2M$, ergo quadrilinea ut $_1E_1B_3B_3E_1E$
et $_1M_1B_3B_3M_1M$ sunt proportionalia. ⟨~~Unde denique sequitur~~
~~BM, vel PN, resectas, esse proportionales temporibus~~
~~progressuum, BE.~~⟩ ⟨~~Unde denique sequitur in omni curva~~
~~projectaria ex centro gravis aut levis, resectas BM (vel~~   440
~~PN) esse proportionales temporibus~~⟩ Adeoque etiam ordinatae BM
vel PN, et BE sunt proportionales. ⟨~~Itaque habemus praeclarum~~

~~theorema: in omni curva projectaria, centrum gravis⟩~~ Ergo
resectae sunt ut incrementa temporum quibus $_1C_2C$ ideo denique
quadrilinea $_1M_1B_3B_3M_1M$ sunt temporibus quibus arcus $_1C_3C$ percurruntur 445
proportionalia. Itaque habemus postremum theorema tale, sane
praeclarum: si sit curva projectaria a gravi aut levi projecto
descripta eiusque sumantur arcus duo $_1C_2C$, et $_2C_3C$ intercepti
inter parallelas $_1B_1C$, $_2B_2C$, $_3B_3C$ et per centrum gravium vel
levium P ducatur ipsis parallela PN, cui tangentes curvam in   450
$_1C$, $_2C$, $_3C$, occurrant in $_1N$, $_2N$, $_3N$ et ipsae PN, transferantur
in BM sumtas in ipsis BC, erit quadrilineum $_1M_1B_2B_2M$ ad
quadrilineum $_2M_2B_3B_3M_2M$, ut (hoc est trilineum $_1CP_2C_1C$ ad
trilin. $_2CP_3C_2C$) ut tempus quo percurritur arcus $_1C_2C$ ad
tempus quo percurritur arcus $_2C_3C$, sive quadrilinea resectarum   455
ex parall⟨el⟩a per centrum a tangentibus factarum $_1M_1B_2B_2M_1M$,
hoc est areae per radios ex curva abscissae P_1C_2C, sunt
temporibus quibus percurruntur arcus $_1C_2C$ proportionales.
Curva CC non potest incidere in P. Ergo curva MM accedens ad
axem AB, ubi infra P descendit ab ipso iterum recedit ibique   460
erit flexus contrarius, et temporis incrementa cum antea
decreverint, nunc crescent.

⟨v.⟩   Jam calculo quaeramus PN. Sit AB, x, et BC, y, et v = PN et
$$AR, r, et RP, p. \; CQ:QN::TB:BC::\overline{dx}:\overline{dy}$$
$$jam \; CQ = r - x \; et \; QN = v - PQ. \; Jam \; PQ = y - p. \quad 465$$

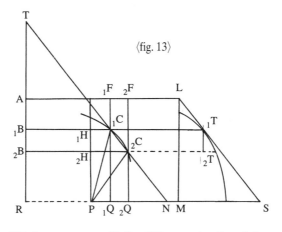

⟨fig. 13⟩

Linea TT sit temporum   Et fiet QN = v − y + p. Ergo habemus:
seu BT tempora integra $\overline{r - x}\langle:\rangle\overline{v - y + p}\langle::\rangle dx:\overline{dy} = \Theta$ seu
impensa. Tangens   fiet: rdy $\langle D\rangle - x\langle d\rangle y = vdx + \overline{p - y}dx \langle\Theta\rangle$ seu
ipsius T occurrat AF $\overline{r - xdy}/dx + \overline{y - p} = v = \overline{dt}$ explicata ergo v, et y
in L, et sit LM   per x, itemque Θ et D, debent r et p sic   470
parallela et aequalis   posse assumi, ut aequatio fiat identica.
ipsi AR, et ipsi RM $t = \int v \overline{dx} = bis \;_1CP_2C_1C\langle = \rangle \int \overline{r - x} \; \overline{dy}$ seu
occurrat LT in S. $_2Q_1C + \overline{y - p} \; dx$ seu $_1H_2C$. Ergo triang.

Debent esse MS
proportionales
ipsis PN.

$_1H_2HL + {_2}Q_1QK$ erit = quadrilatero PLDKP
ut obiter dicam. Alioqui quaesitum   475
impossible est.

Sunt autem v temporum incrementa. Vel
$\overline{r-x}\,dy/dx + \overline{y-p} = v$. Jam tempora tota
seu $\int v\,\overline{dx}$ sunt $ry - \int x\,\overline{dy} = \overline{y}\,dx - px$. Nam
$\int v\,\overline{dx}$ bis $_1CP_2C_1C$ et hoc $= PL._2C + {_1}C.L._2C._1C$,   480

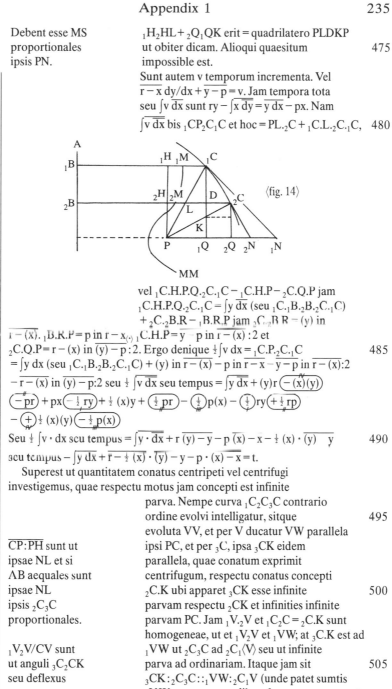

⟨fig. 14⟩

vel $_1C.H.P.Q._2C._1C - {_1}C.H.P - {_2}C.Q.P$ jam
$_1C.H.P.Q._2C._1C = \int y\,\overline{dx}$ (seu $_1C._1B._2B._2C._1C$)
$+ {_2}C._2B.R - {_1}B.R.P$ jam $_3C._2BR - (y)$ in
$\overline{r-(x)}$. $_1B.R.P = p$ in $\overline{r-x}_{(\cdot)}\ _1C.H.P = \overline{y}\ p$ in $\overline{r-(x)}:2$ et
$_2C.Q.P = \overline{r-(x)}$ in $\overline{(y)} - p:2$. Ergo denique $\frac{1}{2}\int v\,dx = {_1}C.P._2C._1C$ 485
$= \int y\,dx$ (seu $_1C._1B._2B._2C._1C$) $+ (y)$ in $\overline{r-(x)} - p$ in $\overline{r-x} - \overline{y-p}$ in $\overline{r-(x)}:2$
$- \overline{r-(x)}$ in $\overline{(y)} - p:2$ seu $\frac{1}{2}\int v\,\overline{dx}$ seu tempus $= \overline{y\,\overline{dx}} + (y)r\,\overline{(-(x)(y))}$
$\overline{(-pr)} + px\,\overline{(-\frac{1}{2}\,ry)} + \frac{1}{2}\,(x)y + \overline{(\frac{1}{2}\,pr)} - \overline{(\frac{1}{2})}p(x) - \overline{(\frac{1}{2})}ry\,\overline{(+\frac{1}{2}\,rp)}$
$- \overline{(+)}\frac{1}{2}\,(x)(y)\,\overline{(-\frac{1}{2}\,p(x))}$
Seu $\frac{1}{2}\int v\cdot dx$ seu tempus $= \int y\cdot\overline{dx} + r\,\overline{(y)} - y - p\,\overline{(x)} - x - \frac{1}{2}\,(x)\cdot\overline{(y)}\ y$ 490
seu tempus $- \int y\,\overline{dx} + \overline{r - \frac{1}{2}\,(x)}\cdot\overline{(y)} - y - p\cdot(x) - x = t$.

Superest ut quantitatem conatus centripeti vel centrifugi
investigemus, quae respectu motus jam concepti est infinite
parva. Nempe curva $_1C_2C_3C$ contrario
ordine evolvi intelligatur, sitque   495
evoluta VV, et per V ducatur VW parallela

$\overline{CP:PH}$ sunt ut
ipsae NL et si
AB aequales sunt
ipsae NL
ipsis $_2C_3C$
proportionales.

ipsi PC, et per $_3C$, ipsa $_3CK$ eidem
parallela, quae conatum exprimit
centrifugum, respectu conatus concepti
$_2C.K$ ubi apparet $_3CK$ esse infinite   500
parvam respectu $_2CK$ et infinities infinite
parvam PC. Jam $_1V._2V$ et $_1C_2C = {_2}C.K$ sunt
homogeneae, ut et $_1V_2V$ et $_1VW$; at $_3C.K$ est ad
$_1VW$ ut $_2C_3C$ ad $_2C_1\langle V\rangle$ seu ut infinite

$_1V_2V/CV$ sunt
ut anguli $_3C_2CK$
seu deflexus

parva ad ordinariam. Itaque jam sit   505
$_3CK:{_2}C_3C::{_1}VW:{_2}C_1V$ (unde patet sumtis
$_1V.W$ semper aequalibus, fore summam conatuum
centripetorum seu ipsarum $_3CK$, ut logarithmos

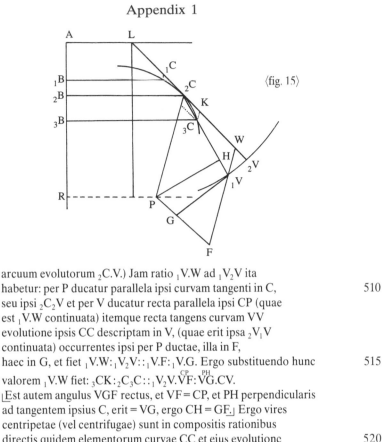

⟨fig. 15⟩

arcuum evolutorum $_2$C.V.) Jam ratio $_1$V.W ad $_1$V$_2$V ita
habetur: per P ducatur parallela ipsi curvam tangenti in C,   510
seu ipsi $_2$C$_2$V et per V ducatur recta parallela ipsi CP (quae
est $_1$V.W continuata) itemque recta tangens curvam VV
evolutione ipsis CC descriptam in V, (quae erit ipsa $_2$V$_1$V
continuata) occurrentes ipsi per P ductae, illa in F,
haec in G, et fiet $_1$V.W:$_1$V$_2$V::$_1$V.F:$_1$V.G. Ergo substituendo hunc   515
valorem $_1$V.W fiet: $_3$CK:$_2$C$_3$C::$_1$V$_2$V.$\overset{CP}{V}$F:$\overset{PH}{V}$G.CV.
⌊Est autem angulus VGF rectus, et VF = CP, et PH perpendicularis
ad tangentem ipsius C, erit = VG, ergo CH = GF.⌋ Ergo vires
centripetae (vel centrifugae) sunt in compositis rationibus
directis quidem elementorum curvae CC et eius evolutione   520
descriptae et radiorum, seu rectarum CP; reciprocis autem
rectarum CV. et PH.

Commentary to Leibniz's *Notes*

⟨Def. 1⟩ The observations in Section 5.2 above can be seen together with
Leibniz's definition of density in *Dynamica*. Leibniz writes that strictly
speaking he does not believe that the same quantity of matter can occupy
a smaller or a larger volume, this being what appears to our senses. He
continues by suggesting that lighter matter is spongier and does not
occupy the whole of its volume, but leaves some space to another subtler
matter which does not pertain to it and should not be considered together
with the body or when calculating its motion. This passage is directly
linked to definition 1 in the *Principia*. In the definition of *moles*—mass or
quantity of matter contained in a moving body—and *pondus*—weight or
quantity of matter of a heavy body—Leibniz repeats that the fluids in the

pores must be excluded.[1] Of course, Leibnizian weight cannot be mass times acceleration; in his system mass and weight are homogeneous magnitudes, the former applying in the general case, the latter in a special one.

Thus Newton used definition 1 as the first step in the attack on the Cartesian philosophy. Leibniz swiftly accepted the proportionality between mass and weight without considering it in relation to subtle fluids—whose existence he considered unquestionable. In *Dynamica* he reinterpreted Newton's definitions and crucial experiments within his own philosophy.

⟨Def. 3 and 4⟩ The definitions of *vis insita* or *inertiae* and *vis impressa* are not commented upon by Leibniz. These were well known to him when the *Principia* appeared, as we have seen in Chapters 1 and 2.

⟨Pag. 20–3⟩ These pages refer to the first part of the Scholium on the third law, where Newton seeks to prove the law of action and reaction for impacts. Leibniz devotes particular attention to collisions in which the relative velocity of the concurring bodies is greater than after the impact, or, as we would say, to 'inelastic' collisions. He discussed this topic in the *Essay de Dynamique*; on this issue compare Section 2.4 above.

⟨Pag. 85⟩ This is the first of a series of page-numbers listed by Leibniz proving that he was reading the entire book and not just some extracts. The striking correlation with the *Marginalia* leads to the following conclusion: the present notes and the great part of the *Marginalia* were drafted at the same time and have to be interpreted together. Page 85 contains lemma ??: 'To change figures into other figures of the same class'.

⟨Pag. 105⟩ Here we find lemma 28: 'There exists no oval figure whose area cut off at will by straight lines might generally be found by means of equations finite in the number of their terms and dimensions.' This lemma is rightly criticized both in the *Marginalia* and in the second set of *Excerpts*, where Leibniz writes: 'est error: tales sunt margaritae Slusii'. The 'pearls' or 'margaritae' appear in the correspondence between René de Sluse and Huygens; their general equation is $y^n = ax^m(a-x)^s$ (n, m and s are positive integers). The problem with these curves is that they are not bounded; the ovals Newton is considering 'are not in contact with conjugate figures proceeding to infinity'. In a letter to Huygens Leibniz proposed a better counter-example: $a^2x^2 = a^2y^2 - y^4$; this, however, is inconvenient in having two loops crossing at the origin, so that, properly

---

[1] Compare *Dynamica, LMG*, 6, pp. 297–8.

speaking, it is not an oval. A counter-example is provided by Tom Whiteside, who discusses the issue in detail.[2]

⟨Pag. 122⟩ On this page in the *Marginalia* Leibniz underlined corollary 2 to proposition 38, which refers to the case of a force proportional to simple distance, such that bodies reach the centre of force from any point in equal times (harmonic motion).

⟨Pag. 139, 141-2⟩ In the *Marginalia* Leibniz underlined the series expansions of the form $(T - X)^n$ in the example following proposition 45.

⟨Pag. 264⟩ In the *Marginalia* Leibniz underlined a passage in the first example of proposition 10, book II, on the geometrical meaning of the fourth and fifth terms of a series expansion.

⟨Pag. 481⟩ There are no *Marginalia* or *Excerpts* relating to this page. Newton's text contains the end of corollary 1, as well as corollaries 2 and 3 to proposition 40, book III, on the orbits described by comets, and the beginning of lemma 5. It is possible that Leibniz was reading the facing page, from which he transcribed a few months later in the first set of *Excerpts* that the heavens are destitute of any resistance.[3]

⟨Pag. 411⟩ This reference is discussed in Section 5.2.

Two passages follow: the former deals with the analogy between light and gravity based on their decreasing in accordance with the inverse-square law (lines 54-6). We have seen in Chapter 7 that Leibniz referred to this analogy in his last letter to Antoine Arnauld of 23 March 1690, in the 'zweite Bearbeitung' of the *Tentamen*, and in the letter intended for Huygens of October 1690. Further, in Chapters 1 and 2 it has been shown that light played an important role in Leibniz's philosophy from a very early stage onwards. One can mention the *Hypothesis Physica Nova*, *Propositiones Quaedam Physicae*, the letters to Fabri and Claude Perrault of 1677. Ismael Boulliau criticized Kepler precisely on this point, for failing to realize that light decreases according to the inverse-square law and not to the simple inverse (*Astronomia Philolaica*, pp. 21-4).

The latter passage regards the causes of cohesion (lines 57-69), which was a standard problem in the natural philosophy of the seventeenth century. In the passage above Leibniz rejects a principle of cohesion, and tries to explain *unio et tenacitas* by means of a qualitative principle of

---

[2] *LMG*, 2, pp. 83-5, 2 March 1691, Leibniz to Huygens; *NNW*, 6, pp. 302-7; an account of the history and properties of pearls is in Loria, *Curve Piane*, vol. 2, pp. 376-83.

[3] First set of *Excerpts* (from page 480): 'Coeli resistentia destituunt alioqui cometae turbarentur ab orbibus planetarum.'

minimum according to which bodies tend to assume a configuration in which their motions least obstruct each other. In *Dynamica* Leibniz states that the cause of cohesion is motion; he deals with this more extensively in the *Animadversiones in Partem Generalem Principiorum Cartesianorum*, where he criticizes Descartes's inference that cohesion arises from rest, and states his own belief that the cause of cohesion is 'concurrent' motion. A brief discussion and references on this problem are given in Section 4.2.

Newton's ideas were very different when he wrote the *Principia*: in the introduction, for example, he expressed his wish to explain phenomena—no doubt, cohesion was a prime concern—in terms of attractions and repulsions. From his early works to the end of his life he pursued the project of explaining cohesion. In *Certain Philosophical Questions* he gave a Boylean account of cohesion resting on air pressure; in the last edition of his *Opticks* he dealt with the problem, also related to alchemy, in Queries 21–2, 28 and especially 31.[4]

⟨Cor. 1⟩ We have seen in Section 3.3 that Leibnizian conatus has two different meanings: the first has been mentioned above in the commentary to the second law and is the differential of velocity, that is, solicitation or an infinitesimal tendency to motion; the other meaning is 'velocity together with direction'. In his first comment on Newton's corollary on composition of forces Leibniz refers to the second meaning, and prefers the term *conatus* to *vis* because he wants to convey the idea of direction (lines 90–1).[5]

Leibniz criticizes Newton's claim that the force along *AC* does not influence the time in which the body in *A* reaches *DB* (lines 102–). Newton's statement is based on the parallelism between *AC* and *DB*. The first corollary can easily be generalized to the case of several forces, but Newton indeed gives no proof of this. In this more general case Leibniz employs the 'centre of gravity', that is, the barycentre of the figure (*ABDC*

[4] See J. E. McGuire and M. Tamny, eds., *Certain Philosophical Questions* (Cambridge, 1983), pp. 292–5 and 349–51. Hall and Hall, *Unpublished Scientific Papers*, part 3, 'De Aere et Aethere', and Part 4, sections 3, 7 and 8 of the preface, conclusion and Scholium Generale to the *Principia*; A. R. Hall, 'Newton's Theory of Matter', *Isis, 51*, 1960, pp. 131–44; 'Newton and the Theory of Matter', in Palter, *Annus Mirabilis*, pp. 54–68. I. B. Cohen, *Papers and Letters*, chapter 3, especially the letter to Robert Boyle of 28 Febr. 1679.

[5] Another definition of 'conatus' is in *Dynamica, LMG, 6*, p. 471. Leibniz's published works on composition of motion appeared in the *Journal des Sçavants* in 1693: 'Règle Générale de la Composition des Mouvemens', pp. 417–19 = *LMG, 6*, pp. 231–3; a modified version is in Costabel, *Dynamique*; 'Deux Problèms Construit par M. de Leibniz, en Employant la Règle Générale de la Composition des Mouvemens', pp. 423–4 = *LMG, 6*, pp. 233–4. Some passages from *Dynamica* are similar in content: *LMG, 6*, pp. 487–8.

in fig. 2), in order to determine the direction in which the body moves: this is along the line between the starting point and the centre of gravity of the figure, and this distance times the number of acting forces gives the intensity of the resultant.[6]

⟨Cor. 3⟩ Newton states the principle of conservation of *quantitas motus*. Whereas for 'quantity of motion' Leibniz means a scalar quantity, 'quantity of progress' carries for him the concept of direction. In the second set of parentheses (lines 123–7) his idea seems to be that living force is conceptually most important, and it should be shown first how to measure it from the mutual recession of the two bodies.[7]

⟨Lem. 1⟩ The rest of this folded folio is devoted to the lemmas in section 1. On the first point, that ratios are also quantities, I cite Leibniz's reply to Samuel Clarke's fourth letter—paragraph 14—:[8] 'As for the objection that space and time are quantities, or rather things endowed with quantity; and that situation and order are not so: I answer, that order also has its quantity; there is in it, that which goes before, and that which follows; there is distance or interval. Relative things have their quantity, as well as absolute ones. For instance, ratios or proportions in mathematics, have their quantities, and are measured by logarithms; and yet they are relations'.[9]

In the *Marginalia* Leibniz writes that the two quantities considered by Newton should tend uniformly to equality, and that asymptotes should be excluded, otherwise there could be no last difference and yet the two quantities would not become equal.[10] Newton, however, considers a finite time, and in the scholium at the end of section 1 makes clear that by 'last

---

[6] Concerning the 'scheda separata' mentioned by Leibniz, see LH 35, 10, 7, f. 41r., *Si trium punctorum quaeratur centrum gravitatis*; the top of the manuscript reads: 'Haec bona sed melius in aliam schedam translatam'. This is probably LH 35, 14, 2, f. 18–19, *Si sint duo conatus corporis*. This manuscript is on a kind of paper used by Leibniz in Vienna in 1688, watermark 510 in the catalogue at the NLB (letters 'M R'). The top right corner of f. 18r. reads: 'Hic de compositione directionum etiam infinitarum et mira de modo describendi lineam centrorum gravitatis arcuum curvae, item data linea centrorum, inveniendi curvam. Ostensum etiam hic vulgarem directionum compositionem non procedere. Et elegantia de decompositione rationum et de earundem additione.'

[7] 'Essay de Dynamique', *LMG, 6*, pp. 215–31, pp. 216–17 and pp. 227–8.

[8] *Leibniz–Clarke Correspondence*, Leibniz's fifth paper, par. 54.

[9] See also Loemker, *Papers*, pp. 704 and 706, and paragraph 47 of the same letter. The theory of relations is an important area of Leibnizian studies. See B. Mates, *The Philosophy of Leibniz. Metaphysics and Language* (Oxford, 1986), chs. 12 and 13; M. Mugnai, 'Bemerkungen zu Leibniz' Theorie der Relationen', *SL, 10*, 1978, pp. 2–21.

[10] *Marginalia*, M 26 A: 'Si ad aequalitatem tendant uniformiter, aut promtius, aut saltem non asymptotos.' M 26 B: 'Si asymptotae sint potest esse nulla differentia ultima, et tamen nunquam fient aequales.'

ratio' he understands that ratio not before the two quantities vanish, nor after, but with which they vanish. On this issue see Sections 3.4 and 7.5.

⟨Lem. 2 and 3⟩ Leibniz is eager to state that the same result proved in lemma 2 on the quadrature of curves by exhaustion can also be obtained by his own method, by which he means the transmutation theorem (see Section 3.2).

Concerning triangle ABC (lines 163–73, Fig. 6), where the angle in B differs infinitesimally from a plane angle, Leibniz's reasoning is not completely rigorous. He draws two infinitesimal arcs of circumference, $\overline{BD}$ with radius AB centred in A, and $\overline{BE}$ with radius BC centred in B. However, ED is twice FD only if ABC is isosceles. Further, since the angles in C and A are first-order infinitesimals, and EF, FD are the versed sines of those angles respectively (namely they are as $1 - \cos$), EF and FD must be second-order infinitesimals.

⟨Lem. 6 and 7, lines 174–84⟩ Leibniz's claim that there is much to be said on the difference between two sides of a triangle enclosing an infinitesimal angle may be related to his idea that this difference is not evanescent, as stated by Newton, but infinitely small. On this issue see paragraph 5 of the *Tentamen*.

⟨Schol., lines 185–205⟩ Leibniz accepts Newton's results, but he criticizes him for not proving that in the sequence AD^2, $AD^{3:2}$, $AD^{4:3}$, etc., the corresponding angles of contact are infinitesimal with respect to the preceding ones. His reasoning in lines 194–7 can be paraphrased as follows: we must find whether subtracting from a 'circular angle of contact'—between a circumference and its tangent an angle homogeneous to it, such as the angle between a parabola and its tangent at the vertex, we are left with an infinitely small angle of contact. I illustrate this problem with the following example; $y = x^2$ is the equation of the parabola, $y = (1:2) - \sqrt{(1:4) - x^2}$ the equation of the semi-circumference osculating it at the origin. The angle of contact obtained by taking their difference is of the same order as that between the curve $y = x^4$ and its tangent at the origin.

In lines 197–205 Leibniz wonders whether he can measure different angles of contact by means of paraboloids and hyperboloids, whose equations are $a^{m-n}x^n = y^m$ and $a^{m+n} = x^n y^m$ respectively, m and n being positive integers and a constant.[11] The problem amounts to finding a measure of the angle of contact between a standard curve, such as $y = |x|^e$, and its tangent at the origin. However, if e is equal to or greater than 1 all possibilities are covered and other cases can be reduced to the

---

[11] *LMG*, 5, p. 103, 'Compendium Quadraturae Arithmeticae'.

following: if e is greater than 2 the angle of contact is incomparably smaller than that between a circumference and its tangent; if $e = 2$ the angles of contact are of the same order; if e ranges between 1 and 2 the angle of contact is incomparably larger; lastly, if $e = 1$ the curve becomes a straight line.

⟨Lem. 9⟩ Leibniz believes lemma 9 to be wrong, because $B(B)$ is not infinitesimal with respect to AB or $A(B)$, and hence to AD or DB; this implies that triangles ADB and $A(D)(B)$ are not similar (see fig. 9 and Section 3.4). Of course, this does not refute Newton's lemma, because they become similar if B, (B) tend to A, or if AD, $A(D)$ are vanishingly small (or infinitesimal). Leibniz's reasoning can be summarized as follows: $AD = x$, $DB = y$, $A(D) = x + dx$, $(D)(B) = y + dy$. He thinks of expressing dx as hx, where h is an infinitesimal constant, and checking whether the areas of triangles ADB and $A(D)(B)$ are proportional to the square of the sides AD and $A(D)$ respectively. Leibniz does not consider that AD, $A(D)$, as well as DB, $(D)(B)$, become infinitesimal, and makes $A(D)$ tend to AD and $(D)(B)$ to DB. Further, he expresses y in terms of x by means of the series $y = a + bx + cx^2 + ex^3 + \ldots$, and fails to realize that $a = 0$, because when $x = 0$ also $y = 0$; he began in the correct way in a part subsequently crossed out for no clear reason (line 218). After some calculations which are the consequence of his two erroneous premises, he arrives at the conclusion that in order to obtain $A(D)(B)$ from ADB it is necessary to substitute $(1 + h)x$, or $x + dx$, for x, taking into consideration the various orders of infinitesimals. Following Leibniz's notation, we ought to have $y = bx + cx^2 + ex^3 + \ldots$; then, $xy = bx^2 + cx^3 + ex^4 + \ldots$ If x becomes infinitesimal, the terms in x^3, x^4, $\ldots$ are negligible, and the area of $ADB(= xy/2)$ is indeed proportional to $AD^2(= x^2)$.

In the *Marginalia* Leibniz marked this lemma with his *distillatur*, meaning that the matter deserves further investigation, and added that Newton's construction in the demonstration could be dispensed with. From Leibniz's perspective Newton's proof was needlessly cumbersome. These observations show the close connection between the present notes and the *Marginalia*: probably Leibniz first wrote *distillatur*, then checked the result without reading the demonstration for lack of time (lines 264–5), and lastly accepted Newton's proof but criticized its lack of directness. Indeed, when using differentials correctly instead of Newton's construction, the proof is immediate. In the first set of *Excerpts* lemma 9 is transcribed without commentary, and seems to be accepted without difficulty.

⟨Lem. 10⟩ Lemma 10 states that a body acted upon by a regular force at the beginning of motion describes spaces proportional to time squared.

This generalization of Galileo's law of fall is presented as a corollary, in mechanical terms, to lemma 9; therefore, Leibniz's rejection of the former is a consequence of his rejection of the latter (see Sections 4.2 and 7.5). Since for him force is proportional to solicitation, at the beginning of motion spaces should be proportional to simple time. Nevertheless, here he believes that in an infinitesimal time the spaces described by a body urged by a regular force need not be proportional to the square of the times, but could also follow a different law. This appears so obvious to him that in his opinion lemma 10 could have made Newton realize his 'mistake' in the preceding lemma. These are probably Leibniz's first impressions on lemma 10.

⟨Lem. 11⟩ Leibniz originally believed that the present lemma was sounder than the two preceding ones, and that there was a contradiction between lemmas 9 and 11. But in that case the angle between the curve $AB(B)$ and the line AD was given (that is, finite), whereas in the present case AC is tangent to the curve $AB(B)$ in A (fig. 10). Later Leibniz added the sentence 'Imo forte et hoc nihil est', thus rejecting lemma 11 as well (lines 274–5). Referring to his own fig. 10, he writes that according to Newton CB would be proportional to AC^2 (equivalent to DB proportional to AD^2 in Newton's diagram, because in fig. 10 letters C and D are inverted with respect to the *Principia*; see Figure 5.1). Referring to Newton's diagram he claims that AD is proportional to AIG^2, where $I(=J)$ is the extreme opposite to A on the osculating diameter; however, if B tends to A then $AD^2 = AIG \cdot BD$. In corollary 3 Newton states that ultimately the curvilinear areas between a curve and its tangent are two thirds of the areas of the corresponding triangles for the known property of parabolas. Further Leibniz seems to infer that following Newton's reasoning a parabola would have six points of contact with a circumference in its vertex! This statement appears to be related to Newton's construction: considering that b and B approach right up to A, and the corresponding three points on the tangent, Leibniz draws his paradoxical conclusion about six points.

In the *Marginalia* lemma 11 is referred to in three instances. On page 33 Newton claimed that since BD, in the case of a vanishing subtense, is vanishingly small compared with AD (BC and AC respectively in fig. 10), any given change in the inclination of BD does not alter the conclusion. Leibniz wrote *non videtur sequi* under the words *quae prius*, thus failing to grasp this point. In the other two comments he questioned the generality of lemma 11 and accepted it only when the curve is a circumference.[12] Newton's lemma is valid only for a curve

---

[12] The relevant passage from the *Principia* reads: 'Inclinetur jam BD ad AD in angulo quovis dato, et eadem semper erit ratio ultima BD ad bd quae prius, adeoque eadem ac AB

with a finite curvature, when its infinitesimal arcs can be approximated by the corresponding arcs of the osculating circumferences, as he explicitly wrote in the subsequent scholium. The extent of Leibniz's failure here is stunning: obviously he read the text superficially and went as far as mocking Newton's 'exaggerated subtlety', probably referring to their correspondence in 1676.

⟨Prop. 1 and Cor. 1⟩ Here Leibniz follows the text very closely. By *motus conceptus* Leibniz means the rectilinear inertial motion which has been acquired by the body (line 295). In the marginal note (lines 306–15), Leibniz introduces the notion of *linea projectitia*, that is, the curve described by a body with respect to a centre of attraction or repulsion, such that the areas are proportional to the times.

There is a manuscript fragment where Leibniz states that Newton did not explain how one or the other ellipse, and in particular a circumference, are obtained; this would be clear in Leibniz's own account. Considering also the second sketch on the opposite side, which seems to represent the polygon described by a body acted upon by central impulsions, I propose that Leibniz may have had in mind proposition 1, where Newton does not explain the nature of the curve described; see Fig. (c).[13] Leibniz was possibly referring to paragraph 30 of the *Tentamen*, which provides a classification of planetary orbits.

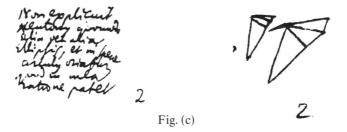

Fig. (c)

⟨Cor. 2⟩ In the first part of the commentary Leibniz considers an arbitrary curve and an arbitrary point *S*, claiming that it is possible to cut the curve so that all triangles centred in *S* have equal areas.

The following comment refers to proposition 1 (lines 329–): if the areas swept out by the radius from the centre of force are proportional

---

quad. ad Ab quad. Q.E.D.' See *Marginalia*, M 41 B: 'Suspectum hoc Lemma generale. In circulo tamen res vera speciali ratione, quia abscissae in circulo sunt ut quadrata chordarum'; and M 42 A: 'Quia Lemma 11 generale nondum admitto, . . .' See also *NMW 6*, p. 117, n. 55.

[13] LH 35, 10, 1, f. 2: 'Non explicuit Neutonus quomodo alia vel alia ellipsis, et in specie circulus oriatur, quod in mea ratione patet.' Compare also *NMW, 6*, pp. 35–7, n. 19 and proposition 17 in book I.

to the times, force is not necessarily attractive, but can be also repulsive because triangles SBC and $SB(C)$ have equal areas (fig. 11). Leibniz makes the same point in the *Marginalia* to proposition 2, still referring to centripetal force.[14]

In the following lines (332–60) Leibniz lists some properties of motion under central forces. For example, a body moving with recti-linear and centripetal motion, or with rectilinear and centrifugal motion, cannot move along the same curve. Here Leibniz is referring to a motion resulting from a repulsive force rather than an attractive one. However, as proposition 12 of the *Principia* and paragraph 30 of the *Tentamen* show, the same hyperbolic branch can indeed be traversed under the action of opposite forces, attractive towards the internal focus and repulsive from the external one. The words *centrum respicere* are underlined by Leibniz: this expression is used as a definition for a body moving of its own inertia under the action of central solicitation (line 358).

⟨Prop. 3⟩ Leibniz's commentary after the transcription of proposition 3 still refers to proposition 1. His reformulation of Newton's theorem states that curves described by a body urged by central forces are planar, and the areas swept out by the radii are proportional to times regardless of whether force is attractive or repulsive, or of its dependence on the distance from the centre, and vice versa. Notice the reference to Galilean parabolas (lines 371–2), which is developed in several later manuscripts such as *Galilaeus* and *De Motu Gravium*, and the claim that Newton generalized Kepler's area law (lines 373–5). Leibniz repeated a similar observation in *De Motu Gravis*, as we have seen in Section 5.4.

Below (lines 385–9) Leibniz states the following problem: given a curve and a relation between times and areas swept out by a radius drawn from a given arbitrary point, to find whether there exists another point with respect to which the areas are as the times. In the *Tentamen* he identified a physicomathematical criterion related to the existence of vortices, equivalent to Kepler's area law, and which characterizes the composite class of curves described by bodies urged by central forces.

The remaining part of the manuscript can be divided into four sections. The first two sections (lines 391–420 and 421–62 respect-ively) consist of as many attempts relevant to the problem we have just seen; in the third Leibniz seeks a procedure for determining the centre of force, and goes on to provide conditions for the solution to this problem (lines 463–91). In these attempts he employed the transmuta-

---

[14] *Marginalia* M 38: 'Imo etiam a centrifuga.'

tion theorem for an obvious reason: in motion under central forces it is convenient to calculate the areas to the centre of force, and his theorem relies precisely on such areas concurrent to a centre. This idea, however, turned out to be less fertile than Leibniz had hoped. Hence the first three sections contain cumbersome calculations which are difficult to follow and are not directly linked to the rest of the work. None the less, they are an interesting document of Leibniz's struggle with the *Principia*. In the fourth section he tried to find an expression for centripetal or centrifugal conatus (lines 492–522); those calculations are definitely more interesting and ought to be compared with the following manuscript *De Conatu*.

In the first section CCC is the curve described by the body and CN its tangent, which is the prolongation of the chord (see fig. 12; I indicate a missing segment in angled brackets. The sketch at the top might be the beginning of a diagram for the transmutation theorem). P is the centre with respect to which the areas are in the given ratio to the times, that is, the areas P_1C_2C, P_2C_3C are in the ratio to the times given by the curve EE; in fact, Leibniz sets BE proportional to the times necessary to traverse the corresponding arcs of the curve CC. He intends to transform the curvilinear figure $P_1C_2C_3CP$ into one of double its area, namely $_1B_1M_2M_3M_3B$. The *resectae PN*, namely the segments cut by the tangents on the ordinate through P, are equal to the corresponding segments BM.[15] In Leibniz's figure the distances PN are not drawn equal to BM and the curve MMM is not consistent with EEE, since both ought to be proportional to time (line 393). I explain this point with the help of Fig. (d), which reproduces the relevant portion of fig. 12. P_1N is equal to G_2B; the letter G is not indicated in the corresponding portion of fig. 12

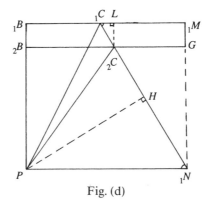

Fig. (d)

15 The notion of 'resecta' is defined in *LMG*, 5, p. 101, 'Compendium Quadraturae Arithmeticae'; see also *LMG*, 5, pp. 113–16.

and can be approximated by $_2M$ if the angle in P is infinitesimal; the axis AB has been taken through P; PH is perpendicular to $_1N_2C$ and $_2CL$ to $_1B_1M$. On the basis of the transmutation theorem, and from the similarity between triangles L_1C_2C and H_1NP, the area of triangle P_1C_2C equals half the area of $_1M_1B_2BG$, because $_1C_2C:P_1N::L_2C:PH$ and $PH \cdot _1C_2C = $ twice the area of triangle $P_1C_2C = P_1N \cdot L_2C$. In order for this to be applicable to the case studied by Leibniz, triangle P_1C_2C must be infinitesimal, so that the chord and the arc $_1C_2C$ can be represented by the tangent $_1C_1N$, as he specifies in lines 426–8. Before analysing Leibniz's solution, I outline a possible procedure starting from the same premisses.

   P must be placed on a straight line through $_2C$ and must satisfy the condition that the areas of triangles P_1C_2C and P_2C_3C are proportional to the times required to traverse $_1C_2C$ and $_2C_3C$. It is easy to show that if a point Z satisfies this condition, any other point along $_2CZ$ satisfies it as well. Therefore, in order to determine P on this straight line, it is necessary to repeat the same reasoning for another portion of the curve, obtaining another straight line which intersects the former in P: this is the common centre of attraction. This reasoning is valid if we already know that there exists a centre of force, otherwise it is necessary to verify that the point P satisfies the same conditions for the whole curve.

   We examine now Leibniz's procedure, which is affected by many mistakes. In the crossed-out lines 396–420 times are taken proportional to PN, but this is true only if $_1B_2B$, $_2B_3B$ are equal (lines 403–4; see also lines 422–3). Since PN is determined apart from a constant factor ('auctae forsitan certa quantitate vel diminutae', line 405), Leibniz tries to consider the differences between the *resectae*, $_1N_2N$, $_2N_3N$, thinking that these could be uniquely determined. But this is obviously wrong: as we have seen, PN, as well as NN, depends indeed on a parameter, which is the proportionality constant between the increments of the times and the areas of the triangles P_1C_2C, P_2C_3C. Leibniz was mistaken in thinking that he could determine an abscissa or an ordinate; what he could determine by his method with three points on the curve CCC was a ratio between abscissae and ordinates. His disappointing conclusion is that the centre P can be placed everywhere on the line RP, and the same would happen if P were lower or higher than R, therefore this arbitrariness appears to be unavoidable. This seems to be absurd to Leibniz (lines 419–20). He crosses out his first attempt and tries again.

   In the second section (lines 421–62) Leibniz sets $_1B_2B = _2B_3B$ and proves that if there is a point P such that the areas swept out by the radii PC are as the times, then the following proportion holds:

$$_1E_1B_2B_2E : _2E_2B_3B_3E :: _1M_1B_2B_2M : _2M_2B_3B_3M :: _1CP_2C : _2CP_3C.$$

This proportion, which follows from the transmutation theorem (line 433), expresses the times needed to traverse the arcs $_1C_2C$ and $_2C_3C$ by means of the areas of three different figures. Further, Leibniz states that the curve CC cannot fall upon P, probably implying that P is a singular point where velocity would become infinite (line 459).[16] In the description of the curve MM Leibniz refers to a *flexus contrarius* in correspondence with R along the axis AB (lines 459–62).

In the third section Leibniz seeks to determine PN by means of a calculation (fig. 13, lines 463–91). In the end, however, he finds that the area of the quadrilateral $PLDKP$ must be equal to the sum of the areas of triangles $_1H_2HL$ and $_1Q_2QK$ (lines 473–4, fig. 14). This result can be easily verified by means of elementary geometrical methods. Notice that Leibniz tries to introduce an abbreviated notation (lines 467–70). In the same fig. 14 the curve MM is incorrectly drawn, because the *resectae PN* are not equal—or even proportional—to BM. With regard to the following calculation (lines 477–91 and fig. 13), I try to clarify Leibniz's notation and point out the slips, leaving the details to the reader: $(x) = A_2B$, $(y) = _2B_2C$, $x = A_1B$, $y = _1B_1C$, $RP = p$, $AR = r$. Since $_1C_2C$ is infinitesimal there is no need to integrate xdy. The calculation is almost correct, apart from the area of triangle $_1C_1HP$ being $(y - p) \cdot (r - x):2$— Leibniz writes (x) instead of x—; in line 488 Leibniz writes $-1:2\ ry$ instead of $-1:2\ r(y)$. Taking into account these remarks and corrections the result is:

$$\text{area } _1CP_2C = (1:2)(y - p)((x) - x) + (1:2)(r - (x))((y) - y).$$

In fig. 13 the curve $_1T_2T$ represents time. Leibniz named a third point with the same letter T. Notice that the tangent CN appears to be the prolongation of the chord, whereas L_1TS does not touch $_2T$. In the marginal note between lines 466–76 MS is proportional to PN because the areas $PC(C)$ are proportional to the times $T(T)$. Thus curves CC and TT ought to be similar.

Lastly, in lines 492–522 Leibniz tries to determine centripetal or centrifugal conatus. His attempt is probably related to corollary 7 to proposition 4, which is examined in the commentary to *De Conatu*. Leibniz's idea of using the curve VV *descripta ex evolutione* from the evolute CC is not useful in this context because it does not help in finding any of the unknowns. In fig. 15 he seems to take the tangent $_3C_1V$ to be the prolongation of the chord $_2C_3C$, but the other tangent $_2C_2V$ is not the prolongation of $_1C_2C$; this introduces an inconsistency into the analysis of curvilinear motion, as in fig. 13. It is possible that Leibniz

---

[16] Newton dealt with a similar case in book I, proposition 7. In the *Marginalia*, M 45, Leibniz writes and crosses out: 'nescio an hoc possibile'.

thought of using his construction remembering his conversations with Huygens some 15 years earlier (see sections 2.3, 4.3, 5.2, and 5.3). The following lines (494–) are similar to *De Conatu*, and will be examined in the commentary to that essay.

In line 492 and in general in the whole calculation Leibniz seems to imply that centripetal and centrifugal endeavours are equal and opposite. In the *Tentamen*, however, they are different in general. In the present manuscript his analysis of curvilinear motion is altogether different and more primitive than the ideas he adopted from 1689 onwards. Moreover, I find it surprising that he wrote such lengthy notes relating to the area law and to the mathematics of planetary motion without ever mentioning any of the results and concepts of the *Tentamen*. In the following essays we shall witness all the steps leading from the present stage to the *Tentamen* including mathematical technicalities as well as physical and mechanical interpretations.

2   *De Conatu*

LH 35, 10, 7, f. 29–30 (compare Figs. (e) and (f))

De Conatu consists of one sheet folded in quarto. Leibniz develops the calculation of centripetal or centrifugal conatus based on the evolute and evolvent first introduced at the end of his notes to the *Principia Mathematica*. This can be inferred from the contents of the two manuscripts—notice the similarity of figs. 15 and 16—and from the edge of sheets 34–5 matching the edge of sheets 29–30. Therefore, *De Conatu* dates after Leibniz's reading of the *Principia*. Further, a series of features indicate a date of composition preceding the *Tentamen*. The most important among these features are the lack of any reference to the *circulatio harmonica*, the notion of *conatus paracentricus*, which is either centrifugal or centripetal and is represented by a one-term expression, the observation that centripetal and centrifugal endeavours are always equal and opposite, and the faulty remark about the order of infinitesimal of the conatus along the rotating ruler.

Initially Leibniz's writing went directly from the first to the fourth side, which was marked as *pag. 2*. Here, soon after having written the first few lines (152–6), Leibniz crossed them out and started again on the second side with the same words: 'Sed ut definiamus' (line 38). Later he crossed out the whole of the second side and began anew with the same words on the third. In the left margin of the second side he noted the letters A to N, which I do not reproduce because they are unrelated to the text. A marginal note at the bottom of the second side (lines 77–81) and a series

Fig. (e) LH 35, 10, 7, f. 29r.–30v., *De Conatu Centripeto aut Centrifugo.*

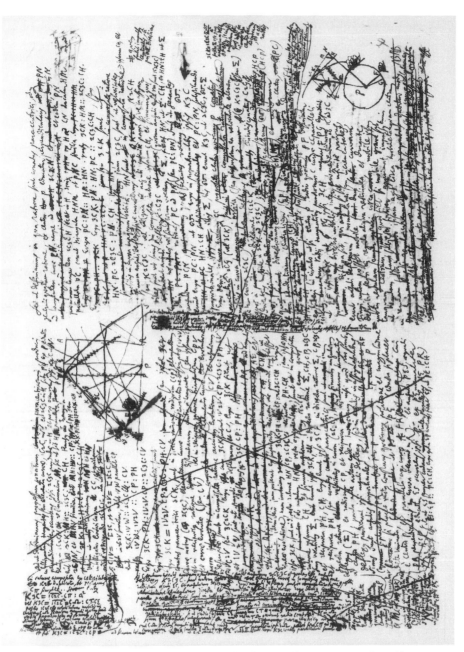

Fig. (f)   LH 35, 10, 7, f. 29v.–30r., *De Conatu Centripeto aut Centrifugo.*

of calculations in the margin of the second and third sides have been inserted at the appropriate places in the running text (lines 113–34). The reader can see the original structure of the manuscript in the enclosed reproduction.

De conatu centripeto aut centrifugo ⌊(vel generaliter p̲a̲r̲a̲c̲e̲n̲t̲r̲i̲c̲o̲)⌋ ⟨pag. 1⟩ mobilis in aliqua curva incedentis

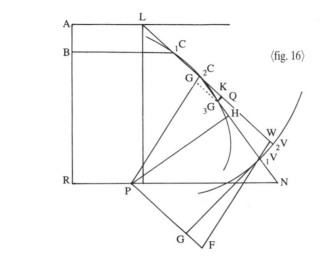

⟨fig. 16⟩

Sit linea ⌊curva quaecunque⌋ C.C.C. ⟨aestimandum est quantus sit continuus accessus mobilis ad B, vel ab eo recessus, vel si mobile ponatur ferri⟩ datumque punctum quodcunque P, aestimandum est, quantus sit ⟨vis⟩ ⌊conatus⌋   5
⟨centripetus aut⟩ centrifugus mobilis in curva lati, seu qua velocitate inter
movendum conetur recedere a puncto P, sumto tanquam centro; vel quod eodem
redit qua conatus ad centrum P tendentis vi opus sit ad mobile in orbita sua
retinendum, ne per tangentem abeat. Unde idem est ⟨vis⟩ ⌊conatus⌋ centrifugus
mobilis ex natura curvae, cum ⟨vi⟩ ⌊conatu⌋ centripeto in orbita retinente.   10
⟨Vis⟩ ⌊Conatus⌋ autem mobilis a centro aliquo ⟨intra concavita⟩ ex concava
parte posito recedendi per tangentem, est ⟨eadem cum vi a centro⟩ ⟨vis⟩
conatus ejusdem mobilis ad centrum aliquod a convexa parte positum accedendi
per tangentem. Et longe differt conatus iste centrifugus vel centripetus, a
conatu in regula circa P mota, qui alteri motu compositus lineam CC   15
describere potest, nam conatus centrifugus est infinite parvus respectu
⟨motus⟩ conatus in regula PC motum in linea componentis; praeterea motus in
⟨linea componens⟩ regula motum in curva componens varius intelligi potest,
prout alius assumitur alter motus simul cum ipso componens; sed ⟨motus
centrifugus⟩ conatus centrifugus semper est idem. ⌊Figuram vide pag. 3⌋   20
    Sit elementum curvae $_1C_2C$ et aliud elementum sequens $_2C_3C$. Productae $_1C_2C$
ultra $_2C$ ⌊in K ut sit $_2CK = {_1C_2C}$⌋ occurrat $_3CK$ parallela P_2C eductae ex centro P,
eique in P_2C sumatur aequalis et parallela G_2C ut compleatur parallelogrammum

G_2CK_3C, patet mobile ex $_2C$ in $_3C$ ferri ⟨m̶o̶t̶u̶⟩ ⌊conatu⌋ composito ex priore $_1C_2K$ 25
seu $_2CK$, et ad centrum P tendente $_2CG$ seu K_3C. Itaque conatus centrifugus vel
magneticus ad centrum attrahens, (centrifugum destruens, eique aequalis,
⌊generaliter paracentricus⌋) est $_2CG$ vel K_3C. Patet elementum conatus
centrifugi esse infinite parvum respectu elementi curvae, seu velocitatis in
curva, atque adeo infinities infinite parvum respectu ⟨r̶e̶c̶t̶a̶e̶⟩ communis
lineae ut PC ⟨n̶a̶m̶ ̶P̶H̶⟩⟨a̶n̶g̶u̶l̶u̶s̶ ̶$_2CK_3C$̶ ̶n̶o̶n̶ ̶s̶i̶t̶ ̶r̶e̶c̶t̶u̶s̶ ̶t̶a̶m̶e̶n̶ ̶s̶u̶b̶t̶e̶n̶s̶a̶⟩. Si 30
angulus $_2CK_3C$ esset rectus foret K_3C sinus anguli cuius radius $_2C_3C$ est autem
angulus K_2C_3C infinite parvus et sinus anguli infinite parvi est ad radium
infinitesimus. Ubi obiter annoto si K_3C esset non parallela PC, sed tendens ad
P, ut $_3CQ$ differentia inter has duas ob angulum K_3CQ erit infinities infinite
parva. Idque etiam praeviideri potest quia sine errore fingi licet conatum 35
centralem non esse perpetuum, sed intervallis interruptum, et durante
intervallo, considerari ut parallelum.

⟨v.⟩ Sed ut definiamus progressum conatuum ⟨c̶e̶n̶t̶r̶a̶l̶i̶u̶m̶⟩ paracentricorum, produci
intelligantur tangentes seu elementa curvae, $_1C_2C$ in N, et $_2C_3C$ in H sitque PH
ex P educta perpendiculariter occurrens ipsi $_2C_3C$ productae in H, et porro 40
occurrens ipsi $_1C_2C$ in N. Denique in $_1C_2CN$ recta sumatur

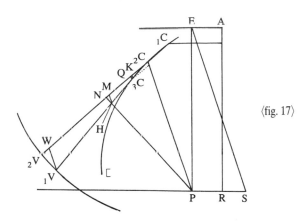

⟨fig. 17⟩

M, sic ut sit HM parallela PC, erit $_3CK:MH::_2C_3C:CH$. Rursus
ob triangula $_2CPN$, et MH⟨N⟩ similia, erit $MH:HN::CP:PN$ vel
PH seu $_3CK \cdot PH:HN \cdot CP::_2C_3C:CH$. Verum ut aliter inveniamus
MH et HN intelligatur evolvi curva CC, et quidem ordine 45
retrogrado $_3C_2C_1C$, ita ut $_3C_1V=\mathsf{L}_3C$, et $_2C_2V=\mathsf{L}_3C_2C$, in $_2C_2V$ sumatur, ut sit
$_1VW$ parallela CP erit $_3CK:_1VW::_2C_3C:CV$ rursus $_1VW:_1V_2V::CP:PH$. Ergo
$_3CK \cdot PH:_1V_2V \cdot CP::_2C_3C:CV$ ⟨E̶r̶g̶o̶ ̶i̶p̶s̶a̶e̶ ̶$_3CK$̶ ̶c̶o̶n̶a̶t̶u̶s̶ ̶p̶a̶r̶a̶c̶e̶n̶t̶r̶i̶c̶i̶,̶ ̶s̶u̶n̶t̶⟩ Ergo
$_3CK=_1V_2V \cdot CP \cdot _2C_3C:PH \cdot CV$ seu conatus paracentrici $_3CK$, sunt in directa
ratione, radiorum $(CP)_{(,)}$ elementorum curvae motus $(_2C_3C)$, elementorum 50
curvae evolutione descriptae $(_1V_2V)$, et in reciproca ratione distantiarum centri
a tangente (seu ipsarum CP), et curvae evolutae (seu CV) vel in reciproca ratione
triangulorum CPV. Porro ipsae $_1V_2V$ elementa curvae evolutione descriptae sunt
in composita ratione arcuum evolutorum (CV) et angulorum deflexus $_1VC_2V$ seu

$_3C_2CK$ ergo cum $_3CK$ sint ut $_1V_2V\cdot CP\cdot_2C_3C{:}PH\cdot CV$ et $_1V_2V$ sint ut anguli deflexuum   55
et CV, ergo $_3CK$ fient ut anguli deflexus et $_2C_3C\cdot CP{:}PH$. ⟨Assumitur directrix
aliqua vel axis AR cui ex P occurrat PR ad angulos rectos, et ex A educatur ad
AR perpendicularis AE et ipsis AE, RP si opus productae occurrat tangens CH,
in E et in S⟩ Idem vero jam poterat ex superioribus derivari nec proinde opus
fuisset lineae evolutione, nam invenimus $_3CK\cdot PH{:}HN\cdot CP{::}_2C_3C{:}CH$. Ergo ipsae $_3CK$,   60
sunt in directa ratione HN, CP, $_2C_3C$, et reciproca $PH\cdot CH$, jam ipsae HN (ob
angulum HNC recto aequivalentem) sunt in composita ratione ipsarum CH, et
sinuum angulorum flexus N_2CH, quos sinus vocabimus Σ; erunt ipsae $_3CK$, in
directa ratione Σ, CH, CP, $_2C_3C$ et reciproca PH, CH, ergo erunt ipsae $_3CK$ in
directa ratione Σ, CP, $_2C_3C$ et reciproca PH seu conatus ⟨centra⟩ paracentrici   65
sunt in directa ratione velocitatis in curva ($_2C_3C$), sinuum angulorum
deflexus (Σ), et radiorum CP, et reciproca distantiarum tangentis a centro PH;
paulo ante autem eadem inveneramus, nisi quod pro sinubus deflexus prodierant
ipsi anguli deflexus, sed in angulis inassignabilibus sinus sunt angulis
proportionales. ⟨Ducantur duae parallelae AE, et RF ⟨quarum si placet una PR   70
transeat per P centrum, at⟩ ⟨Ex centro P ducatur⟩ Sumatur si placet recta
constans quaecunque positione et magnitudine determinata, ut PE (educta si
placet e centro P, et parallela axi curvae AR cui PR perpendicularis ex curva
occurrit in R) et angulo PES, aequali ipsi HCP, educatur ES occurrens ipsi PS
et PE normaliter eductae (ita ut ES semper cadant inter parallelas AE, PR).   75
Patet esse ES{:}EP{::}PC{:}CH, ergo quia EP constans fient ES, ut PC{:}CH.
⌊Sint autem ES ut secantes anguli PES vel CPH ergo denique data curva
quacunque CC assumto centro quocunque P conatus paracentrici $_2CG$ (vel $_3CK$)
sunt in composita ratione velocitatum in curva, ($_2C_3C$) angulorum deflexus
curvedinis et contactus, (Σ) et secantium pro angulis tendentiae centralis ad   80
tendentiam in curva seu radii PC, ad tangentem curvae CH.⌋

⟨r.⟩   Sed ut definiamus in qua ratione sint conatus paracentrici, idque generali
quadam ratione, ex centro P educatur perpendicularis PN ad tangentem curvae
CN, nempe ad $_1C_2CN$ et eadem PN proximam tangentem $_2C_3CH$ secet in H, denique
ex H ad CN ducatur HM parallela PC, erunt triangula HMN et PNC similia. Ergo   85
PC{:}PN{::}HM{:}HN. Rursus $_3CK{:}HM{::}_2C_3C{:}CH$. Ergo $_3CK\cdot PN{:}HN\cdot PC{::}_2C_3C{:}CH$, seu
$_3CK$ sunt ut: $HN\cdot PC\cdot_2C_3C{:}PN\cdot CH$. ⟨Jam HN, sunt in composita ratione sinuum anguli
curvedinis seu deflexus⟩ Jam $_2C_3C$ sunt ut velocitates mobilis in curva tanquam orbita.
HN sunt in composita ratione ipsarum CH, et sinuum anguli deflexus ⌊quem facit
impulsus centralis, a priore directione,⌋ sive curvedinis N_2CH seu K_2C_3C; vel   90
(quia anguli infinite parvi sive inassignabiles sunt ut sinus,) in composita
ratione ipsarum CH, et sinuum quos habent anguli deflexus K_2C_3C (quorum
anguli externi sunt $_1C_2C_3C$). Hos angulos vel sinus eorum vocemus Σ, fient HN,
ut Σ CH, et HN{:}CH ut Σ. Denique rationes PC ad PN, (seu PC{:}PN) sunt ut secantes
angulorum PCN, quos secantes vocemus ϖ. Erunt ergo PC{:}PN ut ϖ. Ergo in   95
proportionalibus ipsis K_3C, HN{:}CH et PC{:}PN, substituendo proportionales illis
Σ, his ϖ erunt K_3C, ut $_2C_3C\cdot\varpi\cdot\Sigma$.

Ergo denique in curva quacunque CC, proposito $_3CK = {}_1C_2C\,\Sigma\,\varpi/a^2$ posito
centro quocunque P conatus paracentrici $_2CG$ Σ esse sinus angulorum
(vel $_3CK$) sunt in composita ratione ⌊1mo⌋   deflexus, ϖ esse   100
velocitatum mobilis in curva tanquam in   secantes angulorum quos

orbita (seu percursorum ea velocitate elementorum curvae, $_2C_3C$) ⌊2do⌋ angulorum deflexus (K_2C_3C seu Σ) quos a curvae directione $_1C_2CK$, facit centralis conatus $_2CG$ divertens mobile a $_2CK$ ad $_2C_3C$); ⟨et angulorum secantium σ͝ pro angulis HCP quos faciunt tendentiae centrales⟩ ⌊3tio⌋ et secantium (σ͝) quos habent anguli (HCP) directionum ⌊seu tendentiarum⌋ curvae (CH) ⟨et radiorum⟩ et tendentiarum centralium (CP) sive anguli radiorum ex centro, (PC) ⌊seu radiorum ex centro ad curvam,⌋ et tangentium curvae (CH). ⌊Hinc eadem manente curva et velocitate, mutatis centris conatus paracentrici sunt ut secantes angulorum quos radii ex diversis centris in eodem puncto ad curvam faciunt.⌋

radii ex centro faciunt ad curvam, at a esse radios.

⌊Sit E centrum circuli curvam osculantis in CC, patet angulum $_2CE_3C$ aequalem esse angulo deflexus K_2C_3C. Sunt autem ⟨velocitates⟩ elementa curvae in composita ratione radiorum osculi EC, et angulorum deflexus. Ergo anguli deflexus sunt in ratione composita ⟨velocitatum⟩ elementorum directa et radiorum osculi reciproca. Ergo denique conatus paracentrici sunt in ratione duplicata velocitatum composita cum simplicibus directa quidem secantium quos habent anguli radiorum ex centro conatus ad curvam facti reciproca vero radiorum osculi. Si in radio osculi CE sumatur CT aequalis sinu totiangulo mensuranti, a, unde recta TΩ occurrat radio centrali PC in Ω, erit CΩ secans anguli quem facit radius paracentricus ad curvam, denique juncta EΩ, ducatur Tπ parallela ipsi EΩ, patet fore Cπ ad sinum totum CT, ut CΩ secans ad CE radium osculi. Ergo K_3C conatus paracentrici sunt in ratione composita ex celeritatis $_1C_2C$ duplicata, et rectarum Cπ simplice. Eritque $K_3C - _1C_2C^2Cπ : a^2$ vel $K_3C = _1C_2C^2 \cdot CΩ : CT \cdot CE$. Posito $_1C_2C$ esse velocitatem mobilis in curva, CΩ secantem anguli quem facit radius paracentricus ad curvam, CT sinum totum, CE radium circuli curvam osculantis. In circulo si centrum conatus sit in circuli centro, coincidunt E et P ct Ω ct fit $K_1C = _1C_3C^2 : CP$. Concludendo ⟨conatus paracentricus aequatur⟩ progressus conatus paracentrici (K_3C) aequatur quadrato elementi curvae ($_1C_2C$) seu progressus mobilis in curva ducto in secantem anguli quem facit radius paracentricus (PC) ad curvam, et applicato ad radium circuli osculantis, (CE) et sinum totum.⌋

Possunt etiam anguli deflexuum exhiberi in circulo, ut si centro P radio quocunque PF describatur circulus, ducanturque radii tangentibus punctorum C curvae CC paralleli, ut PF ipsi $_1CN$, et PG ipsi $_3CH$, angulus FPG exprimit quantum directio mobilis inter progrediendum in curva ab $_1C$ ad $_3C$ deflexa sit, qui angulus si sit infinite parvus, erunt anguli deflexuum momentanei seu incrementales. Fingi autem posset mobile dum curvam describit ferri in regula aliqua rectilinea, et durante motu regulam semper

⟨fig. 18⟩

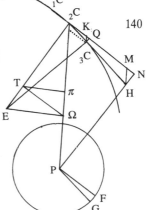

105

110

115

120

125

130

135

140

gyrari circa mobile ipsum tanquam centrum, quod fiet si mobile grave sit, ut
regulam plano subjecto fortiter apprimat, impulsus enim in regulam factus
ipsam aget circa mobile tanquam centrum.   145

Ope huius motus, si regulae circumactio ponatur aequabilis, dabitur modus   NB NB
assignandi angulos laterum curvae aequales inter se, quo casu curvae
evolutione descriptae elementa sunt ut arcus evoluta, et hinc sequitur quod
areae evolutione descriptae posita rectificatione curvae semper sunt
mensurabiles, seu ut arcuum evolutorum ⟨quadrata⟩ cubi; quia autem fictio   150
haec in potestate est, patet rem semper veram esse.

⟨v.⟩   ⟨pag. 2⟩

huius paginae   ‖ Sed ut definiamus ⟨quantitatem⟩ progressum conatuum centralium
contenta sunt   ⟨Evolvi intelligatur curva⟩ ⟨Considerandum quid fiat si
resumenda   inter evolvendum filum ponatur in versus extremitatem   155
            alicubi regidum, vel rigido attigatum, et ex rigido exeat⟩

     Considerandum diligenter quae prodeat linea, si ⟨in orbe aliquo plano
ponatur⟩ planum (horizontale) agatur circa aliquod centrum, per sua transiens
vestigia ⟨(instar orbis moti circa⟩ ⌊seu circa⌋ axem sibi perpendicularem (wie
ein Teller;) et corpus libere alicubi in plano illo positum, extra centrum,   160
⟨vi⟩ conatu centrifugo seu recedendi a centro per tangentem ex ipsa
circumactione concepto continue incitetur. Nempe sit centrum P mobile M, quod
rotatione plani P_1M_2M circa P incitatum transeat ex $_1M$ in $_2M$. In P_1M sumatur

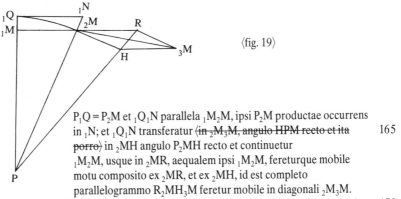

⟨fig. 19⟩

$P_1Q = P_2M$ et $_1Q_1N$ parallela $_1M_2M$, ipsi P_2M productae occurrens
in $_1N$; et $_1Q_1N$ transferatur ⟨in $_2M_3M$, angulo HPM recto et ita   165
porro⟩ in $_2MH$ angulo P_2MH recto et continuetur
$_1M_2M$, usque in $_2MR$, aequalem ipsi $_1M_2M$, fereturque mobile
motu composito ex $_2MR$, et ex $_2MH$, id est completo
parallelogrammo R_2MH_3M feretur mobile in diagonali $_2M_3M$.

Hae lineae variabuntur si planum non ponatur esse unum continuum gyrans, sed   170
interscissum in orbes concentricos inaequaliter gyrantes per quos dum
transit mobile, diversos per tangentem accedendi impetus concipit. Differunt
autem lineae istae ⟨vi⟩ conatu tangentem prosequendi descriptae, a lineis
projectitiis paulo ante descriptis, nam illic R_3M parallela fuisset ipsi P_2M,
hic est ipsi $_2MH$, adeoque perpendicularis ad radium PM, cum in projectitiis   175
sumatur in radio. An dicemus potius grave duplici motu ferri, uno orbis cui
insistit; altero quem descripsimus, illaque linea composita apparebit
spectanti extra orbem posito, linea autem a nobis designata, describetur in
ipso orbe. Idem igitur videtur observandum et in motu projectitio, et si enim
nobis in orbe nostro circumeunte positis linea projecti gravis appareat   180

parabolica, tamen si quis spectaret eam ex alio orbe, erit ipsi composita ex
~~(parabolica)~~ motu parabolico et circulari, et si gyratio orbis in locis a centro
recedentibus minus crescat, quam pro distantia a centro, non quidem motus
compositus erit ex lineis ~~(parabolica et cir)~~ projectitia et circulari,
sed ex conatibus in projectitia et in aliquo circulo. Has compositiones non   185
videtur Neutonus in rationes revocasse, quae tamen videntur habere locum. ⌊Et
hoc modo in orbe ubi non aequabilis gyratio non videntur gravia in ipso orbe
spectata recta ad centrum ire.⌋ Considerandum an fluidi partibus recedere
conantibus per tangentem, solidae quae minus recedunt eo ipso in sola recta
linea detrudantur, an potius retineant aliquid et a suo proprio impetu, et hoc   190
videtur verius. Itaque si gravia ubique in recta linea feruntur ad centrum
terrae, non poterit derivari vis gravium a ~~(tali circulatione)~~ rejectione per
tangentem. Sed motus varii in omnes partes in fluido erunt assumendi
fortiores tamen in majore a centro distantia, vel saltem majoris materiae, ubi
revera est, et ita grave recta detrudetur.   195

Commentary to *De Conatu*

De Conatu can be divided into four sections, which approximately
correspond to the four sides of the manuscript. The first side contains
some introductory remarks on centrifugal and centripetal conatus, which
is called *paracentricus*. The second section comprises an attempt to
determine paracentric conatus for a body moving along a curve. This
attempt is based on the construction of the evolvent of the curve
described by the body. Half-way down the second side Leibniz realizes
that his geometrical construction is redundant, and after a few more lines
he crosses out the whole of the second side and starts again on the third.
There Leibniz succeeds in finding the most general expression for the
paracentric conatus of a body moving along an arbitrary curve with
respect to an arbitrary centre. The fourth side is largely devoted to some
reflections on motion studied by different observers in motion or at rest.
It is not until the very end of the essay that mathematical calculations
give way to physical conjectures on the action of aethereal fluids and on
gravity.

    Within the manuscripts I found, and according to my dating, this is the
first time Leibniz uses the term *paracentricus*, meaning 'towards or from
a centre' (line 1). As we have seen in the *Notes*, Leibniz realized that the
area law is valid for all central forces regardless of their sign. Now his
coinage designates a concept which corresponds to this property. From
now onwards paracentric conatus will be a distinctive element of his
theory of planetary motion. However, from the title of the essay it is
evident that by 'paracentric' Leibniz understands 'either centrifugal or

centripetal', not both at the same time, and this is a notable difference with respect to the *Tentamen*.

Leibniz considers an arbitrary curve *CCC* and an arbitrary point *P*, and tries to determine the centrifugal conatus with respect to *P* with which a body moving along *CCC* tends to escape. He claims that this leads to the same result as the calculation of centripetal conatus (lines 5–10). This and other similar remarks in the essay cast some light on his views concerning curvilinear motion. Referring to fig. 16, Leibniz seems to identify centrifugal conatus with K_3C, in the concave portion of the plane delimited by the curve *CCC*; $_2CG$, which represents centripetal conatus, is in the convex part of the plane (in fig. 16 there are two letters *G*). From 1689 onwards Leibniz believed that centrifugal and centripetal endeavours were different in general. In the *Tentamen* he calculated centrifugal conatus in a different way, and also employed a different expression for the outward tendency in an arbitrary curve, namely *conatus excussorius*, which in the present case would correspond to q_3C (fig. 18). In the particular case of the circumference, Leibniz retains the Huygensian term *centrifugus*.

In the first paragraph Leibniz's concepts and terminology are very uncertain. He claims that centrifugal or centripetal conatus is infinitely smaller than the conatus along a rotating ruler *PC*, which is a component of the orbital motion of the body (lines 16–17). In more detail, he seems to imagine motion along *CCC* to result from the component along the ruler *PC*, and the component due to the rotatory motion of the ruler; these ideas are developed in *Inventum a me est*, *De Motu Gravium*, and appear in print in paragraph 3 of the *Tentamen*. Here Leibniz states that the conatus along the ruler is very different from paracentric conatus, which is infinitesimal with respect to it (lines 14–16). Indeed, the conatus measured in a Newtonian way by the distance from the tangent does differ from the conatus measured along the rotating ruler. In *De Motu Gravium* Leibniz began to realize that the conatus measuring the deviation from the tangent is only attractive, whereas the conatus along the rotating radius is the difference between two terms, infinitesimal of the same order, one repulsive and the other attractive. Motion along the ruler varies according to its degree of rotation; the combination of these motions gives the resultant *CCC*; centrifugal conatus, on the other hand, would always be represented by the same segment ('conatus centrifugus semper est idem', line 20). Leibniz probably believes that the rotatory motion of the ruler can vary, therefore radial motion along it must also vary accordingly in order to produce the given curve *CCC*. In the case of a component along the tangent to the curve and a component along the radius, paracentric conatus is uniquely determined. Leibniz does not explain why centrifugal conatus should be infinitely smaller than the conatus along the rotating ruler. Furthermore, although he is talking about infinitesimals of different orders, he always uses the term 'conatus' and does not distinguish between impetus and conatus or solicitation, as he does in paragraph 5 of the *Tentamen*. The figure referred to in line 20 probably corresponds to fig. 18.

In the second paragraph Leibniz states that the element of conatus is infinitely smaller than the velocity in the curve (lines 27–9); this shows how equivocally the term 'conatus' is used, meaning first- or second-order differential interchangeably. He considers two infinitesimal arcs of curve, $_1C_2C$ and $_2C_3C$. From $_2C$ he prolongs the chord $_1C_2C$ to K, so that $_2CK$ is equal to the chord $_1C_2C$. K_3C is parallel and equal to G_2C and $_3CG$ is equal and parallel to $_2CK$. Leibniz imagines that the motion of the body is decomposed into $_2CK$ and K_3C, which is called 'conatus centrifugus vel magneticus'. $_3CQ$ is the prolongation of the radius P_3C. Leibniz correctly claims that since the angle $K_3\hat{C}Q$ is infinitesimal, KQ is infinitesimal with respect to $_2CK$. This follows because the central conatus can be represented as interrupted in infinitesimal intervals, and during each interval it is parallel to itself (lines 35–7). Leibniz considered curvilinear motion as composed of infinitesimal rectilinear segments, which are the chords $_1C_2C$, $_2C_3C$, etc. During the time in which the body traverses each of them, the central conatus remains parallel to itself and deviates the body acting like an infinitesimal velocity. Therefore, Leibniz does not consider accelerations, but merely instantaneous infinitesimal changes of intensity of the central conatus at each vertex $_1C$, $_2C$, $_3C$, together with their directions. This is confirmed by what Leibniz says on the second side of the manuscript, where he makes clear that the tangents are elements of the curve, or prolongations of the chords (line 39).

The ensuing calculations result from the parallelism between P_2C and HM, K_3C and $_1VW$, and from the similarity between triangles $_2CPN$ and MHN (see fig. 17; lines 42–). He establishes the proportion $K_3C \cdot PH : HN \cdot CP :: _2C_3C : CH$, and this gives K_3C.

Leibniz's reasoning is based on the evolvent $_1V_2V$ of the evolute $[_3C_2C_1C$. The evolvent is constructed by unfolding a well-stretched string from the position $[_3C_2C_1C$ to $_1V_3C_2V$ and $_2V_2C_1C$. In figs 16 and 17 $_1V_2V$ should meet $_3C_2C_1C$ at right angles. It is not until half-way down the second manuscript side that Leibniz realizes that his geometrical construction is redundant. From the equation $K_3C = _1V_2V \cdot CP \cdot _2C_3C : PH \cdot CV$, since the angle of deflection is proportional to $_1V_2V : CV$ and can be expressed without the evolvent, we have that K_3C is equal to the angle of deflection times $CP \cdot _2C_3C : PH$ (lines 59–60, see Sections 2.3 and 5.3).

On the third side Leibniz repeats *mutatis mutandis* the same arguments without using the evolvent. He obtains the equation:

$$K_3C = {_2}C_3C \cdot HN \cdot PC : PN \cdot CH = {_2}C_3C \cdot \hat{\Sigma} \cdot \omega \, ;$$

$HN:CH$ is equal to $\sin K_2\hat{C}_3C$, where $K_2\hat{C}_3C = \hat{\Sigma}$, or, since the angle is infinitesimal, to $\hat{\Sigma}$ itself, which is the angle of deflection; $PC:PN = 1:\sin P\hat{C}N = \omega$, which is the secant of the angle $P\hat{C}N$. The equation above for K_3C is an important result which expresses paracentric conatus (Newton would say 'centripetal force') for a given curve CCC and a given centre P.

The following paragraph corresponds to a series of marginal notes later inserted in the text by Leibniz (lines 113–34). He points out that the angle of deflection $K_2\hat{C}_3C$ between two tangents to infinitesimally distant points is equal to the angle ${_2}C\hat{E}_3C$, where E is the centre of the osculating circumference (see fig. 18). This can be easily proved both when ${_2}CK$ is the prolongation of the chord or is the standard tangent. Since the angle ${_2}C\hat{E}_3C$ is equal to ${_2}C_3C$ over the radius EC, we have that paracentric conatus is directly proportional to the square of ${_2}C_3C$ (which represents velocity) and to the secant $\omega = PC:PN$, and inversely proportional to the radius of the osculating circumference CE. Thus $K_3C = ({_2}C_3C)^2 \, \omega : EC$.

In the same paragraph (lines 120–) Leibniz writes his equation in a slightly different fashion. He fixes an arbitrary constant $CT = a$ along the radius CE (see fig. 18); $T\Omega$ is perpendicular to CT; $C\Omega$ is CT times the secant of the angle $P\hat{C}K$ between the paracentric radius CP and the curve CCC; $E\Omega$ and $T\pi$ are parallel. We have $C\pi : CT :: C\Omega : CE$ and, since CE and PN are parallel, paracentric conatus can be expressed as $K_3C = ({_1}C_2C)^2 C\pi : a^2$ and $K_3C = ({_1}C_2C)^2 C\Omega : CT \cdot CE$. In the special case when the curve CCC is a circumference and its centre is the centre of attraction, we have the well-known result $K_3C = ({_1}C_2C)^2 : CP$.

We return to the text after the marginal notes (line 135). The attempt to represent the angle of deflection in a circumference centred in P is followed by an obscure passage in which Leibniz seems to adumbrate an explanation for gravity. He claims that while the body describes the curve, it is also driven by a rectilinear ruler rotating round the body taken as a moving centre. Interpreting Leibniz's thought, I believe that the ruler must have two motions, one around the body and the other along the trajectory of the body. This passage seems to echo some pages by Descartes, *Principia Philosophiae*, part 3, propositions 59–61. Leibniz goes on to claim that this would be the case if the body were heavy, therefore pressing the ruler more vigorously, and that the impulse exerted on the ruler sets it in motion around the body taken as a centre.

At the bottom of the side (lines 146–51), in a portion crossed out later, Leibniz states that if the motion of the ruler is uniform, it is possible to assign equal angles of deflection between the infinitesimal sides of the curve. I try to explain his reasoning by means of the auxiliary fig. (g). DB is the evolvent of the evolute AB, namely DB is produced by unrolling a stretched thread from B to D; hence AD is equal to the evolute AB. The elements FZ of the evolvent are equal to the arcs $BS = SZ$ of the evolute times the angle of deflection $\widehat{FEZ}$.

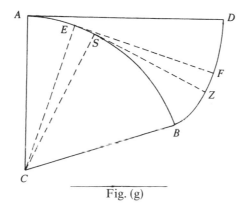

Fig. (g)

According to Leibniz the area between the evolute AB, the evolvent BD and its radius AD is proportional to the third power of AB. Calling the infinitesimal arc of the evolute $ES = ds$, the arc of the evolute $BS = SZ = s$, the infinitesimal angle of deflection $\widehat{FEZ} = d\theta$, we have that $FZ = sd\theta$; since $d\theta$ is constant, Leibniz can say that the elements of the evolvent are as the arc of the evolute. The area of the triangle $FEZ = s^2 d\theta : 2$. Further, $d\theta = ds : CE$, where $CE = CS$ is the osculating radius of the evolute. In conclusion, the infinitesimal element of the area is equal to $s^2 ds : 2CE$, and this is proportional to s^3 only if CE is constant, or if the evolute AB is a circumference.

We have now reached the last side of the essay; the first crossed-out lines (152–6) were marked in the margin as 'page 2', and were originally the continuation from the first side. A later note in the margin shows that Leibniz thought the content of this side worth reconsidering. He studies the case of a horizontal plane rotating along its axis; the remark in German enclosed in brackets reads 'wie ein Teller', like a plate (lines 159–60). He considers the case of a body moving freely on the rotating plane under the action of centrifugal conatus due to the rotation. Leibniz imagines that under the initial conditions the body moves from

$_1M$ to $_2M$ (see fig. 19), perpendicularly to the radius PQ. $_1M_2M$ fixes the initial velocity; in the successive interval of time the body would move from $_2M$ to R because of its inertia. During this interval of time, however, the body, while moving from $_2M$, also has a tangential velocity $_2MH$, perpendicular to P_2M, due to the rotation of the plane from $_1Q$ to $_2M$. Therefore, the resultant will be the diagonal $_2M_3M$. Leibniz states that the curves so described would be different if the rotating plane were divided into concentric orbs, each rotating with a different velocity (lines 170–2). He is beginning to look for mechanical explanations.

Leibniz claims that a heavy body has a twofold motion: the first is common to the motion of the orb in which it is situated; the second is the 'motus projectitius' described above (lines 176–9). Their composition will be observed from outside the orb. He also gives an example of this composition: a line described by a falling body appears parabolic to us, but if it were observed from outside our orb it would be the composition of a parabolic path and a circular motion (lines 179–82). If the circular motion increases in a proportion which is inferior to that of the radius, or, as we would say, if angular velocity decreases in some relation with distance, Leibniz believes that it would be necessary to use endeavours, that is, differentials. In his opinion these remarks were neglected by Newton (lines 185–6). Further, in an orb in which rotation is not uniform, heavy bodies seen from the same orb in which they are situated would not move towards the centre.

With these qualitative remarks on relativity of motion Leibniz seems to be looking for a complement or an alternative to Newton (lines 185–). The very end of this manuscript is devoted to some physical considerations on the actions of fluids with respect to gravity. Two alternatives are considered: either the particles of the rotating fluid receding along the tangent push more solid bodies, with a smaller tendency to recede along the tangent, in a straight line towards the centre; or the solid bodies tend to keep some of their impetus. The second alternative appears to be more reasonable to him. Therefore, if heavy bodies tend everywhere on the surface of the Earth along straight lines towards the centre of the Earth, this cannot be explained by means of their rejection along the tangent, but it is necessary to assume motions in all directions in the fluid, which are swifter the further away from the centre, where there is more matter; this would explain vertical descent.

The reference to a motion which is swifter the further away from the centre implies a law opposite to the harmonic circulation; hence this explanation of gravity cannot be easily extended to planetary motion. Soon we shall see Leibniz shifting his interest from vertical descent to Kepler's laws.

3   *Inventum a me est*

LH 35, 10, 7, f. 36–7

This manuscript consists of one sheet folded in quarto. The last side is blank. The essay is on a kind of paper used by Leibniz in Vienna in 1688. The watermark is identical with that numbered 510 in the catalogue at the NLB, letters 'M R' of a slightly different type from those of f. 34–5. The edge matches perfectly that of f. 16–7, *De Motu Gravis in Linea Projectitia*, on which compare Section 5.4, and this suggests a close date of composition for the two essays.

The first paragraph contains a reference to a text which can be easily identified as *De Conatu*: the analysis of paracentric conatus in the two essays is identical. Leibniz's investigations are purely mathematical. As in the previous manuscript, Leibniz does not mention the *circulatio harmonica*, and this is even more striking because here he develops ideas leading to that concept. Although Leibniz begins to analyse motion along a rotating ruler, paracentric conatus is still calculated with respect to the deviation from the tangent, as Newton had done. Thus conceptually this essay is a hybrid between two approaches. *Inventum a me est* was composed in Vienna in autumn 1688, shortly after *De Conatu* and before the *Tentamen*.

> Inventum a me est et alia scheda explicatum, mobilis in orbita curva
> incedentis conatus ad punctum aliquod certum tanquam centrum respicientes,
> sive paracentricos, esse in composita (1) ratione velocitatum mobilis in
> orbita, (2) angulorum deflexus, et (3) secantium quos habent anguli radiorum
> ex centro eductorum cum curva concursu facti, vel esse in quadrata ratione   5
> celeritatum composita cum directa secantium ipsorum angulorum quos faciunt
> radii paracentrici ad curvam, et reciproca radiorum quos habent circuli
> osculantes.
>
> Pergamus jam ad lineas projectitias nempe gravis aut levis alicuius
> corporis, quod impetu aliquo jam impresso fertur, et interim vi gravitatis aut   10
> levitatis aut magnetismi, ad centrum aliquod tendit
> vel ab ipso recedit. In his enim peculiares debent
> esse respectu conatus paracentrici proprietates.
> Ponamus igitur mobile C moveri in curva $_1C_2C_3C$ etc.
> et cum esset in $_1C$ habere velocitatem et   15
> directionem $_1C_2C$, ita ut dato temporis incremento

⟨fig. 20⟩

aequabili absolvat $_1C_2C$, ergo si nihil mutaret, continuata eadem directione
aequali tempore percurreret $_2CK$ aequalem ipsi $_1C_2C$, et cum ea in directum
jacentem. Itaque in $_2C$ positum, habet ab impetu concepto conatum et
directionem $_2CK$, ponamus jam a gravitatis impulsu accipere conatum　　　　　20
paracentricum ut $_2CG$. Itaque ducta K_3C parallela et aequali ipsi $_2CG$, ad
partes P (nam si conatus esset levitatis ad contrarias duci deberet) et
jungatur $_2C_3C$, erit utique $_2C_3C$ velocitas et directio composita ex ambabus
qua mobile perget in curva projectitia.

　　　Cogitari potest eadem linea describi motu composito ex circulari regulae　　25
indefinitae P⅃ circa centrum P, et rectilineo mobilis C in regula versus
centrum tendentis, quanquam ad ipsum pervenire ⌊in linea huiusmodi projectitia⌋
nunquam possit. Ex $_3C$ in P_2C demittatur perpendicularis $_3C_2L$, manifestum est
in angulo $_2CP_3C$ inassignabili, haberi posse pro arcu, et perinde esse ac si
mobile ex $_2C$ in $_3C$ feratur motu composito ex $_2L_3C$, et ex $_2C_2L$. Sunt ergo ipsae　30
$_2L_3C$ progressus circulares in composita ratione angulorum progressus
circularis, et radiorum, seu distantiarum a centro PC. Jungatur PK. Triangulum
$_1C_2CP$ aequale est triangulo $_2CKP$, (quia idem est vertex P, et bases aequales
$_1C_2C$, et $_2CK$, in eadem recta $_1C_2CK$) et triangulum $_2CKP$ aequale est triangulo
$_2C_3CP$ (sunt enim super eadem basi P_2C, inter parallelas easdem P_2C, et K_3C).　35
Itaque triangula $_1C_2CP$ et $_2C_3CP$, seu areae radiis abscissae incrementa,
aequalibus existentibus temporum incrementis (eo ipso quia $_1C_2C$ et $_2CK$
sumsimus aequales) sunt aequalia, areae igitur radiis ⌊ex centro emissis⌋
abscissae, (quae ex his triangulis componuntur) sunt ut tempora motu
projectitis gravis insumta. Porro triangulum $_2C_3CP$ fit ex ductu radii P_2C in　　40
progressum circularem $_2L_3C$, et similiter triang. $_1C_2CP$ fit ex P_1C in $_1L_2C$. Et
generaliter triangulum C(C)P quod est incrementum areae radiis abscissae, fit
ex ductu radii PC in progressum circularem LC. Itaque incrementa arearum
radiis abscissarum sunt in composita ratione radiorum et progressuum
circularium. Sunt autem incrementa arearum aequalia, aequalibus temporum　　45
incrementis ⌊imo sunt incrementa arearum ut temporum⌋ ⟨Ergo sumtis aequalibus
temporum incrementis in motu projectitio, erunt radii paracentrici reciproce ut⟩

⟨v.⟩ ⌊Itaque generaliter temporum incrementa sunt in composita ratione radiorum
paracentricorum, et progressuum circularium.⌋ Ergo sumtis aequalibus temporum
incrementis in motu projectorum, erunt radii paracentrici PC reciproce ut　　　50
　　　　　　　　　　progressus mobilis circulares L(C).

　　　　　　　　　　Sunt autem progressus circulares L(C) in
　　　　　　　　　　composita ratione radiorum P(C) et angulorum
　　　　　　　　　　circulationis CP(C). ⟨Si aequabilis sit motus
$$L(C) = \frac{1}{PC} = PC \cdot ang$$
　　　　　　　　　　projecti gravis in sua orbita⟩ ⌊Ergo temporum　　55
　　　　　　　　　　incrementa sunt in composita ratione duplicata
$$ergo\ PC \cdot PC \cdot ang = 1$$
　　　　　　　　　　radiorum paracentricorum, et simplice angulorum
　　　　　　　　　　circulationis incrementalium.⌋ Ergo aequalibus
　　　　　　　　　　sumtis temporum incrementis in motu projectitio
　　　　　　　　　　gravis alicujus vel levis centrum certum　　　　60
　　　　　　　　　　respicientis, erunt anguli circulationum circa
　　　　　　　　　　centrum in ratione duplicata reciproca radiorum ⟨ex
　　　　　　　　　　centro emisso⟩ paracentricorum. Ut autem

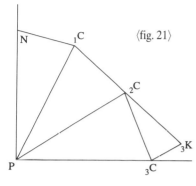

⟨fig. 21⟩

exprimantur integra tempora, sumatur recta ex A
puncto assumto infinite producta AB inque ea   65
sumantur ipsae AB aequales ipsis PC, et ⌊ad has
ordinate⌋ ipsae BE proportionales areis PCN, N
assumto pro motus initio, itaque si A⌋= PN habebimus
lineam ⟨⌐⟩EE temporum seu arearum; per A ducatur AF
parallela ipsis BE, et porro ducantur EF parallelae   70
et aequales ipsis AB. Patet temporibus ⌊vel areis⌋
existentibus AF, radios ⌊paracentricos⌋ fore AB vel
FE, in AF sumatur AQ acqualis ipsi A⌋ et per Q
ducatur QH parallela ipsi AB, et ET curvam tangens
in E, occurrat axi AB in T, et ipsi AH in H. Sumatur   75
in QH ipsa QR aequalis AT et jungatur TR
perpendicularis ad QH. Erunt ipsae RH ut radiorum

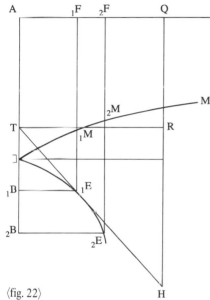

⟨fig. 22⟩

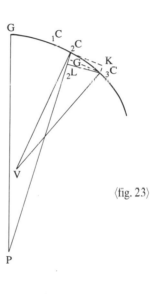

⟨fig. 23⟩

incrementa LC seu ut progressus rectilinei. Progressus autem circulares cum
sint reciproce ut radii, et angulorum seu circulationis incrementa reciproce
ut radiorum quadrata, itaque si ducatur linea ⌐MM talis, ut FM sint ad A⌋   80
reciproce ut quadratum A⌋est ad quadratum FE, erunt ordinatae FM ut
circulationum seu angulorum incrementa, et spatia ⌐AFM⌋ut anguli integri
NPC.

Nunc istis repertis adjungamus quod ex alia scheda repetivimus, ⟨conatus⟩
⌊progressus⌋ paracentricos ut $_3$CK esse in duplicata ratione progressuum in   85
orbita $_2$C$_3$C composita cum directa secantium angulorum PCK inter radios et
curvam et reciproca semidiametrorum VC quos habent circuli osculantes ⌊seu
angulorum contactus sive curvedinum⌋. Itaque nunc demum, omnia ad calculum

revocemus. Radius paracentricus PC sit x, circulatio seu angulus sit y et LC
erit dx, et incrementum circulationis seu angulus CP(C) erit dy, (sumtus pro   90
elemento arcus circuli dati radii a⟨⟩⟩, est autem progressus circularis
$L(C) = dy \cdot x : a$ et areae incrementum $L(C)$ in x erit dy x = dt, quia repraesentat
tempus; seu $t = \int \overline{dy}$ x. Et porro elementum curvae $_2C_3C = \sqrt{dx^2 + \overline{dy}^2} = \overline{dc}$. Sunt autem
elementa curvae in composita ratione incrementorum temporalium $\overline{dt}$,

⟨r.⟩   et velocitatum in orbita quas vocabimus w, seu $\sqrt{dx^2 + dy^2} = dy$ xxw:aa. Jam   95
velocitates mobilis in sua orbita, seu conatus pergendi in tangente, sunt ad
velocitatem circulationis $\overline{dy}$ seu ad conatum rectae mobilis in circulo moti
cuius arcus sunt mensurae angulorum, ut $\sqrt{dx^2 + \overline{dy}^2}$ ad $\overline{dy}$. Ergo w = ⟨interrupted⟩
et velocitatum in orbita, quas vocabimus w, fiet: $\sqrt{dx^2 + \overline{dy}^2} = \overline{dy}$ xw:a. Et
$w = a\sqrt{dx^2 + x^2/a^2 \ \overline{dy}^2} : x \cdot \overline{dy}$. Seu w velocitates sunt in composita ratione directa   100
quidem $\sqrt{dx^2 + dy^2}$ seu $C(C)$ ad dy seu $L(C)$ et inversa radiorum x. Jam $C(C)$ est ad
$L(C)$ ut secans anguli sub curva ⟨et⟩ radio paracentrico ad sinum totum, ergo
velocitates in orbita sunt in composita ratione directa secantium quos habent
anguli curvae ad radios paracentricos et reciproca radiorum paracentricorum.

Et quia secans anguli ad curvam est ad sinum totum ut $\sqrt{dx^2 + \overline{dy}^2}$ ad dy seu ut   105
$x\sqrt{dx^2 + \overline{dy}^2}$ ad xdy fiet sec. in $x\overline{dy}/a$ = sin. tot. in $x\sqrt{dx^2 + \overline{dy}^2}$. Seu erunt
incrementa temporum in ratione ⌊composita⌋ progressuum orbitae, et radiorum, et
reciproca secantium quos habent anguli curvae ad radios. Et progressus
paracentrici ⌊seu incrementa descensus vel ascensus⌋ in projectitia linea,
erunt in triplicata progressuum in orbita et reciproca composita progressuum   110
circularium atque ⟨radiorum⟩ ⟨semidiametrorum osculi seu angulorum contactus
seu⟩ curvedinum. Curvedines autem seu angulos contactus intelligo qui sunt ut
semidiametri circulorum curvam osculantium. Superest ergo ut radii isti seu
semidiametri inveniantur. Ubi tamen nullum inde novum compendium generale
prima fonte deprehendo tantum enim deprehendo semidiametrum osculi seu   115
mensuram curvedinis CV esse = $\overline{dx} : d \ \overline{dy} : \overline{dc} + \overline{wdx} : \overline{adc}$. Posito w esse elementum anguli
sive circulationis sive elementum arcus circuli cuius radius a, quo casu
$\overline{dy}$ = xw:a. Jam d $\overline{dy} : \overline{dc}$ = dy $\overline{ddc}$ – dc $\overline{ddy}$, : $\overline{dc}^2$. Ergo in linea projectitia elementum
descensus vel ascensus gravis aut levis projecti seu K_3C est

$$= \overline{dc}^3 \, d \ \overline{dy} : \overline{dc} + \overline{wdx} : \overline{adc} : dx \cdot dy \text{ seu}$$ 120
$$K_3C = \overline{dc}^2 \cdot \overline{dy \ ddc} - dc \ \overline{ddy} : dc + \overline{wdx} : a : dx \cdot dy.$$

4   *Investigatio Semidiametri Circuli Osculantis*

LH 35, 10, 7, f. 31

The essay reproduced here covers one side of a quarto manuscript sheet.
The verso is blank. The contents are directly connected with *Inventum a
me est*, because the equation for the radius of the osculating circumference attained in the present text is used in *Inventum a me est* in order
to determine paracentric conatus.

Investigatio semidiametri circuli curvam in proposito
puncto osculantis, si pro ordinatis adhibeantur convergentes
ad certum punctum

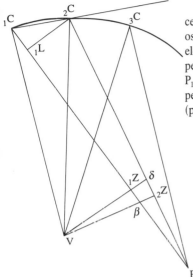

Sit curva $_1C_2C_3C$, unde radii convergentes P ad
centrum radiorum P; quaeritur V centrum circuli 5
osculantis, adeoque CV eius semidiameter, sint
elementa arcuum $_1C_2C$, $_2C_3C$, unde ex $_2C$ et $_3C$ extremis
perpendiculariter eductae concurrent in V. Ex $_2C$ in
P_1C perpendicularis $_2C_1L$ et ex V in P_1C ct P_2C,
perpendiculares V_1Z, V_2Z ex quibus V_1Z secat 10
(producta) P_2C in δ, et V_2Z secat P_1C in β.

⟨fig. 24⟩

In triangulo characteristico $_1C_1L_2C$, vocetur: $_1C_1L$, dx; $_1L_2C$, dy; et $_1C_2C$,
dc et PC, x; ob triangula $_2C_1L_1C$, et CZV similia fit $\overline{CZ : CV} \overset{(1)}{::} \overline{dy} : dc$. Rursus ob
triangula $V_2Z\delta$ et $P_1Z\delta$ similia fit $\delta_2Z : \delta_1Z \overset{(2)}{::} \overline{VZ : PZ}$. Denique ob triangula
$P_1Z\delta$ et P_1L_2C similia fit $_1Z\delta : _1L_2C$ seu dy $\overset{(3)}{::}$ PZ : PC. Tollendo δ_1Z ex analogia 2 15
per analogiam 3 fit $\delta_2Z \cdot PC : dy \cdot PZ \overset{(4)}{::} VZ : PZ$ seu $\delta_2Z : dy \overset{(5)}{::} VZ : PC$. Si $_1Z\delta$ et β_2Z
considerentur ut elementa arcuum centro P radiis P_1Z, P_2Z descriptorum, patet,
sumi posse δ_2Z pro $= \beta_1Z$ seu δ_2Z esse differentiam inter P_1Z et P_2Z, adeoque
δ_2Z esse $\overset{(6)}{=} \overline{dPZ} \overset{(7)}{=} dy \cdot VZ : PC$ per 5. Jam PZ $\overset{(8)}{=}$ PC − CZ. Ergo $\overline{dPZ} \overset{(9)}{=} \overline{dPC} - \overline{dCZ}$ jam
$\overline{dPC} = dx$, et $\overline{dCZ} \overset{(10)}{=} CV \overline{d \, \overline{dy} : dc}$ per 1 quos valores substituendo in aeq. 9 fit 20
$\overline{dPZ} \overset{(11)}{=} dx - CV \cdot d \, \overline{dy : dc} \overset{(12)}{=} dy \cdot VZ : PC$ per 7. Jam VZ : VC $\overset{(13)}{::} dx : dc$ et dy $\overset{(14)}{=} xw :$ a posito w
esse anguli seu circulationis circa P elementum, radio a; nam $_1L_2C$ seu dy sunt in
ratione composita radiorum $\overline{PC}$ seu x, et angulorum $_1CP_2C$. Ergo dy : PC $\overset{(14)(bis)}{=} w : a$. Ergo
denique ex aeq. 12. 13. 14 fit dx − CV $\overline{d \, \overline{dy} : dc} \overset{(15)}{=} CP \cdot w \cdot dx \, CV : \langle x \rangle \cdot a \cdot dc$.
Ergo CV $= \overline{dx : d \, \overline{dy} : dc} + wdx : adc$. Jam $\overline{d \, \overline{dy} : dc} = dy \, \overline{ddc} - dc \, \overline{ddy}, : \overline{dc^2}$. Ergo 25
CV $= dx : \overline{dy \cdot \overline{ddc} - dc \cdot \overline{ddy} : dc^2} + wdx : adc$.

Commentary to *Inventum a me est* and *Investigatio Semidiametri Circuli Osculantis*

Inventum a me est investigates the properties of the *lineae projectitiae,* that is, the lines described by a body whose trajectory is the resultant of its own impetus (inertial motion) and of the action of gravity, levity or a magnetic force. The opening sentences in lines 1–8, and lines 25–32 and 49–52 are translated and discussed in Section 5.3.

In lines 28–63 (see fig. 20) Leibniz studies the properties of the 'motus projectitius': $_2C_1L$ and $_3C_2L$ are perpendicular to the radii P_1C and P_2C respectively and are taken as approximations of the arcs described by the radii P_2C and P_3C with angles $_2C\hat{P}_1C$ and $_3C\hat{P}_2C$. This approximation is correct when the angles are infinitesimal, as in our case. $_2C_1L$ and $_3C_2L$ are named 'circular progressions', and are proportional to their respective angles and radii. Leibniz shows that if the time intervals are equal, triangles $_1CP_2C$ and $_2CP_3C$ have equal areas. These areas are proportional to $_1CP$ times $_2C_1L$ and to $_2CP$ times $_3C_2L$ respectively, or more generally the increments of the areas are proportional to the radii and circular progressions. Taking equal increments of the areas, Leibniz draws two conclusions: first, the paracentric radii PC are inversely proportional to the circular progressions $L(C)$; second, the angles of circulation are inversely proportional to the square of the radii. The first of these properties expresses the law of harmonic circulation. The beginning of the second side (lines 50–2) marks the moment of birth of the most famous concept in Leibniz's theory of planetary motion. At this stage his argument is purely mathematical and descends directly from proposition 1 of the *Principia Mathematica*. Vortices do not enter the stage yet, nor does Leibniz christen the property he has just found 'circulatio harmonica'.

Next (lines 63–) Leibniz seeks to determine the times and the angles $N\hat{P}C$ (see figs. 21 and 22). AB ought to be equal to PC, and in particular $A]=PN$, where N represents the starting point. BE are taken to be proportional to the areas PCN. The curve $]EE$ represents times or equivalently areas with respect to the radii AB. AF are parallel to BE, AQ is taken equal to $A]$ and QH is parallel to AB; TH is the tangent in $_1E$; RH is proportional to the rectilinear progression along the radius LC (this refers to fig. 20) or, as we would say, to radial velocity. If time is given, circular progression is inversely proportional to the radius and the increments of the angles of circulation are inversely as the square of the radii. Therefore, if we draw a curve $]MM$ such that FM is inversely proportional to the square of FE, that is, inversely proportional to the square of the radii, FM will represent the increments of the angles, and the area $AFM]A$ will be proportional to the angle $N\hat{C}P$. In fig. 22 the curve $]MM$ is not consistent with the curve EE. The area of $AFM]A$ is proportional to the integral of dt/r^2, where t is proportional to AF and $1/r^2$ is proportional to FM.

From the following paragraph onwards, Leibniz employs the differential calculus (line 84–). The ensuing remarks are affected by the following problem: he is uncertain whether to fix circular progression $L(C)$ equal to $dy\,x:a$, or simply equal to dy. In the first case dy would measure the arc of a given circumference with radius a, whereas x represents the radius PC; in the second case dy represents directly the arc or its approximation $L(C)$. Once this problem has been pointed out, Leibniz's reasoning becomes straightforward. Since he obtains the result $PC:PN = C(C):(C)L$, the expression for the paracentric conatus K_3C from *De Conatu* can be written as $K_3C = (_1C_2C)^3/CV \cdot _1L_2C$, meaning that paracentric conatus is equal to the third power of orbital velocity $C(C)$ over the radius CV of the osculating circumference and circular progression $L(C)$ (lines 108–12; I introduce brackets for convenience).

At this point Leibniz probably interrupted writing *Inventum a me est* in order to find a suitable expression for the radius of the osculating circumference, and composed a new essay on a separate sheet of paper, *Inquisitio in Semidiametrum Circuli Osculantis si pro Ordinatis Convergentes Adhibeantur Ope Calcui.*[1] This draft was followed by the *Investigatio Semidiametri Circuli Curvam in Proposito Puncto Osculantis*, which is very clear and surprisingly correct. The only mistake (line 25) is trivial; at the right member of the second equation one should have the opposite sign.

Rather than going through Leibniz's reasoning in detail—this task can be easily carried out by the reader—I prefer to draw a comparison between the present manuscript and a similar calculation in an essay published by Leibniz a few years later in the *Acta Eruditorum*,

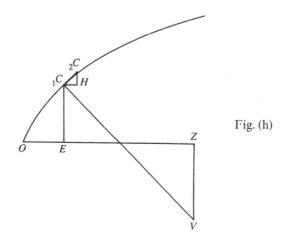

Fig. (h)

[1] LH 35, 10, 7, f. 41v. The manuscript is not reproduced.

Constructio Propria Problematis de Curva Isochrona Paracentrica.[2] In this essay Leibniz uses a similar method to determine the osculating radius of the curve, and finds the expression $-dx:d(dy:dc)$. It is easy to obtain this expression for the osculating radius CV. In the enclosed Fig. (h), $OE = x$, $_1CE = y$, $OZ = f$, $_1C_2C = dc$, $_1CH = dx$, $_2CH = dy$, and $_1C_2CH$ is the characteristic triangle. We have

$$CV:EZ::_1C_2C:_2CH,$$

hence $(dy:dc)CV = f - x$. If we differentiate this equation taking CV and OZ constant (compare equation 10, line 20 in the text), we find the equation $CV = -dx:d(dy:dc)$.

In our case Leibniz uses as abscissae the radii $_1CP$, $_2CP$, etc.—which are not parallel among themselves—and as ordinates their perpendiculars $L(C)$. We have $CZ:CV::_1L_2C:_1C_2C$ (see figure 24). If we differentiate this proportion with CV constant, but CZ varying according to the inclination of CP, we have

$$CVd(L_1C:_1C_2C) = d(_1Z_1C) = d(P_1C) - d(P_1Z) = dx - w\,CV\,dx:adc;$$

after some elementary operations this equation becomes:

$$CV = dx:[d(dy:dc) + wdx:adc].$$

This example shows how the difference in the coordinates generates the additional term $+wdx:adc$ at the denominator and a difference in sign. Going back to *Inventum a me est* (lines 116–21), we see that Leibniz substitutes the expression for CV he has found in *Investigatio*. From the equation $K_3C = (dc)^3:CVdy$ one has

$$K_3C = (dc)^2[(dc\,d^2y - dy\,d^2c):dc + wdx:a]:dxdy,$$

where I have corrected the sign error mentioned above.

5   *Repraesentatio Aliqua*

LH 35, 10, 7, f. 12

The manuscript reproduced here is on a kind of paper used by Leibniz in Vienna in 1688—watermark number 695 in the catalogue at the NLB, letters 'M R'. On the basis of comparison with other similar texts, *Repraesentatio Aliqua* can be shown to date from autumn 1688. These

  ² *AE* Aug. 1694, pp. 364–75 = *LMG, 5*, pp. 309–18, on p. 310. See also Bos, 'Differentials', pp. 40–2. In the copies of the 1694 *Acta* I have consulted, after page 392 pagination starts again from page 312.

texts are briefly discussed in Section 5.4, but one of them, *Galilaeus*, is considered in more detail and extracted in the commentary.

In the title Leibniz had originally written *impetus* instead of *conatus*. Other variant readings consist of trivial stylistic changes and are omitted. The letters 'λ' and 'L' in the text correspond to the same points 'λ' in the figure. In lines 46, 52 and 59 (inside the integral sign) I have substituted $C(\lambda)$ for $C\lambda$.

⌊error⌋

Repraesentatio aliqua curvae quae describitur a gravi projecto in lineis ad centrum concurrentibus conante et ubique aequales conatus novos accipiente

Ponatur grave in A positum projici versus t   5
directione et celeritate A_1t. Dum vero progreditur
ex A in $_1t$, descendet velocitate $A_1\lambda$, et completo
parallelogrammo $A_1t_1C_1\lambda$ ibit composita celeritate
et directione A_1t, seu dum sine gravitate
pervenisset in $_1t$, nunc accedente primo gravitatis   10
impulsu perveniet in $_1C$. Rursus positum in $_1C$

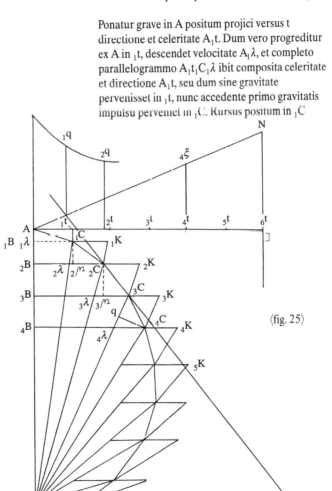

⟨fig. 25⟩

habebit pristinam directionem impressam $_1C_1K$, aequalem ipsi A_1t, seu $_1t_2t$, et
directionem descensus $_1C_2\lambda$ compositam ex aequali pristinae $A_1\lambda$, et novae.
Ponantur esse aequales fiet $C_2\lambda$ dupla $A_1\lambda$, et completo parallelogrammo
$_1C_1K_2C_2\lambda$, eodem tempore quo antea mobile venit ex A in $_1C$, nunc percurret   15
$_1C_2C$; tempore scilicet $_1t_2t$, vel $_1C_1K = A_1t$ et ita porro. Hinc patet velocitates
seu impetus gravitate quaesitos esse temporibus proportionales; patet etiam
pro indesignabilibus, triangulum $P\lambda C$ coincidere triangulo $PC(C)_{(,)}$ est autem
triangulum $P\lambda C = \lambda C$ in PB. Ergo ipsis PB vel earum dimidiis aut tertiis
translatis in tq, tunc posito angulum PAt esse rectum, erunt spatia Aq   20
proportionalia areis APCA. Et proinde in omni linea projectitia si areae
elementa dividantur per altitudinis a distantia centri et verticis detractae
residuum, (posito rectam per centrum et verticem directioni impressae
normalem esse) prodibunt temporum incrementa. Si differentia inter duas
quantitates ut hoc loco duo triangula P_1C_2C et P_2C_3C sit infinite parva,   25
respectu ipsorum non debet negligi quando de summa ex omnibus P_1C_2C, P_2C_3C,
P_3C_4C agitur; summa enim ex summis est summa summarum ex differentiis; sed si
negligi potest, debet esse infinities infinite parva. Potest etiam tempus ita
in unum colligi. Ex C in (L)(C) demittatur $C(\mu)$ erit $\lambda C = \lambda \mu + \mu\, C$. Jam $\int \overline{\lambda C} = At$
et $\int \overline{\mu\, C} = BC$ ergo $At = \int \lambda \mu + BC$. Jam ob triangula $C(\mu)(\lambda)$ et PLC similia fit: $C(\mu)$ 30
seu dx (posito $AB = x$) : $(\lambda)(\mu)$: $C(\lambda)$ :: PB : BC : CP. Ergo: $(\lambda)(\mu)$: dx :: BC : PB seu sit
$BC = y$ et $AP = h$ fiet $PB = h - x$, et fiet $\lambda \mu = dx \cdot y : \overline{h - x}$. Seu $\int \lambda \mu = \int \overline{dx \cdot y : h - x}$. Ergo
fiet: $At = \int \overline{dx \cdot y : h - x} + y = t$. Idem aliter investigemus. Areae APCA quaeramus
incrementum, hoc dividamus per PB, habebimus temporis incrementum, qualium
summa dabit $\frac{1}{2}$ t. Jam area $APCA = \int ydx + y \cdot \overline{h - x} : 2$. Huius differentia est: 35
$ydx + \frac{1}{2} hdy - \frac{1}{2} ydx - \frac{1}{2} xdy$ seu $\frac{1}{2} ydx + \frac{1}{2} dy \cdot \overline{h - x}$, quae divisa per $h - x$ dabit
$\frac{1}{2} ydx : \overline{h - x} + \frac{1}{2} dy$; unde summando fit $\frac{1}{2} \int \overline{ydx : h - x} + \frac{1}{2} y = \frac{1}{2} t$ quod succedit.
Pergamus nunc ad aliquam collectionem descensuum. Ducatur $P\theta$ parallela
At cui tangens C(C) occurrat in θ, manifestum est triangula $C(\lambda)(C)$ et $CP\theta$ esse
similia, eritque $C(\lambda) : (\lambda)(C) : C(C) :: CP : P\theta : C\theta$ seu est ut elementum descensus (id   40
est iisdem temporis incrementis velocitas); elementum progressus paralleli seu
temporis; elementum orbitae seu curvae. Quia autem tempus assumi potest constans
quaeramus $\langle an \rangle$ in curva data projectitia, PC, $P\theta$ sint aequales, quae $P\theta$ sit
Ω, eaque $P\theta \langle = \rangle \Omega$ (vel eius proportionalis $\langle \rangle$) transferatur in $\exists N$ seu $_6tN$.
Atque haec erit mensura seu unitas, facile exhibentur $t\xi$ proportionales ipsis   45
C(λ) elementis descensuum, seu fiat $t\xi : \exists N$, ut PC quaecunque ad constantem Ω,

ubi in casu simplicissimo, quando conatus gravitatis impressus semper est
aequalis, fit $A\xi N$ recta, quia $t\xi$ sunt ipsis At proportionales, sicut est si
constet, quomodo ipsae $t\xi$ seu ipsis proportionales C(λ), crescunt cum
tempore, id est aucta vicinia centro, nam ex relatione ad distantiam a centro,   50
potest haberi relatio ad tempus; licet interdum non nisi transcendenter.
Summa autem ipsarum $t\xi$ seu figura $At\xi$ dabit summan ipsarum C(λ). Sit $_4C_3q$
perpendicularis ex $_4C$ in $_3C_4\lambda$ seu (C)q perp. ex (C) in C(λ). Patet triang.
$C(\lambda)(C)$ esse $= B(B)$ seu dx in dt, seu $(\lambda)(C) = C(\lambda)$ (seu elementum descensus) in
(C)q jam (C)$\langle q \rangle$ est progressus circularis. Ergo elementum descensus in   55
progress. circ. $=$ elem. abscissarum dx in dt elem. temp. seu elementum descensus
convergentis, est ad elementum temporis seu progressus paralleli, ut
elementum abscissarum sive paralleli descensus, seu abscissarum y est ad

elementum progressus circularis. Porro $C(\lambda) = Cq + q(\lambda)$. Ergo $\overline{\int C(\lambda)}$ seu $\underline{\text{summa}}$
$\underline{\text{descensuum est}} \int Cq$ seu $AP - PC + \int q\lambda$. Est autem $q\lambda = \sqrt{(C)(\lambda)^2 - C(C)^2 + Cq^2}$.   60
Seu $q\lambda = \sqrt{\overline{dt^2} - \overline{dx^2} - \overline{dy^2} + \overline{dv^2}}$, posito radium PC vocari v. Supra autem habuimus $\overline{dt}$
quem valorem tollendo habebitur $q\lambda$ per x et y nam et v haberi potest per x et y, et
ita per haec et eorum differentias habebitur et $q\lambda$.

⌊Jam adeo quae sit vera ratio hanc methodum instituendi, in praecedentibus
enim erratum fuit.⌋   65

Commentary to *Repraesentatio Aliqua*

This manuscript is marked at the beginning with the word *error* and ends
with an admission of failure in which Leibniz wonders how the problem
can be solved. Both comments are later additions.

Repraesentatio Aliqua is similar (see Section 5.4) and complementary
to another manuscript essay, *Galilaeus tractare incipit de linea gravium
projectorum*; since their contents are to some extent mutually clarifying, I
discuss the two essays together. *Galilaeus* starts with the claim that
Galileo began dealing with curves described by bodies moving with
uniform rectilinear motion composed with acceleration of gravity acting
along lines parallel to one another. In the attempt to generalize Galileo's
reasoning, Leibniz studies curves described by a body tending towards a
centre at a finite distance, with a conatus depending on the distance from
the centre. This is the inverse problem of central forces.[1]

In *Galilaeus* and in the present manuscript Leibniz studies the motion
of a body tending towards a centre and acquiring a constant conatus.
The relevant portion of the figure in *Galilaeus* corresponds to that of the
present essay. Notice that in fig. 25 $C(C)$ is a rectilinear segment.
Leibniz's mistake is obvious from the figures, and in *Repreaesentatio*

---

[1] LH 35, 10, 7, f. 18v.–19r. The whole essay has been crossed out by Leibniz. The watermark
(number 695 in the catalogue at the NLB, letters 'M R') is identical to that of *Repraesentatio
Aliqua*. We have seen in Section 5.5 that the edge of *Galilaeus* matches perfectly that of *Si
mobile aliquod ita moveatur*, LH 35, 10, 7, f. 38–9. The opening passage reads: 'Galilaeus
tractare incipit de linea gravium projectorum, quae motu composito ex semel impresso
aequabili, et gravitatis accelerato feruntur. Posuit autem quod ipsi ad scopum praefixum
sufficiebat, gravia omnia descendere conari in rectis inter se parallelis, et quolibet temporis
incremento aequalem a gravitate ipsi imprimi conatum descendendi novum, prioribus
addendum. Verum prosequendo hanc speculationem, considerare porro licet, quae sit linea
projectorum gravium, si gravia, ut revera fit ponantur tendere ad unum aliquod gravium
centrum, et praeterea quod fiat, si ponatur conatus novus impressus aequalibus temporibus
non esse aequales, sed crescens vel decrescens, pro distantiis scilicet a centro, vel aliis causis.
Ponamus primo gravia ad centrum aliquod tendere, conatus autem a gravitate impressos
semper esse aequales.' Leibniz seems to believe that the case of a constant conatus is very
simple; in *Repraesentatio Aliqua* this is called 'casus simplicissimus' (line 47).

Aliqua from the first lines of the text (lines 11–13); Leibniz tries to consider rectilinear inertial motion on the one hand, and the sum of the impressions of gravity on the other, and then tries to find their composition. I have called this 'pseudo-Galilean' motion. From this representation it follows that at each point C we have to consider the sum of all previous impressions of gravity, which in the present case, unlike Galileo's, are not parallel.

Let us consider the text. In fig. 25 time is proportional to the segments CK, or equivalently $t(t)$. Motion from A to $_1C$ entails no problems: the body has an initial velocity A_1t and is acted upon by gravity $A_1\lambda$. Once in $_1C$, the body would be acted upon by gravity $_1C_2\lambda$, and keep its velocity $_1C_1K$ in the direction previously impressed At, which is not tangent to the curve in $_1C$. The action of gravity $_1C_2\lambda$ is the composition of the previous action $A_1\lambda$ with the new one acquired in $_1C$. If the new and the old actions are equal, he states that $_1C_2\lambda$ is equal to twice $A_1\lambda$ (line 14). This decomposition is not correct: first, it does not take into proper consideration the changing directions of gravity;[2] secondly, if $C(\lambda)$ is the sum of all previous impressions of gravity, it ought to start from a point t on the tangent rather than on the curve. In Fig. (i) I have adopted a representation whereby KC represents the deviation from the tangent or the instantaneous impression of gravity, whilst the curve tKC results from the sum of all previous impressions. At each point C the body

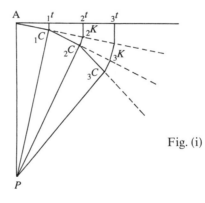

Fig. (i)

[2] This problem is clearly stated in *Galilaeus*: 'Similiter aggregatum ex omnibus $C(\lambda)$, seu $_1C_{\langle2\rangle}\lambda$, $_2C_{\langle3\rangle}\lambda$, constituit aggregatum spatii descensu percursi. Quamvis nec omnes $C(\lambda)$, nec omnes λC hic facile in unam rectam componi possit, quod in eo casu facile est quem Galilaeus tractavit, quando descensus sunt paralleli.' In *Galilaeus* the letters L and λ are used indifferently: for convenience I use only the latter; further, its indices are shifted with respect to those in *Repraesentatio Aliqua*. In order to make the comparison of the two essays easier I have rearranged the notation; altered indices are enclosed in angled brackets.

tends to continue its rectilinear uniform motion along $C(K)$. If the curve $PC(C)$ is a circumference, in the limit the deviations tKC from the tangent At to the curve $C(C)$ coincide with its evolvent, as we have seen in Section 2.3 on Huygens.

Below Leibniz explains that the curve $_1q_2q$ represents the decreasing heights PB, hence the areas between the curves $q(q)$ and $t(t)$ are proportional to the areas of the triangles $PC(C)$. In fig. 25 the perpendiculars from q do not follow on to the points $_1t$ and $_2t$. The provisional conclusion is that in every *linea projectitia* the increments of time are proportional to the infinitesimal areas of triangles $PC(C)$ over PB, namely the difference between AP and AB (lines 21–4). In fig. 25 the letters $_1q_2q$ are not related to the letter q along $_4\lambda_3C$.

The text continues with a passage (lines 24–8) which can be better understood in connection with *Galilaeus*, where Leibniz used both the area law and his own construction according to which the segments CK measure time. In *Galilaeus* Leibniz had originally taken the difference between triangles P_1C_2C and $P_{(2)}\lambda_2C$ to be infinitesimal, and therefore negligible. But later he added the following point in a marginal note: triangles $_1C_2CP$ and $_{(2)}\lambda_2CP$ are different, and *a fortiori* triangles $_1C_2CP$ and $_2C_3CP$. Since each triangle $PC(C)$ differs from the preceding one, when taking their sum we have to consider the progressively larger differences between them. This means that it is necessary to perform two summations: first, a summation over the difference between each successive pair of triangles, in order to determine the area of the nth triangle; second, a summation over all triangles $PC(C)$. Unless the difference between the areas of two contiguous triangles is at least a second-order infinitesimal with respect to the areas of the triangles, this difference cannot be neglected. In fact, the difference between the areas of $_1C_2CP$ and $_nC_{n+1}CP$ is the sum of $n-1$ terms. The tacit assumption is that since the number of triangles is infinite, the difference between their areas cannot be neglected if it is a first-order infinitesimal with respect to the areas of the triangles. According to the marginal note in *Galilaeus* this mistake had been made by Newton; possibly Leibniz had in mind proposition 1 of the *Principia*, but his criticism is not justified because Newton did not decompose motion in the same way as Leibniz.[3]

---

[3] The text from *Galilaeus* reads: 'Nam triangulum $_1C_2$CP differt indesignabiliter a triangulo $_{(2)}\lambda_2$CP [(id est differentiarum ratio ad ipsa est minor quavis assignabili) imo error; hic necesse esset ipsam differentiam esse infinities infinite parvam respectu differentium, in hoc lapsus est Neutonus, nam repetuntur semper omnes errores priores et error sit summa summarum, ideoque assignabilis]'. I have inserted the marginal note within square brackets. From the words 'imo error' onwards the hand is slightly different and the ink darker; this argues for a later addition.

In *Galilaeus* Newton's representation of the area law was first accepted and successively criticized in the marginal note. The contents of that note are now included in the present manuscript. This may suggest that *Repraesentatio Aliqua* follows *Galilaeus* and Leibniz's first reading of the *Principia*. With respect to the dating, notice also the expression *progressus circularis* (line 55) for the perpendicular $_4C$ to the radius P_3C—the same expression is used in *Inventum a me est*—the lack of any reference to the mathematical notions of the *Tentamen* and to the action of a vortex.

The following lines (52–63) are affected by the initial mistakes. Once these have been clarified, the reader can follow them without difficulty. Leibniz wants to determine the total time (lines 28–37), and the collection of descents (lines 38–63); in the similar triangles $C(\lambda)(C)$ and $CP\theta$, $C(\lambda)$ and CP are homologous. Transferring $P\theta$ on to $_6tN$, which is taken to be constant, he finds (line 46): $t\xi :]N :: PC : \Omega$; Ω is fixed and is equal to a given radius PC; $]N$ is also fixed; $t\xi$ and PC are the only variables and they are proportional. Further, they are also proportional to $C(\lambda)$. However, Leibniz overlooks that the proportionality constant is different for each triangle $PC\theta$, and this introduces an inconsistency in his representation. According to him, the descents $C(\lambda)$ grow uniformly because attractive endeavours are constant; therefore $t\xi$, which is proportional to them, is represented along a straight line $A\xi N$. This reasoning is based on the wrong assumptions that motion can be represented in this way and that descents remain parallel among themselves.

At first sight it seems almost unbelievable that Leibniz adopted such an unsatisfactory decomposition of motion and ignored that although the impressions of gravity are equal, they are not parallel. We can conjecture that he was groping his way to both an understanding of the *Principia*, and the development of a different approach to orbital motion.

6 *De Motu Gravium*

LH 35, 10, 7, f. 1, 2, 3, 25 (compare Figs. (j)–(l))

The manuscripts reproduced here are on two separate folded folii, numbered 1–25 and 2–3 respectively. They are on a kind of paper used by Leibniz in Vienna in 1688, watermark 'cross the cavalier', number 564 in the catalogue at the NLB. The first three sides of the first folded folio have been written in continuous succession and form the proper

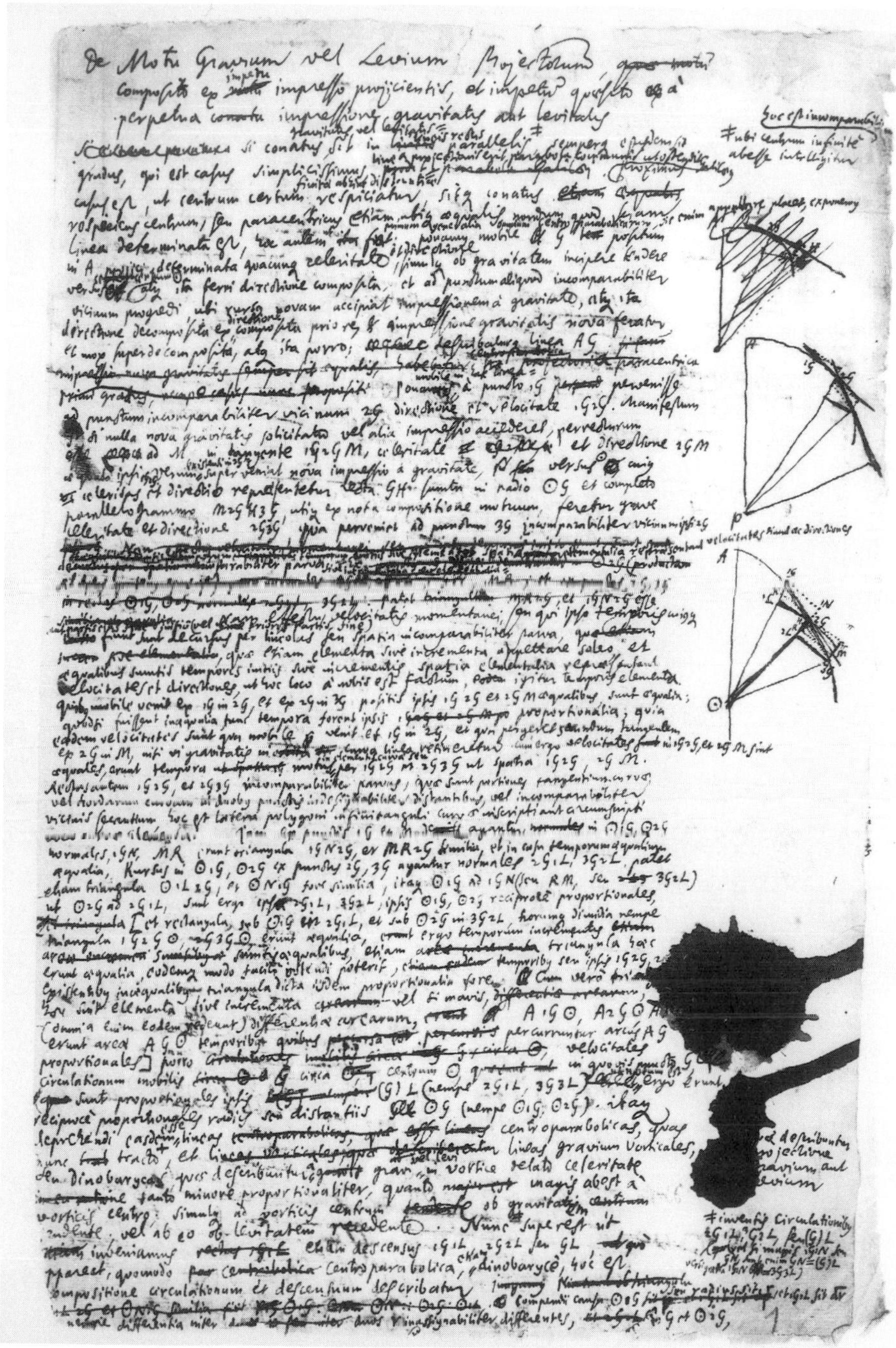

Fig. (j) LH 35, 10, 7, f. 1r., *De Motu Gravium vel Levium Projectorum.*

Fig. (k) LH 35, 10, 7, f. 25r., *De Motu Gravium vel Levium Projectorum.*

Fig. (l) LH 35, 10, 7, f. 2r., *De Motu Gravium vel Levium Projectorum.*

essay *De Motu Gravium*. The fourth side contains some complementary remarks not directly connected with the preceding text.

The second folded folio contains three clearly separate sections. The first one, f. 2, presents a development of some calculations on paracentric conatus from the preceding essay. The second section, f. 3 recto, consists of two attempts to prove lemma 12 in book 1 of the *Principia*. The last section, f. 3 verso, is an independent essay on planetary motion. I reproduce the texts in the same order as they appear in the manuscripts, taking the complementary sections as Additions. The main essay and the appendices are drafts and contain several truncated lines.

In the essay we find some calculations in which Leibniz begins to see paracentric conatus as the difference between a centrifugal and a centripetal term. The terminology employed by Leibniz, while different from that of the *Tentamen*, is more precise than in the preceding essays. He introduces a vortex rotating with a speed inversely proportional to the distance from the centre, and calls the curves described by a body carried by such a vortex *lineae vorticales* or *dinobarycae*. The manuscripts appear to date from the same period; Addition 2 is the continuation of the essay and in both folii Leibniz does not make the mistake by a factor of two in the term representing centrifugal conatus. The texts are influenced by Newton, who is referred to twice on folii 2 recto and 3 recto. The present manuscripts date from autumn 1688.

De motu gravium vel levium projectorum,
composito ex ⟨motu⟩ ⌊impetu⌋ impresso projicientis, et impeto quaesito a
⟨conatu⟩ perpetua impressione gravitatis aut levitatis

Si conatus ⌊gravitatis vel levitatis⌋ sit in lineis rectis parallelis ⌊ubi
centrum infinite ⌊hoc est incomparabiliter⌋ abesse intelligitur⌋, semperque   5
ejusdem sit gradus, qui est casus simplicissimus, linea projectionis erit
parabola communis ut ostendit Galilaeus. Proximus casus est, ut centrum
certum ⌊finitae abhinc distantiae⌋ respiciatur, sitque conatus respiciens
centrum, seu paracentricus etiam ubique aequalis, nondum quod sciam linea
determinata est, haec autem ⟨ita⟩ ⌊ut⌋ fiat ⌊primum generalia omnium centro   10
parabolicarum, sic enim appellare placet, exponemus⌋. Ponamus mobile G positum
in A projici determinata quacunque celeritate ⌊et directione⌋, simulque ob
gravitatem incipere tendere versus centrum Θ atque ita ferri directione
composita, et ad punctum aliquod incomparabiliter vicinum progredi, ubi
rursus novam accipiat impressionem a gravitate, atque ita directione   15
decomposita ex ⌊directione⌋ composita priore, et impressione gravitatis nova
feratur, et mox super decomposita, atque ita porro; describaturque linea AG;
si jam impressio nova gravitatis semper sit aequalis, habebitur ⟨projectaria⟩
⟨centroparabolica⟩ paracentrica primi gradus, nempe casus nunc propositi.
Ponamus ⌊mobile in hac linea⌋ a puncto ₁G pervenisse ad punctum   20

incomparabiliter vicinum $_2$G directione et velocitate $_1$G$_2$G. Manifestum est si
nulla nova gravitatis solicitatio vel alia
impressio accederet, pervecturum esse ad M, in

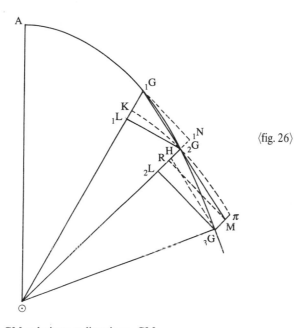

⟨fig. 26⟩

tangente $_1$G$_2$GM, celeritate et directione $_2$GM
aequali ipsi $_1$G$_2$G. Verum existenti in $_2$G   25
superveniat nova impressio a gravitate, versus Θ,
cuius celeritas et directio repraesentetur recta GH
sumta in radio ΘG et completo parallelogrammo
M$_2$GH$_3$G, utique ex nota compositione motuum, feretur
grave celeritate et directione $_2$G$_3$G, qua perveniet ad punctum $_3$G   30
incomparabiliter vicinum ipsi $_2$G. Nam effectus velocitatis momentanei, seu
qui ipso temporis cuiusque initio (vel prioris partis fine) fiunt$_{(,)}$ sunt
decursus per lineolas seu spatia incomparabiliter parva, quae etiam elementa
sive incrementa appellare soleo, et aequalibus sumtis temporis initiis sive
incrementis, spatia elementalia repraesentant velocitates et directiones, ut   35
hoc loco a nobis est factum, igitur temporis elementa, quibus mobile venit ex
$_1$G in $_2$G, et ex $_2$G in $_3$G, positis ipsis $_1$G$_2$G et $_2$GM aequalibus, sunt aequalia;
quodsi fuissent inaequalia, tunc tempora forent ipsis proportionalia; quia
eaedem velocitates sunt qua mobile venit ex $_1$G in $_2$G, et qua pergeret secundum
tangentem ex $_2$G in M, nisi vi gravitatis in curva linea retineretur. Cum ergo   40
velocitates in $_1$G$_2$G, et $_2$GM sint aequales, erunt tempora motus ⌊in elementis
curvae seu⌋ per $_1$G$_2$G et $_2$G$_3$G ut spatia $_1$G$_2$G, $_2$GM. Rectas autem $_1$G$_2$G, et $_2$G$_3$G
incomparabiliter parvas, quae sunt portiones tangentium curvae, vel chordarum
curvam in duobus punctis indesignabiliter distantibus, vel incomparabiliter
vicinis secantium, hoc est latera polygoni infinitanguli curvae inscripti,   45

aut circumscripti voco curvae elementa. Jam ex punctis $_1$G et M agantur in
Θ_1G, Θ_2G normales, $_1$GN, MR$\langle;\rangle$ erunt triangula $_1$GN$_2$G, et MR$_2$G similia, et in casu
temporum aequalium, aequalia. Rursus in Θ_1G, Θ_2G ex punctis $_2$G, $_3$G agantur
normales $_2$G$_1$L, $_3$G$_2$L. Patet etiam triangula Θ_1L$_2$G, et ΘN$_1$G fore similia, itaque
Θ_1G ad $_1$GN (seu RM, seu $_3$G$_2$L) ut Θ_2G ad $_2$G$_1$L, sunt ergo ipsae $_2$G$_1$L, $_3$G$_2$L, ipsis　50
Θ_1G, Θ_2G reciproce proportionales, [et rectangula sub Θ_1G in $_2$G$_1$L, et sub Θ_2G
in $_3$G$_2$L, horumque dimidia nempe triangula $_1$G$_2$GΘ, $_2$G$_3$GΘ, erunt aequalia, ergo
temporum incrementis sumtis aequalibus, etiam triangula haec erunt aequalia,
eodemque modo facile ostendi poterit, temporibus seu ipsis $_1$G$_2$G, $_2$GM
existentibus inaequalibus, triangula dicta iisdem proportionalia fore. Cum　55
vero triangula haec sint elementa sive incrementa vel si mavis, (omnia enim
eodem redeunt) differentiae arearum, A$_1$GΘ, A$_2$GΘ, A$_3$GΘ, erunt areae AGΘ
temporibus quibus percurruntur arcus AG proportionales] porro, velocitates
circulationum mobilis circa centrum Θ, in quovis puncto G, sunt
proportionales ipsis (G)L (nempe $_2$G$_1$L, $_3$G$_{(2)}$L) ⌊ut notum est,⌋ ergo erunt　60
reciproce proportionales radiis seu distantiis ΘG (nempe $\Theta_1\overline{\text{G}}$, Θ_2G). Itaque
deprehendi easdem esse lineas centroparabolicas, quas nunc tracto ⌊aequae
describuntur projectione gravium aut levium,⌋ et lineas gravium vorticales,
seu dinobarycas, quae describuntur a gravi ⌊vel levi⌋ in vortice delato
celeritate tanto minore proportionaliter, quanto magis abest a vorticis　65
centro; simulque ad vorticis centrum ob gravitatem tendente; vel ab eo ob
levitatem recedente. Nunc ⌊inventis circulationibus $_2$G$_1$L, $_3$G$_2$L, seu (G)L (vel
si mavis $_1$G$_1$N seu GN, sunt enim GN = (G)L verbigratia $_1$G$_1$N = $_3$G$_{(2)}$L),⌋ superest ut
inveniamus etiam descensus $_1$G$_1$L, $_2$G$_2$L seu GL, quo appareat, quomodo
centroparabolica, etiam dinobaryce, hoc est compositione circulationum et　70
descensuum describatur. Compendii causa Θ_1G seu radius sit r, et $_1$G$_1$L
sit $\overline{\text{dr}}$ nempe differentia inter duos inassignabiliter differentes, Θ_1G et Θ_2G,

⟨v.⟩ et $_2$G$_2$L erit differentia ⌊proxime sequens⌋ inter duos radios Θ_2G, et Θ_3G, quae
proinde erit dr + ddr, decrescentibus r, ut in figura, vel dr − ddr crescentibus.
Nempe ut dr mihi est differentia inter radios, ita mihi ddr est differentia　75
differentiarum inter radios. Itaque Θ_2G est r − dr. Jam M$_3$G vel R$_2$L vocemus m,
qua detracta a $_2$G$_2$L restat $_2$GR, seu $_2$GN, quae proinde erit dr + ddr − m.
ΘN = Θ_2G + $_2$GN$\langle = \rangle$r − dr + dr + ddr − m. Habemus ergo ob triangula Θ_1L$_2$G, ΘN$_1$G
similia Θ_1G,r:Θ_2G,r − dr::ΘN,r + ddr − m$\langle:\rangle\Theta_1$L,r − dr. Unde necesse est esse ddr = m,
seu m sive $_2$GM ⌊quae repraesentat impressionem novam gravitatis⌋ esse differentiam　80
inter differentias duas proximas proximorum radiorum, seu inter duos descensus,
$_1$G$_1$L, $_2$G$_2$L, sive quod eodem redit inter ⌊duos proximos⌋ impetus integros
descendendi, jam acquisitos, quod etiam praevideri poterat ex ipsa natura
impetus descendendi totius acquisiti, quippe qui ex omnibus conatibus novis
in quovis loco curvae impressis colligitur; ⌊addito forte ⌊aut ademto⌋ primo　85
aliquo impetu versus centrum aut a centro, si qua in ipsius projicientis
prima impressione continebatur, quod si projectionis impressus impetus
computari ponatur a puncto A, ubi radius ad curvam AG angulum facit rectum,
⌊quale utique in curva assumi potest,⌋ cessat primi alicuius impetus versus
centrum additio aut detractio utcunque autem computes, cum commune additum　90
vel detractum differentias non minuat, manet⌋ conatus a gravitate, esse
impetuum seu gravitationum incrementa seu differentias. Unde etiam

intelligitur ⟨vim⟩ ⌊conatum⌋ gravitatis ad gravitationem seu ⟨vim⟩ collectum
aliquandiutino descensu impetum esse infinite seu incomparabiliter parvum.
Unde soleo conatus appellationem magis tribuere infinite parvis   95
celeritatibus, quarum vim appellare suevi mortuam, qualis est initio
gravitatis, at ⟨vim⟩ ⌊celeritatem⌋ accelerando acquisitam ex infinitis
conatibus compositam, magis appellare amo impetum, et vim huius impetus voco
vivam. Ostendi etiam vires mortuas esse in composita ratione celeritatum et
corporum, sed vires vivas in composita ratione corporum simplice, et   100
celeritatum duplicata. Quorum confusio plerosque in errorem induxit. Unde et
cum de viribus mortuis servandis agitur, eadem servatur quantitas motus ⌊seu
summa motuum⌋, ⟨cum de vivis eadem servatur⟩ quod in vivis locum non habet,
etsi eadem servetur quantitas progressus, id est modo summa modo differentia
motus. Ne quis autem miretur, quod $_1L_2G$ sumsimus pro circulatione circa   105
centrum, quasi coincideret arcui circulari $_2GK$ centro Θ descripto, dabimus
regulas aliquas quae pro filo erunt in labyrintho incomparabilium sive
infinitorum aut infinite parvorum; nempe in omni triangulo, cuius duo latera
sunt comparabilia, si tertium sit ipsis incomparabile, erit et angulus cui
subtenditur recto incomparabilis, ⌊vel contra, et porro⌋ erit differentia   110
duorum laterum, inter se comparabilium, ipsis incomparabilis. Hinc fit etiam
ut dentur non tantum infinite parva, sed etiam infinities infinite parva, et
infinitesies infinities parva, et sic porro sine fine. Quae tamen in geometria
et motuum computatione locum habere possunt, ut vel hinc patet. Sic ob angulum
$_1G\Theta_2G$ infinite parvum utique in triangulo $_1G_2G\Theta$ cum radii seu latera inter se   115
comparari possint, seu homogenea sint, et quidem hic pro aequalibus habenda,
subtensa $_1G_2G$ erit respectu radiorum incomparabilis, ac proinde cum radii
sint quantitates ordinariae, subtensa $_1G_2G$ quae velocitatem mobilis in sua
orbita repraesentat, erit infinite parva. In ipso autem triangulo $_1G_1L_2G$
licet latera sint infinite parva, erunt tamen comparabilia inter se, et   120
assignari potest triangulum ordinarium, tali (quod characteristicum curvae
olim appellabam) simile sed si angulus ad L sit rectus, et jungatur recta K_2G,
patet ob arcum circularem K_2G, fore K_1L, $_1L_2G$, $_1L\Theta$ continue proportionales,
ergo, ut est $_1L_2G$ incomparabiliter ⌊minor⌋ ipsa $_1L\Theta$ fore et K_1L
incomparabiliter minorem ipsa $_1L_2G$, ⟨ergo differentia inter rectam $_2CK$, et   125
$_2C_1L$ ipsis incomparabilis, imo et inter $_1L_2C$ et arcum K_2C, ut facile
demonstrari potest, unde $_1L_2C$ assumi potest pro circulatione. Et differentia⟩
Habemus ergo infinities infinite parvam K_1L, et talis infinities infinite
parva est etiam $_2GH$, cum enim velocitas viva seu impetus exprimatur per
quantitatem infinite parvam, ut per latera trianguli $_2G_2L_3G$, utique velocitas   130
mortua seu infinite parva, sive primus conatus descendendi exprimetur per
infinite parvam. Sed ecce aliam infinitesies infinities parvam, nempe
differentiam inter $_2GK$, et $_2G_1L$, nam si centro $_2G$ radio $_2GK$ describatur radius
secans ipsam G_1L ultra $_1L$ productam, distantia inter $_1L$ et punctum
intersectionis, erit haec differentia, quae ipsis $_2G_1L$ infinite parvae, $_1LK$ 135
infinities infinite parvae tertia proportionalis, adeoque uno adhuc gradu
inferior infinita parvitate secundi gradus. Unde patet tanto magis $_2G_1L$ pro
aequivalente ipsi $_2GK$ imo et arcui K_2G adeoque pro circulatione haberi posse.
Caeterum assumamus tamen K_1L quippe ipsi m comparabilem, et vocemus k, et $_1G_1L$

erit dr − k, et $_2G_2L$ erit dr + ddr − (k) et Θ_1N erit Θ_2G + $_2GN$, et $_2GN$ est $_2G_2L − R_2L$ 140

seu m, ergo $\Theta_1N = r\overline{(-dr + dr)} + \overline{ddr} − (k) − m$. Ergo correximus tandem calculos

nostros, et ⟨descensus est aggregatum⟩ una cum conatu descendendi a gravitate,

conjungenda est vis centrifuga a circulatione. Nempe vis centrifuga a

circulatione, ut K_1L vel ΩQ est ut $\overline{_1L_2G^2}:\Theta_1L$, jam $_1L_2G$ sunt ut $1:\Theta_1L$, vel

$1:\Theta_2G$, ergo si mobile ⟨moveatur vorticaliter⟩ circuletur in vortice   145

celeritate reciproca radiorum, erunt vires centrifugae in triplicata ratione

reciproca radiorum, vel triplicata directa circulationum. Idem est si pro vi

centrifuga sumas ipsam $_1G\lambda$, sumto descripto ex Θ arcu $_1G\pi$ et sumta $\pi\lambda =$ et

parall. $_1L_2G$.

Hinc patet ⟨ad⟩ ab elementis descensibus elementaribus ⟨ddr⟩ dr   150

⟨detrahendam⟩ ⟨addendam⟩ detrahendam esse summam conatuum centrifugorum seu

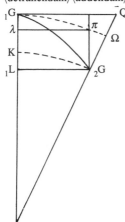

Hinc patet si addantur in unum omnes conatus a gravitate m, et ab hinc detrahantur omnes conatus centrifugi k, habitum iri impetum descensus $_1G_1L$, seu dr = ∫m − ∫k, seu ddr = m − k. Jam k sunt semper   155 ba²θ:r³. Posita θ constante infinities infinite parva, fiet ddr = m − ba²θ:r³. Hinc si m = θ fiet $\overline{dr} \cdot \overline{ddr} = \theta dr − ba^2\theta\overline{dr}:r^3$. Ergo $\overline{dr^2}:2 = \theta r − ba^2\theta:3r^2$.

⟨fig. 27⟩

⟨r.⟩   Ab hac aequatione | differentiali ad absolutam ita veniri. Aequatio differentialis est

$\overline{dr^2}:2 = \theta r − ba^2\theta:3r^2$. Pro θ infinities infinite parva scribamus $\overline{dt^2}$:a et fiet: 160

$adr \cdot r\sqrt{3} = \overline{dt}\sqrt{ar^3 − ba^3} \sqrt{2}\langle;\rangle a\sqrt{3}\int dr \cdot r:\sqrt{ar^3 − ba^3} = t\sqrt{2}$, quae est relatio inter tempora

seu areas, et radios. Si ponatur $m = r^4\overline{dt^2}:a^5 + ba \cdot \overline{dt^2}:3r^2$ fiet

$a^2\sqrt{a:2} \, dr:r^2 = dt$ et $a^2\sqrt{a:2:2r} = t$ et hoc demum casu erunt tempora radiis ⟨reciproce⟩

proportionalia. Unde eo casu quo temporum et radiorum relatio haberi potest,

patet aream curvae esse quadrabilem. Sed quando m seu vis gravitationis est   165

simplex, non aeque facile radiorum et temporum relatio habetur.

    Una adhuc aequatio est ex figura, nempe $\overline{_2G_1N^2} + \overline{_1N_1G^2} = \overline{_1G_1L^2} + \overline{_1L_2G^2}$. Est autem

$_1G_1L = \overline{dp} + k$ et $_1N_1G = adt:\overline{p − dp}$ et $_2G_1N = (dp) + (k) − m$ seu dp + ddp + (k) − m et

$_1L_2G = adt:p$ seu $_1N_1G^2 − _1L_2G^2 = \overline{_2G_1N^2} − _1G_1L^{\langle 2\rangle}$ seu $a^2\overline{dt^2}:\overline{r − dr^2} − a^2\overline{dt^2}:r^2$ et fiet:

$a^2\overline{dt^2}, \overline{(r^2 − r^2)} + 2rdr\overline{(−dr^2)}, :r^4\overline{(−2r^3dr + r^2dr^2)} = \overline{(dp^2)} \overline{(+dp\,ddp)} + \overline{ddp^2} + 2\overline{dp}(k) +$ 170

$2\overline{ddp}(k) \overline{(−2dp\,m)} − 2\overline{ddp}\,m \overline{(+(k)^2)} − 2(k)m + m^2\overline{(−dp^2)} \overline{(−2dp\,k)} \overline{(−k^2)}$.

Generaliter pro omni ⟨aequatione⟩ curva sit AB x, BG y, AC, h, CG, r erit

rr = yy + h² − 2hx + x² ergo rdr = ydy − hdx + xdx. Seu rursus sit circulatio e, fiet

$\overline{dx^2} + \overline{dy^2} = \overline{dr^2} + e^2$. Sit jam $e^2 = a^2\overline{dt^2}:r^2$ fiet $\overline{dx^2} + \overline{dy^2} = \overline{dr^2} + a^2\overline{dt^2}:r^2$.

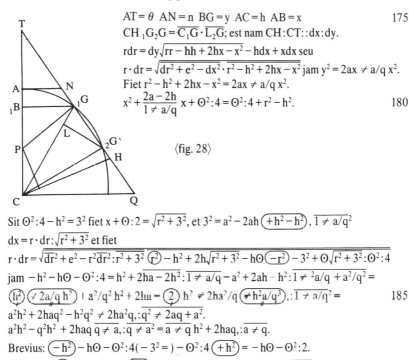

$AT = \theta \;\; AN = n \;\; BG = y \;\; AC = h \;\; AB = x$ 175

$CH \,_1G_2G = \overline{C_1G} \cdot \overline{L_2G}$; est nam $CH:CT::dx:dy$.

$rdr = dy\sqrt{rr - hh + 2hx - x^2} - hdx + xdx$ seu

$r \cdot dr = \sqrt{dr^2 + e^2 - dx^2} \cdot r^2 - h^2 + 2hx - x^2$ jam $y^2 = 2ax \neq a/q \, x^2$.

Fiet $r^2 - h^2 + 2hx - x^2 = 2ax \neq a/q \, x^2$.

$$x^2 + \frac{2a - 2h}{1 \neq a/q} \, x + \Theta^2 : 4 = \Theta^2 : 4 + r^2 - h^2.$$ 180

⟨fig. 28⟩

Sit $\Theta^2 : 4 - h^2 = 3^2$ fiet $x + \Theta : 2 = \sqrt{r^2 + 3^2}$, et $3^2 = a^2 - 2ah \left(+ h^2 - h^2\right), \overline{1 \neq a/q}^2$

$dx = r \cdot dr : \sqrt{r^2 + 3^2}$ et fiet

$r \cdot dr = \sqrt{\overline{dr^2 + e^2 - r^2 dr^2 : r^2 + 3^2}} \, \textcircled{r^2} - h^2 + 2h\sqrt{r^2 + 3^2} - h\Theta \left(-r^2\right) - 3^2 + \Theta\sqrt{r^2 + 3^2} : \Theta^2 : 4$

jam $- h^2 - h\Theta - \Theta^2 : 4 = h^2 + 2ha - 2h^2 : \overline{1 \neq a/q} - a^2 + 2ah - h^2 : \overline{1 \neq}{}^2 a/q + a^2/q^2 =$

$\textcircled{h^2} \, \textcircled{\not 2a/q \, h^2} \; | \; a^2/q^2 \, h^2 + 2 l h u - \textcircled{2} \, h^2 \neq 2ha^2/q \, \textcircled{\not+ h^2 a/q^3}, : \overline{1 \neq a/q}^2 =$ 185

$a^2 h^2 + 2haq^2 - h^2 q^2 \neq 2ha^2 q, : q^2 \neq 2aq + a^2$.

$a^2 h^2 - q^2 h^2 + 2haq \; \overline{q \neq a}, : q \neq a^2 = \overline{a \neq q} \, h^2 + 2haq, : \overline{a \neq q}$.

Brevius: $\textcircled{-h^2} - h\Theta - \Theta^2 : 4 (-3^2 =) - \Theta^2 : 4 \textcircled{+ h^2} = - h\Theta - \Theta^2 : 2$.

$h + \Theta : 2 = \textcircled{h} \neq a/q \, h + a \textcircled{-h}, : , 1 \neq a/q, \; \mathbb{C} = + ah \neq qa, : \neq q + a.$

Jam si C sit umbilicus ellipseos et q distantia centri a vertice primario, f   190
fieri distantia focorum. Fiet $h \neq q = f$, seu fiet $af : \overline{a \neq q} = h + \Theta : 2$.
Jam $a = \overline{f^2 - q^2}, : q$.

$r \, dr = \sqrt{, dr^2 + e^2 - r^2 dr^2 : r^2 + 3^2}, \overline{h + \Theta : 2} \; 2\sqrt{r^2 + 3^2} - \Theta$,

seu $r^4 dr^2 + 3^2 r^2 = \textcircled{r^2 dr^2} + \overline{3^2 dr^2 + e^2} \textcircled{- r^2 dr^2} \, \overline{af : \overline{a \neq q}} \cdot 2\sqrt{r^2 + 3^2} + \dfrac{2a - 2h}{1 \neq a/q} \overset{\Theta}{\overline{}}$

seu $\overline{dr} = e\sqrt{, \mathbb{D} \, 2\sqrt{r^2 + 3^2} + \Theta, : , r^4 + 3^2 r^2 \; \mathbb{D} \cdot 2\sqrt{r^2 + 3^2} + \Theta}$, ⟨interrupted⟩   195

$x \cdot \overline{2q - x} : y^2 (::) q^2 : b^2$ fiet $2qx \mp x^2 = q^2/b^2 \, y^2$ multiplicatur per a/q fit

$2ax - a/q \, x^2 = aq/b^2 \, y^2$.

$q + f$ vel $q - f$. Ergo $h - q = f; f^2 = q^2 - b^2$ jam $bb = aq$. Ergo $f^2 = q^2 \neq aq$.

⟨Addition 1; f. 25v.⟩

In ellipsi accessus ad focum vel ab eo recessus seu
crementum radiorum ejusdem foci est ad
circulationem circa eundem focum, ut sinus
complementi semianguli inter radios duorum focorum
est ad sinum rectum.   5

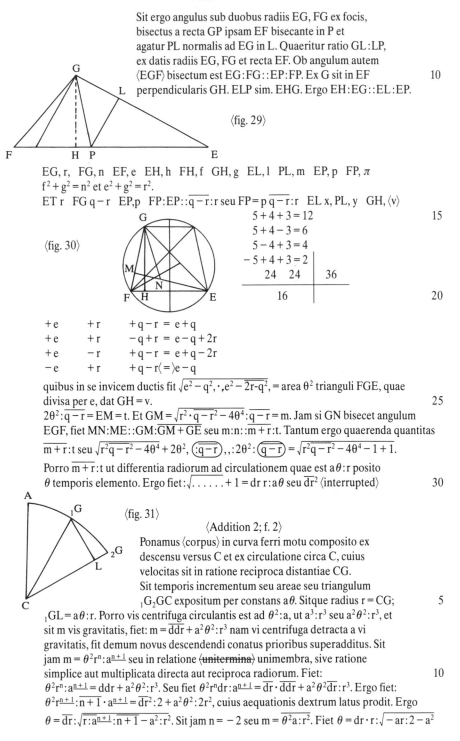

Sit ergo angulus sub duobus radiis EG, FG ex focis,
bisectus a recta GP ipsam EF bisecante in P et
agatur PL normalis ad EG in L. Quaeritur ratio GL:LP,
ex datis radiis EG, FG et recta EF. Ob angulum autem
⟨EGF⟩ bisectum est EG:FG::EP:FP. Ex G sit in EF
perpendicularis GH. ELP sim. EHG. Ergo EH:EG::EL:EP.

⟨fig. 29⟩

EG, r, FG, n EF, e EH, h FH, f GH, g EL, l PL, m EP, p FP, π
$f^2 + g^2 = n^2$ et $e^2 + g^2 = r^2$.
ET r FG q−r EP,p FP:EP::$\overline{q-r}$:r seu FP=p$\overline{q-r}$:r EL x, PL, y GH,⟨v⟩

⟨fig. 30⟩

$5 + 4 + 3 = 12$
$5 + 4 − 3 = 6$
$5 − 4 + 3 = 4$
$− 5 + 4 + 3 = 2$

| | 24 | 24 | 36 |
| 16 | | | |

+e	+r	+q−r = e+q
+e	+r	−q+r = e−q+2r
+e	−r	+q−r = e+q−2r
−e	+r	+q−r⟨=⟩e−q

quibus in se invicem ductis fit $\sqrt{e^2 − q^2}, \cdot, e^2 − \overline{2r\text{-}q^2}$, = area θ^2 trianguli FGE, quae
divisa per e, dat GH = v.

$2\theta^2:\overline{q−r}=EM=t$. Et $GM = \sqrt{r^2 \cdot \overline{q−r^2} − 4\theta^4}:\overline{q−r}=m$. Jam si GN bisecet angulum
EGF, fiet MN:ME::GM:$\overline{GM+GE}$ seu m:n::$\overline{m+r}$:t. Tantum ergo quaerenda quantitas
$\overline{m+r}$:t seu $\sqrt{r^2q^2 − r^2 − 4\theta^4} + 2\theta^2, (\overline{:q−r}),,:2\theta^2:(\overline{q−r}) = \sqrt{r^2q − r^2 − 4\theta^4} − 1 + 1$.
Porro $\overline{m+r}$ ut differentia radiorum ad circulationem quae est aθ:r posito
θ temporis elemento. Ergo fiet:$\sqrt{\ldots\ldots} + 1 = dr$ r:aθ seu $\overline{dr}^2$ ⟨interrupted⟩

⟨fig. 31⟩

⟨Addition 2; f. 2⟩

Ponamus ⟨corpus⟩ in curva ferri motu composito ex
descensu versus C et ex circulatione circa C, cuius
velocitas sit in ratione reciproca distantiae CG.
Sit temporis incrementum seu areae seu triangulum
$_1G_2GC$ expositum per constans aθ. Sitque radius r = CG;
$_1GL = a\theta$:r. Porro vis centrifuga circulantis est ad θ^2:a, ut a^3:r^3 seu $a^2\theta^2$:r^3, et
sit m vis gravitatis, fiet: m = $\overline{ddr} + a^2\theta^2$:$r^3$ nam vi centrifuga detracta a vi
gravitatis, fit demum novus descendendi conatus prioribus superadditus. Sit
jam m = θ^2r^n:a^{n+1} seu in relatione ⟨unitermina⟩ unimembra, sive ratione
simplice aut multiplicata directa aut reciproca radiorum. Fiet:
θ^2r^n:$a^{n+1} = ddr + a^2\theta^2$:$r^3$. Seu fiet θ^2r^ndr:$a^{n+1} = \overline{dr} \cdot \overline{ddr} + a^2\theta^2\overline{dr}$:$r^3$. Ergo fiet:
θ^2r^{n+1}:$\overline{n+1} \cdot a^{n+1} = \overline{dr}^2$:$2 + a^2\theta^2$:$2r^2$, cuius aequationis dextrum latus prodit. Ergo
$\theta = \overline{dr}$:$\sqrt{r:a^{n+1}:\overline{n+1} − a^2:r^2}$. Sit jam n = − 2 seu m = $\overline{\theta^2a}$:r^2. Fiet $\theta = dr \cdot r$:$\sqrt{− ar:2 − a^2}$

10

15

20

25

30

5

10

seu t tempus, vel area integra, $t = \int dr \cdot r : \sqrt{-ar : 2 - a^2}$. Hac ergo res redit, ut inveniamus

aream seu summam talium: $\int \overline{dr} \cdot r : \sqrt{r+1}$. Sit $r+1 = v^2$. Fiet $r = v^2 - 1$ et $dr = 2vdv$. 15

Ergo fiet inde in rationalibus $\int 2v \cdot dv \cdot \overline{v^2 - 1} : v^2 = vv - 2 \int \overline{dv} : v$ quae pendet ex

quadratura hyperbolae, non ergo curva GG potest esse ellipsis, ubi areae ex radiis

per quadraturam hyperbolae non determinantur. Agat gravitas in ratione

distantiarum reciproca triplicata, fiet $n + 1 = -2$. Et fiet $\theta = dr \cdot r \sqrt{\text{const.}}$ Fiet $\int \theta$

seu t ut r^2, seu tempora insumta erunt in duplicata ratione radiorum. 20

Investigemus qualis sit figura, in qua areae A_1GC, A_2GC sint ut quadrata

radiorum C_1G, C_2G. Sit $BG = y$ et $AB = x$ et $AC = h$.

Fient $\int y \cdot dx + \frac{1}{2} y \cdot \overline{h-x} = y^2 + h^2 - 2hx + x^2$ seu $\int y \cdot dx = y^2 + h^2 - 2hx + x^2 - \frac{1}{2} y \overline{h-x}$ et

fiet $\boxed{(y \cdot dx)} + \frac{1}{2} h \, dy \boxed{(-\frac{1}{2} y \cdot dx)} - \frac{1}{2} x \, dx, = 2y \, dy - 2hdx + 2x \, dx.$

Ergo $\int x \cdot dy = hy - 2yy + 4hx - 2x^2$. Ergo $\int y \cdot dx + \int x \cdot dy$ hoc est 25

$\underset{\frac{1}{2}xy}{\boxed{(xy)}}, -\boxed{(y^2)} + h^2 \boxed{(-2hx)} \boxed{(+x^2)} \boxed{(-\frac{1}{2} yh)} \boxed{(+\frac{1}{2} yx)} \overset{+\frac{1}{2}hy}{\boxed{(+hy)}} - \boxed{2} yy + \boxed{4} hx - \boxed{2} x^2$ quae

est aequatio ad hyperbolam. Imo fiat error in calculo. Repetimus

$\int y \cdot dx + \frac{1}{2} y \cdot \overline{h-x} = y^2 + h^2 - 2hx + x^2$. Seu $\int y \cdot dx = y^2 + h^2 - 2hx + x^2 - \frac{1}{2} yh - \frac{1}{2} yx$

cuius aequatio differentialis est

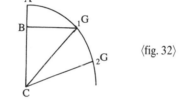

A

B

G

2G

⟨fig. 32⟩

C

$\frac{3}{2} ydx \boxed{(ydx)} - 2ydy - 2hdx + 2xdx - \frac{1}{2} hdy \boxed{(-\frac{1}{2} ydx)} \boxed{(-\frac{1}{2} xdy)}$ อou 30

$4xdx + 4ydy = 3ydx + xdy + 4hdx + hdy$ seu $4xdx - 3ydx - 4hdx = xdy + hdy - 4ydy.$

~~⟨Si quadrata temporum⟩~~ Sit $ldx + mxdx + nydx = pdy + qxdy + rydy$ et fiet

$\overline{4x - 3y - 4h}$ in $\overline{p + qx + xy} = \overline{x + h - 4y}$ in ~~⟨Assumitur⟩~~

$l + mx + ny + pxy + qx^2 + ry^2 = 0$. Fiet $mdx + ndy + pxdy + pydx + 2qxdx + 2rydy =$

⟨interrupted⟩ Seu fiet $mdx + pydx + 2qxdx = -ndy - pxdy - 2rydy$ 35

$y = a + bx + cy + exy + fx^2 + gy^2 + hx^2y + lxy^2 + mx^3 + ny^3$ etc.

Si tempora ⟨interrupted⟩

$h + x \quad -4y$

$m + 2qx + py$

$+ hpy + pxy - 4py^2$ 40

$+ 2hqx \quad - 8qxy \quad + 2qx^2$

$+ hm + mx \quad - 4my$

$- 4h + 4x - 3y$

$n + px + 2ry$

$- 6ry^2$ 45

Ad Neut. p.50 QR vocemus m et CA, q

QR seu $PX : PV \overset{(1)}{::} CA : CP$ porro $\overline{GV \cdot PV} : QV^2 \overset{(2)}{::} PC^2 : \overline{CD}^2$.

$QX^2 : QT^2 \overset{(3)}{::} \overline{AC}^2 : \overline{PF}^2 \overset{(4)}{::} CD^2 : CB^2$. Sit L latus rectum primarium $\overset{(5)}{=} 2\overline{BC}^2 : AC$. Ergo

$L \cdot QR \overset{(6)}{=} 2\overline{BC}^2 : AC$, in PV CA : CP per 1 et 5. Rursus $PV \overset{(7)}{=} QV^2 \ \overline{PC}^2 : \overline{CD}^2 GV$ ex 2.

Et ex aeq. 6 fiet $L \ QR = 2\overline{BC}^2 : \boxed{(AC)}$, in $QV^2 \ PC^\circledast \boxed{(CA)}, : CD^2 \cdot GV \cdot \boxed{(CP)}$ at   50

$\overline{CD}^2 = QX^2 \cdot \overline{CB}^2 : QT^2$ ex 3 et 4; ergo ex aeq. 8 fiet $L \ QR = 2 \ \boxed{(BC^2)} \ QV^2 \ PC \ QT^2, :$

$QX^2 \cdot \boxed{(CB^2)} \cdot GV$. Seu $L \ QR = 2PC \ QV^2 \ QT^2, : GV \cdot QX^2$. QR seu PX vocemus m, et PC, c,

et PV, v, et $GV = 2c - v$; et QX seu $RP = h$, et $XV = n$ et $QV = h + n$ et $QT = k$. Ergo

fit $L \cdot m \cdot \overline{2c - v} \cdot h^2 = \overline{h^2 + 2hn + nn} \cdot 2c \cdot k^2$, ubi ipsius n infinities infinite parvae

quadratum (ut infinitesies infinite infinite parvum) negligere possumus. Porro ob   55

triangula PVX et PCE similia fit PV seu $v : VX$ seu $n, : PX$ seu $m :: PC$ seu $c : CE$ seu

$e : PE$ seu a, et fiet $v = cm : a$ et $n = em : a$. Ergo $L \cdot m \cdot \overline{2c - cm : a} \cdot h^2 = \overline{h^2 + 2hem : a} \cdot 2c \cdot k^2$.

Ergo $2L \ m \ h^2 - L \ m^2 h^2 : a = 2h^2 k^2 + 2ehk^2 m : a$. Si jam rejiciamus $Lm^2 h^2 : a$ itemque

$2ehk^2 m : a$ tanquam caeterarum respectu infinite parvas fit utique $2L \ mh^2 = 2h^2 k^2$

seu m erunt ipsorum k quadratis proportionales, ut conclusit et Neutonus.   60

~~(Verum talis argumentatio //////// potest enim intelligi differentia)~~

⟨v.⟩	Vis centrifuga est $= k^2 : r$. Jam $k = a\theta : r$ posito θ temporis elemento
a seu L	et $a\theta$ area seu triangulo. Ergo vis centrifuga est $a^2\theta^2 : r^3$. Jam
integrum hic	si vis centripeta seu gravitas m sit $k^2 L$ seu $L\theta^2 : r^2$ seu (si
latus rectum	$L = 2a$) $a\theta^2 : 2r^2$, fiet descensus integer seu $ddr = \theta^2 : r^2 - a^2\theta^2 : r^3$. 65
	Et fiet $\int \overline{dr} \cdot ddr = \theta^2 \int dr : r^2 - \theta^2 \int dr : r^3$. Et $\frac{1}{2} \ \overline{dr}^2 = L$.

Sit θ temporis incrementum et $a\theta$ area seu triangulum constans. Jam vis

centrifuga $k^2 : r$. Et in casu arearum aequalium est $k = a\theta : r$ (posito radio r) fiet

vis centrifuga $a^2\theta^2 : r^3$. Rursus vis centripeta seu gravitas m, sit $k^2 : L$, et

$L = 2a$. Posito L esse latus rectum primarium, fiet: $m = a^2\theta^2 : 2ar^2$ seu $a\theta^2 : 2r^2$. Jam   70

ddr nova vis descendendi integra, est differentia vis centripetae et

centrifugae. Ergo fiet $ddr = a\theta^2 : 2r^2 - a^2\theta^2 : r^3$. Et fiet:

$\int dr \cdot \overline{ddr} = \theta^2 a \int dr : 2r^2 - a\int dr : r^3$, et summando $1/2 \ \overline{dr}^2 = \theta^2 a, 1 : 2r - a : r^2 : 2$.

Et fiet $\theta = \overline{dr} \cdot r : \sqrt{ar - aa}$ et $\int \theta$ seu tempus $= \int dr \cdot r : \sqrt{ar - aa}$. Sit $\sqrt{ar - aa} = v$, fit

$r = v^2 + a^2, : a$ et $dr = 2vdv + a^2, : a$, et fit: tempus $= \int 2vdv : v + \int a^2 dv : v$ seu tempus $= 2v +$

$\int a^2 dv : v$. Verum hoc obstat, quod r radius non incipit ab infinite parvo. Ergo pro

r scribamus $z - h$, posita h distantia umbilici a vertice et fiet $dz = dr$ et

$ddz = ddr$, et aequatio ita stabit $dz \cdot ddz = dz \cdot a\theta^2 : 2z^2 - 2zh + 2h^2 - dz \ a^2\theta^2 : \overline{z - h}^3$. Et

jam $\int dz \langle : \rangle \overline{z - h}^2 = -\int dr : r^2 = h : a^2 - 1 : r$ et $\int \overline{dz} \langle : \rangle \overline{z - h}^3 = h : a^2 - 1 : r^2$ et fiet

$1/2 \ a \cdot \overline{dz}^2 \cdot \overline{z - h}^2 = \boxed{(+)} 1/2 \ \theta^2 hr^2 - a^2\theta^2 r \boxed{(-\theta^2 hr^2)} + \theta^2 a^3$ 80

seu $\sqrt{1/2 \ a \cdot dz \cdot \overline{z - h}}, : \sqrt{1/2 \ h \cdot \overline{z - h}^2 - a^2 \overline{z - h} + a^3}, = \theta$.

Si sint temporum quadrata ut cubi radiorum, fiet $t^2 = \overline{h - z}^3$. Ergo ⟨interrupted⟩

Si planetae jovis ferantur circa jovem in circulis sintque temporum quadrata,

ut cubi distantiarum, et tempora sint areis ex centro proportionalia, erunt

sectorum CAB, C(A)(B) quadrata, id est radiorum CA, (C)A quadrato quadrata,   85

proportionalia cubis suis quod est absurdum. Ergo si

moventur in circulis et sint aequales non possunt tempora

habere areis proportionalia, quod si habent, necesse est ut

non in circulis, sed ellipsibus ferantur. Verum si planetae

joviales inaequales, sintque ⟨interrupted⟩   90

⟨fig. 33⟩

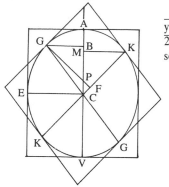

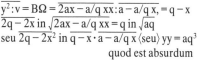

$$\overline{y^2:v} = B\Omega = \overline{2ax - a/q\, xx} : \overline{a - a/q\, x}, = q - x$$
$$\overline{2q - 2x} \text{ in } \sqrt{2ax - a/q\, xx} = q \text{ in } \sqrt{aq}$$
seu $\overline{2q - 2x^2}$ in $\overline{q - x} \cdot \overline{a - a/q\, x}$ ⟨seu⟩ $yy = aq^3$
quod est absurdum

⟨fig. 34⟩

Parallelogramma omnia circa eandem ellipsin conscripta esse inter se   5
aequalia; a conicis scriptoribus dudum demonstratum est, sed placet idem
investigare calculo. Sit AGEV semiellipsis, sit centrum C, vertex primarius A,
oppositus V. AB sit x, BG, y, latus rectum a, transversum seu CA, q. Constat
esse $2ax + a/q\, x^2 - y^2$ et $adx + a/q\, xdx = ydy$. Seu $dy:dx::\overline{a \neq a/q\, x},:y::BP:BG$ seu y.
Ergo $BP = a \neq a/q\, x$ et jam $C\Lambda - q$ fit $PC = q - \langle x\rangle - a \neq a/q\, x$, jam $PF:PC::PB:PG$ et   10
$PG^2 = 2ax \neq a/q\, xx + a^2 \neq 2a^2/q\, x + a^2/q^2\, x^2$. Et sit $q - a = e$. Fiet: et $\neq a/q + a^2/q^2 = 3$
et $2a \neq a^2:q = 5a$ et fiet $PG^2 = a^2 + 5ax + 3x^2$ et $PC = e \neq a/q\, x$. Ergo
$PF = \overline{a \neq a/q\, x} \cdot \overline{e \neq a/q\, x} : \sqrt{a^2 + 5ax + 3x^2}$ et $GF = PG + PF$. Fit autem quartae partis
parallelogrammi circumscripti integri, seu parallelogrammi GCK area ex ductu
ipsius GF in CK. Hanc ergo CK quaeramus. Et primum quidem CM et MK ob   15
triangula similia GBP et CMK sit $CM:MK::GB$ seu $y,\langle:\rangle BP$ seu $a \neq a/q\, x$. Sit AM m,
et MK, k, fiet $2am \neq a/q\, m^2 = k^2$ et $q - m:k::\Theta::\overline{dx}:\overline{dy}$ et $m = q - \Theta K$. Ergo fiet
$2aq - 2\Theta K \neq a/q\, q^2 \quad 2q\Theta K \mid \Theta^2 K^2 \neq K^2$ hoc est quia $\neq = $ fict $\neq aq \mid a/q\, \Theta^2 K^2 - K^2$
seu $K = \sqrt{aq:\overline{1 - a/q\, \Theta^2}}$. Jam $CM = q - m$. Ergo ΘK. Et fiet $CK = \Theta^2 K^2 + K^2 = $
$K\sqrt{1 + \Theta^2}$. Ergo $CK = \sqrt{aq \cdot \overline{1 + \Theta^2}:\overline{1 - \Theta^2}}\, GB:BP::\Theta:1$. Ergo $\overline{GB^2} + \overline{BP^2}$, seu   20
$\overline{PG^2}:\overline{BP^2}::\Theta^2 + 1,:1$. Seu $PG = BP\sqrt{\Theta^2 + 1}$ et $PC:PF::PG:PB:\sqrt{\Theta^2 + 1}:1$. Ergo
$PF = PC:\sqrt{\Theta^2 + 1}$. BC scribatur $v = a \neq a/q\, x$. Erit $\Theta = y:v$ et $PC = v - BP$.
$GF = PG + PF - BP\sqrt{\Theta^2 + 1} + PC:\sqrt{\Theta^2 + 1}$ seu $PG + PF = BP\sqrt{\Theta^2 + 1} + PC:\sqrt{\dots}$ seu
$= \overline{BP \cdot \Theta^2 + BC},:\sqrt{\Theta^2 + 1} = GF$. Restat ut inveniamus. Jam $BP = a \neq a/q\, x = v$ et
$BC = q - x$ et $\Theta^2 = y^2:v^2$. Et $BP\,\Theta^2 = \overline{y^2:v} = \overline{2ax \neq a/q\, x^2},:\overline{a \neq a/q\, x}$ et $BP\,\Theta^2 + BC = $ 25
$\boxed{2ax}\boxed{\neq a/q\, x^2} + aq \boxed{-ax \neq ax}\boxed{\neq a/q\, x^2}:v^2$. $1 - \Theta^2 = a^2 \neq 2a^2/q\, x + x^2 - 2ax \neq a/q\, x^2,:v^2$.

Parallelogramma omnia circa eandem ellipsin descripta esse inter se
aequalia ex scriptoribus constat, placet autem veritatem eius theorematis
investigare calculo.

Sit ellipseos AGEVK vertex A, oppositus V, centrum C, latus rectum a,   30
transversum seu AC, q. Diametri coniugatae GG, KK, ex puncto G et ex puncto K
in axem AV ducantur ordinatae GB, KM, erit AG, x, BG, y, AM, m, MK, k. Ex puncto
G sit perpendicularis GF in conjugatum diametrum CK, secans AC in P. Patet
parallelogrammum ACK, fieri ex ductu GF in CK. Investigemus ergo utramque.

Quia GB seu $y:BP::dx:dy$ seu $\Theta:1$ ⌊quia GP perpendicularis ad tangentem in G⌋ et   35

aequatio ellipseos est $2ax - a/q\ x^2 \overset{(2)}{=} y^2$ erit eius differentialis $adx - a/q\ xdx \overset{(3)}{=} ydy.$
Seu $dx:dy::y:\overline{a-ax:q}$ ergo fiet $BP = a - ax:q \overset{(4)}{=} v.$ Sit $AM = m,$ et $MK = k,$ et $CM,$
$q - m.$ Fiet $q - m : K \overset{(5)}{::} \Theta : 1$ et $2am - am^2 : q \overset{(6)}{=} KK.$ Ex aeq. 5 est $m \overset{(7)}{=} q - \Theta K$ quo
substituto in aeq. 6 fit: $\textcircled{2}$ aq $\overline{\bigcirc -2a\Theta K} - \overline{a:q}\ \textcircled{q^2} \cdot \overline{\bigcirc -2q\Theta K} + \Theta^2 K^2 = K^2.$ Seu
$qK^2 + a\Theta^2 K^2 = aq^2$ seu $K = aq : \sqrt{aq + aa\Theta^2} = CK.$ Rursus ut inveniamus GF, est $\qquad$ 40
$GF = GP + PF$ jam quia $GB:BP::\Theta:1$ fiet $\overline{GP}^2 = \overline{GB^2 + BP^2}, : \overline{BP}^2 :: \overline{\Theta^2 + 1}, :1.$ Sit BP, v.
Fiet $GP = v\sqrt{\Theta^2 + 1} = \sqrt{y^2 + v^2}.$ BC sit c. Erit $PC = c - v.$ Rursus
$PC:PF::GP:BP:: \sqrt{\Theta^2 + 1} : 1$ fiet $PF = \overline{c - v} : \sqrt{\Theta^2 + 1} = v\overline{c - v} : \sqrt{y^2 + v^2}.$ Ergo
$GF = v\sqrt{\Theta^2 + 1} + \overline{c - v} : \sqrt{\Theta^2 + 1}$ seu $v\Theta\ \textcircled{$+v$}) + c\textcircled{$-v$}, : \sqrt{\Theta^2 + 1}$ et explicando erit
$GF = y^2 + vc : \sqrt{y^2 + v^2}$ et $CK = aqv : \sqrt{aqv^2 + aay^2}.$ Seu $GF = GP + PF =$ $\qquad$ 45
$\sqrt{y^2 + v^2} + \overline{c - v}\ v : \sqrt{y^2 + v^2} = y^2\textcircled{$+v^2$}) + cv\textcircled{$-v^2$}, : \sqrt{y^2 + v^2} = GF.$ Seu
$CK = aqv : \sqrt{aqvv + aayy}.$ Est autem $y^2 + v^2 = 2ax - ax^2:q + a^2 - 2a^2x:q + a^2x^2:qq$
seu $a^2 + \overline{2ax - ax^2 : q} : \overline{1 - a : q}.$

$$k = 1 : \sqrt{2};\ m = 1 - 1 : \sqrt{2};\ m + k = 1$$
$$2 - \sqrt{2} - 1 + \sqrt{2} - 1 : 2 = 1 : 2.$$

⟨Addition 4; f. 3v.⟩

Ponamus planetas ita ferri, ut ab orbibus suis ad circulandum impellantur simulque vi eorum centrifuga deprimantur. Et quidem initio ⟨vis centrifuga est⟩ vis circulandi est exigua, quia materia orbis est valde tenuis, ita ut ⟨intra certum tempus circulatio illa⟩ velocitas circulandi primum impressa, habeat ad velocitatem circulandi ordinariam, rationem prope infinite parvam, $\qquad$ 5
ita ut parum admodum progrediatur. ⟨Vis⟩ ⟨Conatus centrifugus autem cum sit infinitesimus conatus circulandi, multo minor adhuc descendet grave et ab orbe suo depelletur; paulatim autem⟩ Conatus autem centrifugus ipsi planetae impressus cum sit infinitesimus circulationis, multo magis pro nullo habebitur, et cum ⟨conatus⟩ ⌊vis⌋ descendendi planetae sit vis centrifuga orbis $\qquad$ 10
demta vi centrifuga planetae; (unde si posterior excedit, degenerat in vim ascendendi) ideo sola initio in rationes venit vis centrifuga orbis, qua planeta deorsum impellitur. ⟨Sunt ergo velocitas circulandi initialis, et conatus descendendi initialis, homogenei. Ergo dant mobili statim directionem.⟩ ⟨Sed tamen ob materiae orbis raritatem⟩ Verum et ipsa ob $\qquad$ 15
materiae obis raritatem parum initio operatur.
Ut res fiat clarior, sit ⟨velocitas⟩ ⟨vis⟩ potentia orbis v imprimenda ad circulandam materiam seu ⟨soliditas⟩ ⟨densitas⟩ crassities m, ⟨vis impressa⟩ planetae corpus p, si jam materiae soliditas orbis sit prope infinitesima planetae, erit vis planetae mv, et potentia eius $mv:p.$ $\qquad$ 20
Et ut res fiat clarior consideretur corpus quo minores habet particulas, etsi solidiores, eo alteri minus impetus ⟨imprimere. Itaque vis circulandi est a partibus supeficiariis, sed vis centrifuga est a⟩ imprimere, nam tanta est vis, quantus est numerus particularum stringentium, ideo vis impellendi eodem tempore impressa est in ratione celeritatum, ob repetitiones particularum. $\qquad$ 25
Sed his missis consideremus planetam cum impetu aliquo jam dato circulandi

⟨fig. 35⟩

ferri PK, intereaque detrudi vi ⟨orbis centrifuga⟩
impressa centripeta, quae sit oriunda a centrifuga orbis.
Ponatur autem sic moveri, ut areae sint temporibus
proportionales, seu ut ipsae PL circulationes ⟨integrae⟩   30
⌊collectae⌋ sint radiis reciproce proportionales. Et si
linea sit ellipsis cuius umbilicus S, erunt ipsae $_2P_2G$
reciproce ut quadrata distantiarum SP. Huius phaenomeni
ut causam inveniamus. Considerandum est novum ⟨descensum
priori super⟩ conatum descendendi priori superadditum,   35

⟨fig. 36⟩

⟨qui est vis centripetas impressio⟩ fieri detrahendo a
conatu gravitatis, conatum centrifugum circulationis.
Porro conatus ⟨centrifugus⟩ novus componitur ex his
conatibus: impresso $_3P_3K$ (aeq. $_2P_3P$) et gravitatis, et
impresso conatu circulandi.   40
    Sit ellipsis PPP, via planetae, in cuius umbilico sol
S, et ponamus aequalibus temporibus aequales describi
areas, sunt ergo ipsae L(P) ipsis PS reciproce
proportionales, et si dicatur (G)((P)) parallela P(P)
secans S(P) in (G) ob naturam ellipsis debent esse ipsae PG in reciproca   45
duplicata ipsarum SP ⌊quae est centripeta impressio nova ex natura gravitatis⌋.
Compleatur parallelogrammum $_2P_2L_3P_2F$, et centro S radio S_2P describatur
arcus, secans $_3P_2F$ in /ϱ, erit /ϱF, ut vis centrifuga ⌊nova⌋ ex natura
circulationis; quodsi a $_{(2)}P_2G$ detrahas /ϱ$_2F$ oritur impressio descendendi nova
seu excessus ipsius $_2P_2L$ super $_1P_1L$. Verum hinc jam patet, si novum addamus   50
conatum circulandi additum a novo orbe, quem ingreditur mobile, ut Pw, cum
ille nihil faciat ad descensum in $_2PS$, atque adeo maneat P_2G, ideo non posse
concipi, ut maneat $_2P_2K$ aequal. $_1P_2P$. Et proinde si novus impetus circulandi

$_2$Pw in quolibet orbe imprimitur, et servatur praeterea $_2$P$_2$K impetus jam
impressus in priori orbe, non posse areas describi temporibus proportionales.
Non video igitur, quomodo excusari possit impressio nova a quolibet vortice,
nisi dicamus mox fieri ut delatus planeta in alium orbem circulationi se
orbis accomodet. Cum enim poni possit alicubi in orbe aliquo habere
celeritatem sui orbis, certe descendens in proximum, cuius paulo minor est
celeritas, parum differentem habet celeritatem, ut cum in eo aliquandiu
movetur, mox acquiret, vitandae perturbationis causa; celerius enim motu
resistit materia orbis. Si quod autem discrimen est, a nobis notari non
poterit. Idem est si recedens magis ab S in ⟨contrarium⟩ alium veniat orbem.
Ibi enim major est celeritas, adeoque continue acceleratur vis in orbe, donec
aequetur celeritas; verum sciendum est vim gravitatis descendere non a motu,
utrum scilicet celerior sit motus planetae, quam sui orbis, sed a soliditate,
ponamus enim planetae corpus habere spatia materiam vorticis excludentia,
quae sint instar bullarum vitrearum in aqua, tunc etsi motus orbis et
planetae sit idem, nihilominus vis centrifuga aequalis materiae quam planeta
loco suo expellit fortior est, quam vis centrifuga ipsius planetae, etsi enim
eadem sit utriusque celeritas, tamen differunt massa.

⟨Sunt autem vires centrifugae⟩ ⟨celeritates centrifu⟩ Conatus autem
centrifugi positi circulantium in ratione reciproca radiorum sunt in ratione
reciproca triplicata radiorum. Sed vis ⟨centrifuga⟩ gravitatis est in ratione
composita conatus centrifugi et ⟨virium⟩ ⟨massae⟩ ⟨gravitatis seu
soliditatis massae⟩ densitatis massae, densitates autem massae sunt in
ratione radiorum directa; seu materia solida ab orbe remotior est. Et ratio
composita ex reciproca triplicata et directa simplice, est reciproca
duplicata. Sunt ergo vires gravitatis in ratione radiorum reciproca
duplicata. ⌊Quando materia homogenea est, et orbes cum celeritates sunt ut
distantiae reciproce, tunc eadem est potentia in quolibet orbe.⌋ Sin orbes ita
ferrentur, ut essent quadrata temporum ut cubi distantiarum, videamus quae
sit ratio potentiae posita massa crescente in ratione distantiarum. ⟨Cum sint
quadrata sit t^2:(t)^2::r^3(r^3). Ergo t:(t)::r$\sqrt{}$r:(r)$\sqrt{}$(r) jam t:(t)::cel. in⟩ circ.
////// celeritatis rad. ad cel. quad. seu vr:(v)(r) v:r:(v)(r)⟩ tempora t,
radii r, veloci ///// t^2 ut r^3, jam t directe ut circuli seu ut radii reciproce
et celeritates //////////////////////
Ut potentiae sint reciproce ut r, si massae ut r^2, debent v^{2}r^2 esse ut 1:r seu
v^2 ut 1:r^3; jam t ut r:v. Ergo t^2 ut r^{2}v^2, ergo t^2 ut 1:r; si t^2 ut r^3, videamus
quod debeat esse m, ut potentiae p seu mv^2 sint ut 1:r. Sunt autem t ut r:v.
Ergo t^2 ut r^2:v^2, jam t^2 ut r^3 ergo r ut 1:v^2 ergo v^2 ut 1:r ergo m ut 1. Jam m
ur rs. Ergo rs ut 1. Ergo s ut 1:r ergo deberent densitates remotiorum orbium
esse minores, in ratione distantiae, et densiora esse prope centrum.
Generaliter p ut mv^2, m ut rs, t ut r:v. Quando densitates ut r, erunt m ut rr,
si jam v$^{(2)}$ ut 1:r, erunt p ut r. Si sit globus g, globorumque medius motus, ut
quadrata temporum sint ut cubi distantiarum t^2 ut r^3, ut r^2:v^2, ergo r ut 1:v^2,
ergo v ut 1:$\sqrt{}$r. Ergo v^2, seu potentiae in reciproca distantiarum quod si
densitates in directa distantiarum, erunt corporum ////////////// ⟨h⟩inc
magnitudines, si massae ut distantiae, erunt aequales ut orbes /////////

Commentary to *De Motu Gravium*

The opening of the present essay is similar to the first lines of *Galilaeus*. Leibniz intends to classify the curves described by a body under the action of central forces, which he calls *lineae centroparabolicae* (lines 10–11). The first instance is the parabola, which results from an infinitely distant centre and a constant conatus. The second case results from a constant conatus, which is called *respiciens centrum seu paracentricus*, and a centre at a finite distance, hence the name 'centroparabolic' for the curve. The denominations *centroparabolica* and *projectitia* or *projectaria* are equivalent, as one can see from lines 18–19. The word *centroparabolica* was used in *Si mobile aliquod ita moveatur*, which I have briefly described in Section 5.5, whereas the expression *conatus centrum respiciens* was defined in the *Notes* to Newton's *Principia*. The study of centroparabolic curves begins at the end of the second manuscript side. Before reaching that point Leibniz's theory of planetary motion has taken a new shape.

The remarks in line 11 and the following lines resemble those which we have seen in *De Conatu* and *Inventum a me est*, and are free from the mistake concerning decomposition of motion typical of the *Repraesentatio Aliqua*. Leibniz considers the curve as composed of infinitesimal chords which he calls 'elements of the curve' and which are traversed in equal times (lines 41–6). Then, following a reasoning which echoes a passage in *Inventum a me est*, he shows that taking equal elements of time, the velocities of circulation are inversely proportional to the radii (lines 50–1).

In lines 62–6 Leibniz discovers a crucial proposition: the *lineae centroparabolicae* are the same as the *lineae vorticales* or *dinobarycae*, which are described by a body carried by a vortex with a speed of rotation inversely proportional to the distance from its centre. Leibniz's proposition states that, with respect to the area law, central forces are equivalent to a special case of vortical motion. This proposition, which was implicitly mentioned in *Inventum a me est* in purely mathematical terms, is now stated with greater clarity and with reference to the vortex. Its role in Leibniz's theory is crucial, because it transforms central forces into a physically acceptable cause. The lack of any reference to the harmonic circulation in this context, and the usage of alternative denominations which have not been employed by Leibniz since 1689, indicate that *De Motu Gravium* precedes the *Tentamen*. The following considerations provide further arguments for the dating.

Leibniz wants to determine the descent $_1L_1G$, which he names dr, representing the difference between the two radii Θ_1G and Θ_2G

(compare fig. 26 and lines 71–9). Since the angles in $_1L$ and $_1N$ are right angles, triangles Θ_1L_2G and Θ_1N_1G are similar and we can establish the proportion:

$$\Theta_1G:\Theta_2G::\Theta_1N:\Theta_1L.$$

For Leibniz $\Theta_1G = r$ is a radius; $\Theta_2G = r - dr$ is another radius infinitesimally distant; $\Theta_1N = \Theta_2G + _2G_1N$ and $_2G_1N = _2GR$ because Leibniz takes the tangent $_2GM$ to be the prolongation of the chord $_1G_2G$; MR is perpendicular to Θ_2G, hence triangles $_1G_2G_1N$ and $_2GMR$ are equal, because $_1G_2G = _2GM$. Further, $_2GR = _2G_2L - R_2L$ and $_2G_2L = dr + ddr$; $R_2L = M_3G = m$ is the deviation from the tangent and represents gravity; therefore $\Theta_1N = r + ddr - m$. Lastly, since Leibniz sets $\Theta_1L = r - dr$, the proportion becomes

$$r:(r - dr)::(r + ddr - m):(r - dr);$$

hence $ddr = m$. Leibniz's immediate reaction is that this result was predictable (line 83): the conatus of gravity is the difference between two descents or between two contiguous impetuses of descent.

At this point the reader has certainly realized the difference between *De Conatu* and *De Motu Gravium*. In *De Conatu* Leibniz calculated the attractive conatus, which was measured by the deviation from the tangent. Here Leibniz is trying to calculate the difference in length between two radii infinitesimally distant, and he compares them by transferring one on to the other by rotation along a circular arc. He begins to change one of the components of motion from uniform rectilinear to uniform circular, and to use a representation along the rotating radius. Motion from $_1G$ to $_2G$ should be composed of the descent $_1GK$ and of the uniform circular motion K_2G. At this stage Leibniz is not yet aware of the consequences of this change for the component of motion along the radius, or for paracentric conatus.

A correct form of the preceding reasoning would be as follows: $K_1L = k = 1:2$ centrifugal conatus; $\Theta_1G = r$; $\Theta_2G = r - dr_1$; $\Theta_1N = \Theta_2G + _2G_2L - R_2L = r - dr_1 + dr_2 + k - m$, where $dr_1 = \Theta_1G - \Theta_2G$ and $dr_2 = \Theta_2G - \Theta_3G$; $\Theta_1L = r - dr_1 - k$. Further, $ddr = dr_1 - dr_2$ and the proportion above becomes:

$$r:(r - dr_1)::(r - ddr + k - m):(r - dr_1 - k).$$

Neglecting third-order infinitesimals, we have $ddr = 2k - m$, or the second-order differential of the radius is equal to the difference between centrifugal conatus and solicitation of gravity. This calculation is modelled on paragraph 15 in the *Tentamen*. Notice that motion along M_3G is uniform because the tangent is the prolongation of the chord; motion along K_1L, however, ought to be accelerated because K_2G is a circle-arc (see Section 4.2). This is the reason why the solicitation of gravity is equal to m, whereas centrifugal conatus is equal to $2k$.

In lines 95–8 we find some remarks on the distinction between conatus and impetus, a problem which plagued several passages in *De Conatu*. Leibniz also refers (lines 96 and 99) to dead and living force, which he claims were confused by the Cartesians. They believed that quantity of motion is conserved, not living force, but for Leibniz only quantity of progress is conserved, that is, velocity with direction.

After a few observations on infinitesimals and on the characteristic triangle (lines 108–25), Leibniz, while trying to determine $\Theta_1 N$, carries out some of the calculations we have seen above. He writes (expressions in square brackets are mine):

$$_1G_1L = dr - k \ [= dr_1 + k],$$

$$_2G_2L = dr + ddr - (k) \ [= dr_2 + k],$$

where (k) denotes centrifugal conatus at the successive instance of time. Since k and (k) differ by a third-order infinitesimal, they can be taken to coincide, as we have seen in Section 8.4. Recalling that

$$_2GN \ [= _2GR] = _2G_2L - R_2L,$$

we have a new expression for $\Theta_1 N$, namely

$$\Theta_1 N = \Theta_2 G + _2GN = \Theta_2 G + _2G_2L - R_2L = r - dr + dr + ddr - (k) - m$$
$$[r - dr_1 + dr_2 + k - m = r - ddr + k - m].$$

Thus the result found by Leibniz is $\Theta_1 N = r + ddr - (k) - m$. On the basis of the previously established proportion involving $\Theta_1 N$, he claims to have found a new expression for *ddr* contradicting what he obtained above, that the second-order differential of the radius is equal to gravity, $ddr = m$. Now Leibniz believes that his former calculations were wrong, and that centrifugal force arising from the circulation should be considered together with the conatus of gravity (lines 141–55). Leibniz is not worried whether gravity and centrifugal conatus have to be added or subtracted: with the word *conjungenda* (line 143) he seems to indicate that gravity and centrifugal conatus act together. In the case under consideration he sets the second-order differential of the radius equal to the difference between gravity and centrifugal conatus, $ddr = m - k$. An expression for *ddr* emerges more convincingly in paragraph 15 of the *Tentamen* as the difference between two successive differentials of the radius. In *De Motu Gravium*, however, Leibniz obtains first the expression $ddr = m + k$ by chance and, indeed, inconsistently: he does not prove, but rather simply infers that *ddr* results from the composition of a centrifugal and an attractive term. The present essay, however, marks a crucial step in the evolution of Leibniz's theory with respect to *De Conatu*, because here he begins to consider a two-term paracentric conatus. On the basis of the elements pointed out so far we can conclude that *De Motu Gravium* dates after *De Conatu* and *Inventum a me est*, and *a fortiori* after Leibniz's notes to Newton's *Principia*. These elements also show the evolution between *De Motu Gravium* and the *Tentamen*. Although the latter essay is more coherent and better structured, the present text comprises all the elements of the final theory.

Concerning the calculation of centrifugal force (see fig. 27 and lines 143–9), Leibniz writes $K_1L = \Omega Q = {}_1L_2G^2/\Theta_1L$. Figure 27 is based on the relevant portion of fig. 26; $\overset{\frown}{K_2G}$ and $\overset{\frown}{{}_1G\Omega}$ are circular arcs with radii Θ_2G and Θ_1G respectively. K_1L ought to be equal to $({}_1L_2G)^2:(\Theta_1L + K\Theta)$. Since the arc $\overset{\frown}{K_2G}$ is infinitesimal, K_1L can be set equal to $({}_1L_2G)^2:2K\Theta$. Thus Leibniz neglects a factor of 2. This mistake is purely mathematical and compensates exactly for the other mistake regarding accelerated motion in the expression of centrifugal conatus. The former was corrected in 1688 and does not appear in the *Tentamen*, the latter was corrected thanks to Varignon in 1706. Thus the lack of the mistake in the expression of centrifugal conatus, far from indicating that Leibniz was already aware of it, appears as the casual result of a less sophisticated theory.

In line 155 Leibniz provides a mathematical formulation of the relation between impetus and conatus, $dr = \int m - \int k$. This equation is written with no reference to time and entails no accelerations (see Section 4.3).

From the bottom of the second side onwards Leibniz deals with the problem formulated at the beginning of the essay, namely the problem of investigating the properties of centroparabolic curves. He takes constant gravity, centrifugal conatus $k = ba^2\theta:r^3$, where θ is proportional to the square of the differential of time dt, a and b are constant factors whose meaning is not explained. He found above that centrifugal force is inversely proportional to the third power of the distance (lines 146–7).

The calculations which we are about to study are fascinating both because they are based on very advanced techniques of integration of differential equations, and because they illustrate an important aspect in Leibniz's dynamics (compare Sections 3.3 and 5.5). Leibniz's study of centroparabolic curves starts from the following equation:

$$ddr = m - ba^2\theta:r^3. \tag{1}$$

Setting $m = \theta$ and multiplying by the differential of the radius he finds

$$\overline{dr} \cdot \overline{ddr} = \theta dr - ba^2\theta\overline{dr}:r^3. \tag{2}$$

Taking the integral of equation (2) Leibniz writes:

$$\overline{dr^2}:2 = \theta r - ba^2\theta:3r^2. \tag{3}$$

Leibniz makes some standard mistakes, neglecting to change sign in the second term at second member and to insert the integration constant; the last constant factor should be 2 and not 3. Setting $\theta = dt^2:a$, we have (this equation is introduced by the editor for convenience):

$$\overline{dr^2}:2 = \overline{dt^2}r:a - ba\overline{dt^2}:3r^2. \tag{4}$$

From this equation Leibniz separates variables in two passages:

$$a \cdot dr \cdot r\sqrt{3} = dt\sqrt{ar^3 - ba^3}\sqrt{2}, \tag{5}$$

$$a\sqrt{3} \cdot dr \cdot r:\sqrt{ar^3 - ba^3} = dt\sqrt{2}, \tag{6}$$

where a factor 3 has been forgotten in the first term of the radicand. The last equation expresses the relation between time and distance. Although Leibniz believes these calculations to be the easiest generalization of Galilean parabolas, the case which he is investigating involves a non-trivial elliptical integral.

On realizing the difficulty in calculating the integral, Leibniz tries again, giving a different value to gravity. He sets it as composed of two terms, one of which cancels out the expression of centrifugal conatus (line 162):

$$m = r^4\overline{dt^2}:a^5 + ba\overline{dt^2}:3r^2; \tag{7}$$

the value of the second term is fixed according to the expression of centrifugal conatus after the integration, in equation (4). Consequently, the first term of m is not integrated. After this remarkable series of errors, the result is that the radius is inversely (not directly!) proportional to time:

$$a^2\sqrt{a:2}\ dr:r^2 = dt \tag{8}$$

and

$$a^2\sqrt{a:2}:2r = t; \tag{9}$$

where the factor 2 and a sign are wrong. Leibniz is aware that this is a mathematical trick for integrating the equation, and adds that when gravity consists of a simple term it is not equally easy to find a relation between time and radius (lines 165–6). This remark and the following calculation on ellipses indicate that Leibniz had in mind Kepler's laws. We shall come back to this problem below and again in Addition 2.

Soon afterwards Leibniz adopts an approach similar to the one we have seen above (line 167). He fixes (see figs. 26 and 28)

$$_1L_1G = \overline{dp} + k,$$

$$_2G_1N = (dp) + (k) - m = dp + ddp + (k) - m,$$

$$_1N_1G = adt : \overline{p - dp},$$

$$_1L_2G = adt : p.$$

From these equations we have the following relation:

$$\overline{_1N_1}G^2 - \overline{_1L_2}G^2 = \overline{_2G_1}N^2 - \overline{_1G_1}L^2,$$

at the right member signs should be inverted. From here Leibniz tries unsuccessfully to find a relation between time and distance. In fig. 28 CH ought to be perpendicular to GH; the role of point P is not explained.

From the equations above we should have:

$$a^2(dt)^2 : (p - dp)^2 - a^2(dt)^2 : p^2 = (dp + k)^2 - (dp - k)^2;$$

neglecting the infinitesimal of higher order we have $2k = a^2(dt)^2/p^3$. This expression, far from giving us a relation between time and distance for the orbital curve, tells us that centrifugal conatus is inversely proportional to p^3.

Lastly, Leibniz tries a different approach based on the equations $r^2 = y^2 + h^2 - 2hx + x^2$ and $\overline{dx^2} + \overline{dy}^2 = \overline{dr}^2 + e^2$, where r is the radius CG, $BG = y$, $h - x = BC$, $e^2 = a^2\overline{dt}^2 : r^2$ (lines 172–). After a series of substitutions which eliminate the variable y from the former equation differentiated, $rdr = ydy - hdx + xdx$, Leibniz continues forgetting the expression $-hdx + xdx$ (lines 177–8). This example is typical of his style of carrying out calculations which become hopelessly involved, also because of his errors.

Throughout the present essay Leibniz deals with arbitrary curves described by a body under the action of central forces. Towards the end (line 178), however, he introduces the equation of the ellipse when the symbol '$\neq$', standing for '+' or '−', is a minus sign: a represents the *latus rectum*, b and q the minor and major axes respectively. Lines 196–7 and 198 are written in the margin to lines 176–8 and 189–94 respectively.

Commentary to addition 1

This addition consists of a series of sketchy reflections and calculations. Leibniz seeks the expressions for the descent and circulation of a body in the specific case of ellipses. The opening sentence is based on the property of ellipses that the perpendicular *GP* to the tangent at any point *G* on the curve bisects the angle between the radii from *G* to the foci *E* and *F*; in fig. 29 the ellipse through *G* is not drawn, but see my

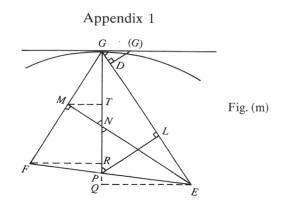

Fig. (m)

enclosed Fig. (m). *FE* is the focal distance; *GP* bisects the angle in *G*. It is easy to prove that the triangle *GPL* is similar to the triangle formed by the velocity in the orbit (homologous to *GP*), the velocity of descent (homologous to *PL*) and the velocity of circulation (homologous to *GL*). In fact, the angle $G\hat{P}L$ is equal to the angle $D\hat{G}(G)$ in Fig. (m), where (G) is the following point on the ellipse, and *D*(*G*) the perpendicular to the radius *GE*. Thus triangles *G*(*G*)*D* and *PGL* are similar: of course, this is an application of the characteristic triangle. Setting $P\hat{G}E = \phi$, we have the proportion $GL:PL::\cos\phi:\sin\phi$; hence in lines 1–5 Leibniz ought to state that the velocity of circulation is to the velocity of descent as $\cos\phi$ to $\sin\phi$. The proportion $EG:FG::EP:FP$ (line 10) can be proved by drawing the perpendiculars *FR* and *EQ* to the prolongation of *GP*, as in Fig. (m), and considering the similar triangles *FPR* and *EPQ*, as well as *GFR* and *GEQ*; $EG:FG::EQ:FR$ and $EQ:FR::EP:FP$, and the result is proved.

In the second part of the text (lines 15–), Leibniz has recourse to the formula for the area of a triangle given the sides *a*, *b* and *c*: the area is equal to $\frac{1}{4}\sqrt{(a+b+c)(a+b-c)(a-b+c)(-a+b+c)}$. In spite of his checking this formula by means of a numerical example, he neglects the constant factor and a sign (line 23). Fig. 30 reproduces a triangle with the same characteristics as *FGE* as in fig. 29. If *EM* is perpendicular to *FG*, we have $GM^2 = GE^2 - EM^2$. If we recall that $GF = q - r$, where *q* is the major axis of the ellipse, and $GE = r$, we have that *EM* is equal to twice the area of the triangle *EFG* over $q - r$. We know that *GN* bisects the angle $E\hat{G}M$; the proportion $GE:GM::NE:NM$ is proved immediately by drawing the perpendiculars *MT* and *EQ* to the prolongation of *GN*, and considering the similar triangles *GMT* and *GEQ*, as well as *MNT* and *ENQ*. Then we have $(GE:GM) + 1 = (NE:NM) + 1$, hence $(GE + GM):GM::(NE + NM):NM$, and therefore $(GM + GE):EM::GM:MN$, which can be written as $(m + r):t::m:n$; namely $GM = m$, $EM = t$, $MN = n$. Since $GM:MN$ is (inversely!) proportional to the descent *dr* over the velocity of circulation $a\theta:r$, Leibniz finds a way of expressing those differential expressions in the case of an elliptical orbit by means of known quantities.

The aim of these calculations is the same as in paragraph 18 of the *Tentamen*, where Leibniz finds an expression for velocity of rotation, orbital velocity and descent in terms of the parameters of the ellipse. The present text appears to be a preliminary draft leading towards the *Tentamen*.

Commentary to addition 2

The main points of this text are discussed in Section 5.5. The calculations on this folio appear to be the direct continuation of those on folio 25r., namely the last side of the essay *De Motu Gravium*; both deal with the same problem virtually with the same notation. Notice in line 6 that $_1G_1L$ ought to be equal to $2a\theta{:}r$, and the lack of the mistake by the factor $1/2$ in the term representing centrifugal conatus.

Leibniz tries to find a relation between time of revolution of an orbiting body and its distance from the centre of motion. The starting point is an equation similar to that which we have seen towards the end of the essay *De Motu Gravium*:

$$m = \overline{ddr} + a^2\theta^2{:}r^3, \tag{1}$$

where m and r have the usual meaning; θ is the constant differential of time which was previously written as dt; a is the constant proportionality factor between times and areas.

The expression for centrifugal force probably stems from the proportion:

$$\text{centrifugal force}{:}\theta^2/a{::}1/r^3{:}1/a^3; \tag{2}$$

the second term seems to represent the centrifugal force of a body moving along a circumference with radius a, θa representing the constant infinitesimal elements of the area. At the end of the *Horologium Oscillatorium* Huygens dealt with centrifugal force using proportions in a similar way (see Section 2.3). Gravity is set proportional to an arbitrary power of the distance:

$$m = \theta^2 r^n{:}a^{n+1}, \tag{3}$$

which has the same dimensions as the expression for centrifugal force. The rest of the calculation down to line 16 can be easily followed. In spite of the remarkable series of mistakes involving mainly signs and constant factors, the procedure adopted by Leibniz for both separation and substitution of variables is correct.

In lines 16–17 we find a passage which clarifies the purpose of the whole calculation. Leibniz says that since the integral he has found depends on the quadrature of the hyperbola, the curve GG cannot be an ellipse. As in the essay *De Motu Gravium*, Leibniz aims at a relation between times and areas for elliptical trajectories satisfying Kepler's laws.

Since the integral does not seem to lead to the result he wanted, Leibniz tries again with gravity inversely proportional to the third power of the distance, $n + 1 = -2$ (line 19). Setting the integration constant equal to zero, time becomes directly proportional to the square of the distance. Leibniz tries (lines 21–37) to determine the figure described by a body under the action of this force. After two unsuccessful attempts in which he tries to use the relation between radii and times rather than the differential equation of the curve, he crosses out his calculations realizing that they are leading nowhere. For a force inversely proportional to the third power of the distance, setting the integration constant equal to zero, the curve is a logarithmic spiral, where the angle of circulation is proportional to the logarithm of the distance. We find a similar problem in the case of uniform paracentric motion with $ddr = 0$ in paragraph 13 of the *Tentamen*; the curve described there is a hyperbolic spiral, where the angle of circulation is inversely proportional to the radius.

The following calculations (lines 67–76) are discussed in Section 5.5. At the end of the text (lines 82–90) we find some remarks on the alleged incompatibility between Kepler's second and third law; these remarks show once again that Leibniz had in mind planetary ellipses. At this point in the manuscript there is a figure which I do not reproduce because it is unrelated to the text. The contents of the last lines are similar to the manuscript *Ponamus duos planetas joviales*, LH 35, 9, 2, f. 72.

Commentary to addition 3

This text consists of two attempts to prove lemma 12 in book I of the *Principia*. Lemma 12 states that all parallelograms circumscribed about any given ellipse have equal areas. In the second edition of the *Principia* Newton makes clear that the parallelograms are described around conjugate diameters of a given ellipse. Newton adds that this result is known from the *Conics*.[1] A further attempt by Leibniz to solve the same problem is in LH 35, 10, 7, f. 15r. (f. 15v. is empty).

[1] See Descartes, *Geometria*, ed. F. van Schooten (Amsterdam, 1659–1661); Newton's copy with marginal annotations is in the Wren Library, Trinity College, Cambridge, classmark

The text and fig. 34 are affected by several inaccuracies. From line 1, for example, we infer that $B\Omega$ is on the prolongation of the major axis AV; more seriously, both BC (line 22) and BP (line 24) are set equal to v. Leibniz's interest in lemma 12 was aroused by the crucial proposition 11, where the lemma is used in the demonstration (see *NMW*, **6**, pp. 44–9). As we have seen, proposition 11 is annotated in the *Marginalia* and transcribed in Addition 2; line 48, point (4) relates to lemma 12.

Commentary to addition 4

Leibniz investigates the mode of action of vortices on planets, the mechanism of gravity, the properties of vortices and their relation to Kepler's laws. The last paragraph, which extends into the margin, has been included in the continuous text.

In the opening lines (1–16) Leibniz draws a series of distinctions in the order of infinity of several variables. At the beginning of its motion a planet has a small *vis circulandi*, or force of circulation; further its *velocitas circulandi*, or circulation, is infinitesimal with respect to ordinary circulation because the matter of the orb pushing it is very subtle. Moreover, the planet's centrifugal conatus is infinitesimal with respect to the circulation, and at the beginning of motion the centrifugal force of the vortex does not act much, because its matter is very subtle. The force of descent of the planet is defined by Leibniz as the difference between the centrifugal forces of the vortex and of the planet; if the latter is bigger than the former, the planet has levity instead of gravity. This mechanism entails a problem, because it requires the density of the planet to be comparable with that of the fluid matter. We have seen in Section 1.2 that an analogous objection was raised by Kepler in the *Epitome*, and that Leibniz was aware of it, although he does not refer to this problem here.

In lines 21–5 we find a reference to Leibniz's theory of motion in a resisting medium. Although the sentence is partially crossed out and unfinished, it is possible to infer what Leibniz wanted to say. First, if the particles of a fluid body (referred to as *corpus*) become smaller, so also does the impetus impressed on a solid body floating in it. In that portion of the sentence which has been crossed out (lines 22–3), Leibniz claims that the force of circulation arises from the surfaces of

---

NQ.16.203, esp. p. 220. M. Galuzzi, 'I *marginalia* di Newton alla seconda edizione latina della *Geometria* di Descartes ed i problemi ad essa collegati', *Quaderno n.15/1988*, Dipartimento di Matematica, Università di Milano, p. 14. See also G. A. Borelli's edition of Apollonius, *Conicarum libri V. VI. VII* (Florence, 1661), VII 31.

the fluid's particles. A description of this kind of action, or equivalently of friction, can be found in the *Schediasma de Resistentia Medii*, dating from the same time and based on Descartes's ideas (see Section 2.2). On the basis of the *Schediasma*, we can infer that in the present text Leibniz intended to show that the force of circulation arises from absolute resistance, because it depends on the surfaces of the particles of the fluid, whereas gravity or levity arise from respective resistance. According to Leibniz, since absolute resistance is the cause of circulation, if the particles of the fluid are smaller, they impress less impetus or velocity on the floating body; even if they have more solidity, this does not influence the impetus of the body in the fluid, because only respective resistance depends on solidity. This Cartesian account is not entirely consistent with the claim that gravity is the difference between the centrifugal forces of the body and of the vortex, because if gravity and levity depend on the centrifugal force of the body, they must also depend on absolute resistance. This may be the reason why Leibniz crossed out these lines and abandoned this project. With regard to the dependence of resistance on velocity compare Section 2.2. The passage we have just examined is the only example I know of the application of the theory of motion in a resisting medium to planetary motion in Leibniz.

In lines 34–40 Leibniz states that the new conatus of descent is the difference between gravity and centrifugal conatus, as he discovered in the essay *De Motu Gravium*, and illustrates the components of the new conatus of a planet: $_3P_3K$ which is equal to the preceding conatus $_2P_3P$, the conatus of gravity $_2P_2G$, and the conatus impressed by the circulation (see fig. 35). This statement highlights Leibniz's belief that the law of inertia holds in his representation of motion too.

In the second paragraph (lines 41–) Leibniz tries to determine the conatus impressed by the circulation. At the beginning he constructs the parallelogram $_2P_2L_3P_2F$ (see fig. 36); $_2$ is the intersection between $_2F_3P$ and the arc drawn from $_2P$ with radius S_2P, and $_2F$ represents centrifugal force. $_3P_2G$ is parallel to $_1P_2P$, therefore $_2G_2P$ represents gravity, and is set inversely proportional to the square of the distance 'because of the nature of the ellipse' (lines 45–6). Further, $_2P_2G - _2F$ is the new impression of descent *ddr*, which is also equal to $_2P_2L - _1P_1L$. Leibniz improves his method developed in the essay *De Motu Gravium*; however, he still makes two compensating mistakes in the term representing centrifugal force. The main point in this paragraph is that if we consider the vortex as acting on the body, and let this action be represented by $_2Pw$, perpendicular to $_2PS$, the areas cannot be proportional to the times. Leibniz's solution is that the difference in velocity between the planet and the fluid must be either zero or so small as to be undetectable. This solution is forced by the acceptance of Kepler's second law and is repeated in a similar fashion in paragraph 8 of the *Tentamen*. According to Leibniz, the equality in velocity between the planet and the fluid orb entails no problem as to the cause of gravity, because this originates from the different densities of the matters of the fluid orb and of the body. Leibniz refers to the example of water and glass as the constituents of the vortex and the planet respectively; we have seen in Section 7.4 that in *De Causa Gravitatis* he chooses mercury and glass.

In line 72 and following lines Leibniz states that centrifugal conatus is inversely proportional to the third power of the distance. Since gravity depends on centrifugal force and density of matter, which is proportional to distance, it follows that gravity is inversely proportional to the square of the distance. Lines 88–99 are devoted to a standard topic in Leibniz's manuscripts on planetary motion, namely the attempt to reconcile Kepler's second and third laws with the idea that gravity is inversely proportional to the square of the distance, and the hypothesis of equal forces for each vortex shell; the only variable is the density of the fluid matter. However, all these constraints cannot be satisfied simultaneously.

Appendix 2
CHRONOLOGICAL OUTLINE

The following chronological outline is based on Leibniz's life and publications. I have also inserted some references to the relevant data on Newton.

1664–6   Newton's 'anni mirabiles'; invention of the fluxional calculus.

1660s   Following the appearance of the 1664 comet Newton takes up astronomy. Preliminary investigations on curvilinear motion in the Waste Book, Vellum Manuscript, and other manuscripts.

1669   Leibniz excerpts Huygens's paper on the impact laws in the *Philosophical Transactions.*
          Newton composes *De Analysi*, which remains unpublished.

1670   Leibniz reads Thomas Hobbes, *De Corpore*, whence he takes the notion of 'conatus'.

1671   The *Hypothesis Physica Nova* appears in Mainz.

1672   Leibniz moves from Mainz to Paris, where he establishes contacts with Huygens, Edme Mariotte, Ismael Boulliau, and Claude Perrault.

1673   Huygens publishes the *Horologium Oscillatorium.* Leibniz reads *Saggi di Naturali Esperienze*, which refers to a Galilean 'peso morto'. Leibniz mentions for the first time a 'force morte' and 'animée' in a letter to Mariotte. First trip to London.

1675   Invention of the differential calculus. Work on motion in a resisting medium.

1676   Leibniz writes a letter to Claude Perrault on the cause of gravity.
          (June and October) Newton's first and second letters to Leibniz.
          (October) Leibniz leaves Paris and visits London for the second time.

1676   (December) Leibniz arrives in Hanover.

1677   Leibniz sends a letter on natural philosophy to Honoré Fabri.

1679   Hooke restarts his correspondence with Newton.

1680–1   Appearance of the big comet; Newton corresponds with Flamsteed.

1680s   Leibniz excerpts part 3 of Descartes's *Principia Philosophiae* and astronomical texts: *Observationes Astronomicae Novissimae*.

1684   Halley visits Newton in Cambridge; Newton drafts *De Motu Corporum*.
            Leibniz publishes *Demonstrationes Novae de Resistentia Solidorum* and *Nova Methodus pro Maximis et Minimis* in the *Acta*.

1686   Leibniz publishes *Brevis Demonstratio Erroris Memorabilis Cartesii* and *De Geometria Recondita et Analysi Infinitorum* in the *Acta*.

1687   Newton publishes the *Principia Mathematica*.
            (November) Leibniz leaves Hanover for the Italian journey.

1688   (May) Leibniz arrives in Vienna.
            (June) Christoph Pfautz reviews Newton's *Principia* in the *Acta*.
            (Autumn) Leibniz reads the *Principia Mathematica* and writes the *Notes* and *Marginalia*. He composes in short succession *De Conatu, Inventum a me est, Investigatio Semidiametri Circuli Osculantis, De Motu Gravis in Linea Projectitia, Galilaeus, Repraesentatio Aliqua, Si mobile aliquod ita moveatur, De Motu Gravium, Calculus Motus Elliptici* and *Tentamen de Systemate Universi*.

1689   (January) Leibniz's *De Lineis Opticis* and *Schediasma de Resistentia Medii* appear in the *Acta*.
            (February) Leibniz leaves Vienna. At the same time the *Tentamen de Motuum Coelestium Causis* appears in the *Acta*.
            (April) Leibniz arrives in Rome.
            Leibniz writes the *Excerpts* from the *Principia Mathematica*, and composes the *Tentamen de Physicis Motuum Coelestium Rationibus*. He starts working on the 'zweite Bearbeitung' of the *Tentamen* and *Dynamica*.
            (November) Leibniz leaves Rome.
            (December) Leibniz is in Florence. After talks with Vincenzo Viviani he writes a memorandum on the censorship of the Copernican system and an accompanying letter to the Jesuit Antonio Baldigiani in Rome.

1690   (March) Leibniz writes from Venice his last letter to Antoine Arnauld summarizing his work on planetary motion.
            (May) Leibniz's *De Causa Gravitatis* appears in the *Acta*.
            (June) Leibniz returns to Hanover.

            (October) Leibniz composes *De Causis Motuum Coelestium*

and a letter for Huygens on planetary motion in which he claims that he first saw the *Principia* in Rome. The letter is not sent.

1691–2 Newton composes *De Quadratura Curvarum.*

1692 Leibniz composes the *Essay de Dynamique* for the Paris Academy.

1695 Leibniz's *Specimen Dynamicum* appears in the *Acta.*

1702 David Gregory publishes *Astronomiae Physicae et Geometricae Elementa,* where he criticizes the *Tentamen.*

1705–6 Correspondence between Leibniz and Pierre Varignon on central forces. Leibniz writes the *Illustratio Tentaminis;* an abridged version appears in the *Acta.*

1706 Leibniz composes the *Antibarbarus Physicus.*

1710 Leibniz publishes the *Essais de Théodicée.*

1713 Second edition of the *Principia Mathematica* with a preface by Roger Cotes.

1710s Newton composes two memoranda on the *Tentamen.*

1714 John Keill attacks the *Tentamen* in the *Journal Litéraire.*

1715 Newton publishes anonymously in the *Transactions* his *Account of the 'Commercium Epistolicum'.*

1715–16 Correspondence between Leibniz and Samuel Clarke.

1716 Death of Leibniz.

Appendix 3
LIST OF MANUSCRIPT SOURCES

The following is a list of Leibniz's manuscripts to be found at the NLB. With the exception of the last three items, the classmark is 'LH 35, *Mathematica*'.

8, 30, f. 25 *Pro curva sumatur linea polygona*
    f. 69 *Keplerus plurimus aliorum inventis principium et occasionem dedit*
9, 2, f. 1–36 *Illustratio Tentaminis* (clear copy)
    f. 37–52 *Illustratio Tentaminis* (draft)
    f. 53 *Sit ellipsis planetaria Kepleriana*
    f. 54–5, 60–7, 74–8 'zweite Bearbeitung' of the *Tentamen*
    f. 56–9 *Tentamen* (draft with later additions)
    f. 68–9 *Massa materiae per se continua et uniformis*
    f. 72 *Ponamus duos planetas joviales*
    f. 80 excerpts from Kepler, *Astronomia Nova*
9, 5, f. 29 excerpts from G. D. Cassini, *Journal des Sçavans* 22 April 1686
    f. 30 *Si motus sit aequabilis*
9, 9, f. 1–2 *Tentamen de Systemate Universi*
    f. 3–4 *Incrementa angulorum circulationis harmonicae*
    f. 5–6 *Calculus Motus Elliptici*
    f. 7–8 *Si mobile ex centro semel emittatur aut repellatur*
    f. 9–10 *Si mobile feratur motu composito*
9, 26, f. 1 *Tentamen de Physicis Motuum Coelestium Rationibus* published in Leibniz, *Vorausedition*, Faszikel *8*
10, 1, f. 1 and 3 *De Causis Motuum Coelestium*
    f. 2 *Non explicuit Neutonus*
    f. 4 *Observationes Astronomicae Novissimae* (excerpts)
    f. 9 excerpts from Kepler, *Harmonice Mundi*
    f. 10–11 *Si mobile feratur motu composito*
    f. 12–13 *Calculus de Elementis Radiorum Ellipseos*
    f. 14r. *Tentamen* (draft of paragraph 18)
    f. 14v. *De Lineis Opticis* (draft)
    f. 15 *Si mobile lineam describat motu composito*
    f. 16–17 *Compositio Motus* (with separate observations)
10, 4, f. 1 *Ad Relationem Actorum* (draft of *De Lineis Opticis*)
    f. 1v.–2 *Tentamen de Legibus Naturae Mundi*

10, 7, f. 1–3 and 25 *De Motu Gravium*
    f. 4v.–5 *Inveniendus est Calculus Differentialium Ellipseos*
    f. 8 *Sit MM linea in qua gravia projecta*
    f. 12 *Repraesentatio Aliqua*
    f. 13–14 excerpts from Pfautz's review of the *Principia*
        Mathematica
    f. 15 *Clarissimus Neutonus* (on *Principia*, book I, lemma 12)
    f. 16–17 *De Motu Gravis in Linea Projectitia*
    f. 18v.–19r. *Galilaeus*
    f. 20 *De Areis Ellipticis*
    f. 23–4 *Constructio calculo inserviens*
    f. 29–30 *De Conatu*
    f. 31 *Investigatio Semidiametri Circuli Osculantis*
    f. 32–5 notes to Newton's *Principia Mathematica*
    f. 36–7 *Inventum a me est*
    f. 38–9 *Si mobile aliquod ita moveatur*
    f. 40 *Nova Methodus Tractandi Lineas Corporum Gravium*
    f. 41r. *Si trium punctorum quaeratur centrum gravitatis*
    f. 41v. *Inquisitio in Semidiametrum Circuli Osculantis*
14, 2, f. 18–19 *Si sint duo conatus corporis*
    f. 37–8 first set of *Excerpts*
    f. 31–6 second set of *Excerpts*
15, 2, f. 1 *De Lineis Opticis* (draft)
15, 6, f. 25–7 excerpts from Descartes, *Principia Philosophiae*
    f. 28–9 excerpts from Kepler, *Epitome Astronomiae*
    Copernicunae
LH 1, 3, 8, d, f. 1–2 *Tentamen de Physicis Motuum Coelestium*
        Rationibus, published in Leibniz, *Vorausedition*,
        Faszikel *8*
LH 38, f. 120–1 *Machina Coelestis* (draft)
    f. 122–7 *Machina Coelestis*, published by Gerland (1906,
    pp. 134–41)

Texts with marginalia in Leibniz's hand (classmark at the NLB: 'Leibn. Marg.')
   63 N. Mercator, *Institutiones Astronomicae* (London, 1676)
   70 C. Huygens, *Horologium Oscillatorium* (Paris, 1673)
   97 J. Kepler, *Epitome Astronomiae Copernicanae* (Frankfurt/M, 1635[2])
124 D. Gregory, *Astronomiae Physicae et Geometricae Elementa* (Oxford, 1702)

Texts with marginalia in Huygens's hand at the NLB, classmark Nm. A
104 G. A. Borelli, *Theoricae Mediceorum Planetarum* (Florence, 1666)

ABBREVIATIONS

AE	*Acta Eruditorum*
AHES	*Archive for History of Exact Sciences*
AIHS	*Archives Internationales d'Histoire des Sciences*
AS	*Annals of Science*
BJHS	*British Journal for the History of Science*
Excerpts	Bertoloni Meli, D., (1988). Leibniz's Excerpts from the *Principia Mathematica*. *AS*, **45**, 477–504.
GLI	*Giornale de' Letterati d'Italia*
GOF	G. Galilei (1890–1909) *Opere* (ed. A. Favaro) Florence. 20 volumes.
HOC	C. Huygens (1888–1950). *Oeuvres Complètes*. La Haye. 22 volumes.
JBB	Johann Bernoulli (1955–). *Briefwechsel*. Basel.
JBO	Johann Bernoulli (1742). *Opera Omnia*. Lausanne, Geneva. 4 volumes.
JHA	*Journal for the History of Astronomy*
JHI	*Journal of the History of Ideas*
KGW	J. Kepler (1937–). *Gesammelte Werke* (ed. by W. von Dyck, M. Caspar, F. Hammer *et al.*). Munich.
LBG	Gerhardt, C. I. (ed.) (1899). *Der Briefwechsel von G. W. Leibniz mit Mathematikern*. Berlin.
LMG	Gerhardt, C. I. (ed.) (1849–63). *Leibnizens mathematische Schriften*. Berlin, Halle. 7 volumes.
LPG	Gerhardt, C. I. (ed.) (1875–90). *Die philosophischen Schriften von G. W. Leibniz*. Berlin. 7 volumes.
LSB	G. W. Leibniz (1923–). *Sämtliche Schriften und Briefe*. Darmstadt, Leipzig, Berlin.
Marginalia	G. W. Leibniz (1973). *Marginalia in Newtoni Principia Mathematica* (ed. E. A. Fellmann). Paris.
MASP	*Mémoires de l'Académie Royale des Sciences de Paris*
NC	H. W. Turnbull, J. F. Scott, A. R. Hall, and L. Tilling (eds.) (1959–77). *The Correspondence of Isaac Newton*. Cambridge. 7 volumes.
NMW	D. T. Whiteside (ed.) (1967–81). *The Mathematical Papers of Isaac Newton*. Cambridge. 8 volumes.
NLB	Niedersächsische Landesbibliothek, Hanover
NRRS	*Notes and Records of the Royal Society of London*
PT	*Philosophical Transactions*
RHS	*Revue d'Histoire des Sciences*
SHPS	*Studies in History and Philosophy of Science*
SL	*Studia Leibnitiana*
ULC	University Library, Cambridge.

INDEX